Applications and Science in Soft Computing

Advances in Soft Computing

Editor-in-chief
Prof. Janusz Kacprzyk
Systems Research Institute
Polish Academy of Sciences
ul. Newelska 6
01-447 Warsaw, Poland
E-mail: kacprzyk@ibspan.waw.pl
http://www.springer.de/cgi-bin/search-bock.pl?series=4240

Mieczysław Kłopotek, Maciej Michalewicz
and Sławomir T. Wierzchoń (Eds.)
Intellligent Information Systems
2000. ISBN 3-7908-1309-5

Peter Sinčák, Ján Vaščák, Vladimír Kvasnička
and Radko Mesiar (Eds.)
The State of the Art in Computational Intelligence
2000. ISBN 3-7908-1322-2

Bernd Reusch and Karl-Heinz Temme (Eds.)
Computational Intelligence in Theory and Practice
2000. ISBN 3-7908-1357-5

Rainer Hampel, Michael Wagenknecht,
Nasredin Chaker (Eds.)
Fuzzy Control
2000. ISBN 3-7908-1327-3

Henrik Larsen, Janusz Kacprzyk,
Sławomir Zadrozny, Troels Andreasen,
Henning Christiansen (Eds.)
Flexible Query Answering Systems
2000. ISBN 3-7908-1347-8

Robert John and Ralph Birkenhead (Eds.)
Developments in Soft Computing
2001. ISBN 3-7908-1361-3

Mieczysław Kłopotek, Maciej Michalewicz
and Sławomir T. Wierzchoń (Eds.)
Intelligent Information Systems 2001
2001. ISBN 3-7908-1407-5

Antonio Di Nola and Giangiacomo Gerla (Eds.)
Lectures on Soft Computing and Fuzzy Logic
2001. ISBN 3-7908-1396-6

Tadeusz Trzaskalik and Jerzy Michnik (Eds.)
Multiple Objective and Goal Programming
2002. ISBN 3-7908-1409-1

James J. Buckley and Esfandiar Eslami
An Introduction to Fuzzy Logic and Fuzzy Sets
2002. ISBN 3-7908-1447-4

Ajith Abraham and Mario Köppen (Eds.)
Hybrid Information Systems
2002. ISBN 3-7908-1480-6

Przemysław Grzegorzewski, Olgierd Hryniewicz,
Maria A. Gil (Eds.)
Soft Methods in Probability, Statistics and Data Analysis
2002. ISBN 3-7908-1526-8

Lech Polkowski
Rough Sets
2002. ISBN 3-7908-1510-1

Mieczysław Kłopotek, Maciej Michalewicz
and Sławomir T. Wierzchoń (Eds.)
Intelligent Information Systems 2002
2002. ISBN 3-7908-1509-8

Andrea Bonarini, Francesco Masulli
and Gabriella Pasi (Eds.)
Soft Computing Applications
2002. ISBN 3-7908-1544-6

Leszek Rutkowski, Janusz Kacprzyk (Eds.)
Neural Networks and Soft Computing
2003. ISBN 3-7908-0005-8

Jürgen Franke, Gholamreza Nakhaeizadeh,
Ingrid Renz (Eds.)
Text Mining
2003. ISBN 3-7908-0041-4

Tetsuzo Tanino, Tamaki Tanaka,
Masahiro Inuiguchi
Multi-Objective Programming and Goal Programming
2003. ISBN 3-540-00653-2

Mieczysław Kłopotek, Sławomir T. Wierzchoń,
Krzysztof Trojanowski (Eds.)
Intelligent Information Processing and Web Mining
2003. ISBN 3-540-00843-8

Ahmad Lotfi, Jonathan M. Garibaldi (Eds.)
Applications and Science in Soft-Computing
2004. ISBN 3-540-40856-8

Ahmad Lotfi
Jonathan Garibaldi (Eds.)

Applications and Science in Soft Computing

With 125 Figures and 56 Tables

Springer

Dr. Ahamad Lotfi
School of Computing and Technology
Nottingham Trent University
Burton Street
Nottingham NG1 4BU
United Kingdom
ahmad.lotfi@ntu.ac.uk

Dr. Jonathan M. Garibaldi
Automated Scheduling, Optimisation and Planning Group
School of Computer Science and IT
Universtiy of Nottingham
Jubilee Campus
Wollaton Road
Nottingham NG8 1BB
United Kingdom
jmg@cs.nott.ac.uk

Cataloging-in-Publication Data applied for
Bibliographic information published by Die Deutsche Bibliothek
Die Deutsche Bibliothek lists this publication in the Deutsche Nationalbibliografie; detailed bibliographic data is available in theInternet at <http://dnb.dd.de>

ISSN 16-15-3871
ISBN 3-540-40856-8 Springer-Verlag Berlin Heidelberg NewYork

Springer-Verlag is a part of Springer Science+Business Media

springeronline.com

Printed in Germany

Cover design: Erich Kirchner, Heidelberg
Typesetting: Digital data supplied by the authors

Printed on acid-free paper 62/3020Rw-5 4 3 2 1 0

Preface

Soft computing techniques have reached a significant level of recognition and acceptance from both the academic and industrial communities. The papers collected in this volume illustrate the depth of the current theoretical research trends and the breadth of the application areas in which soft computing methods are making contributions.

This volume consists of forty six selected papers presented at the Fourth International Conference on Recent Advances in Soft Computing, which was held in Nottingham, United Kingdom on 12th and 13th December 2002 at Nottingham Trent University.

This volume is organized in five parts. The first four parts address mainly the fundamental and theoretical advances in soft computing, namely Artificial Neural Networks, Evolutionary Computing, Fuzzy Systems and Hybrid Systems. The fifth part of this volume presents papers that deal with practical issues and industrial applications of soft computing techniques.

We would like to express our sincere gratitude to all the authors who submitted contributions for inclusion. We are also indebted to Janusz Kacprzyk for his advices related to this volume. We hope you find the volume an interesting reflection of current theoretical and application based soft computing research.

Ahmad Lotfi & Jon Garibaldi
Nottingham, July 2003

Contents

Part I

Artificial Neural Networks

A Recurrent Self-Organizing Map for Temporal Sequence Processing

T. A. McQueen, A. A. Hopgood, J. A. Tepper and T. J. Allen

Department of Computing & Mathematics, The Nottingham Trent University, Burton Street, Nottingham, NG1 4BU, United Kingdom
e-mail:{thomas.mcqueen,adrian.hopgood,jonathan.tepper,tony.allen}@ntu.ac.uk

Abstract. We present a novel approach to unsupervised temporal sequence processing in the form of an unsupervised, recurrent neural network based on a self-organizing map (SOM). A standard SOM clusters each input vector irrespective of context, whereas the recurrent SOM presented here clusters each input based on an input vector and a context vector. The latter acts as a recurrent conduit feeding back a 2-D representation of the previous winning neuron. This recurrency allows the network to operate on temporal sequence processing tasks. The network has been applied to the difficult natural language processing problem of position variant recognition, e.g. recognising a noun phrase regardless of its position within a sentence.

1 Introduction

Temporal sequence processing (TSP) is an increasingly important field for neural networks, with applications ranging from weather forecasting to speech recognition [1]. TSP involves the processing of signals that vary over time. Problems such as predicting the weather generally cannot be solved by just examining a set of current inputs from the dynamic system in question, e.g. a satellite image showing today's cloud cover. Rather, any prediction must be based on the current input in the context of a number of previous inputs, e.g. a satellite image for today along with satellite images from the previous five days, showing how the weather has changed so far over the week.

Neural network models for TSP outperform alternative methods, such as NARMAX [9], mainly due to their ability to learn and generalize when operating on large amounts of data [9]. Supervised learning is usually used to solve TSP problems, i.e. the recurrent neural network must be explicitly trained by providing a desired target signal for each training exemplar. Current supervised learning methods are computationally inefficient [8] and are unable to solve certain types of problems [6].

A number of unsupervised neural networks for TSP have been proposed [6], mostly based on the self-organizing map (SOM) [5]. These models use a variety of external and internal memory mechanisms to capture information concerning past inputs, e.g. tapped delay lines and leaky integrators. Unsupervised learning has advantages over equivalent supervised techniques in that it makes fewer assumptions about the data it processes, being driven solely by the principles of self-organization, as opposed to an external target signal.

We present a novel, unsupervised, recurrent neural network based on a SOM to identify temporal sequences that occur in natural language, such as syntactic groupings. The network uses both an input vector and a context vector, the latter of which provides a 2-D representation of the previous winning neuron. The proposed network is applied to the difficult natural language processing (NLP) problem of position variant recognition, e.g. recognizing a noun phrase regardless of its position within a sentence.

2 Architecture and algorithm

The network has a 28-bit input vector that provides a binary representation of the input tag being processed. In addition to this input vector, the network also uses a second context vector. The size of this context vector can be varied depending on the size of the network, but in experiments detailed below the context vector was set to 10 bits (Fig. 1). Both the input and the context vector are used in the Euclidean distance calculation to determine the winning neuron in a similar manner to a standard SOM.

The context vector represents the previous winning neuron using a 10-bit coordinate vector. The most significant five bits of this vector (i.e. the five bits on the left) represent the binary number of the winning neuron's column, while the least significant five bits (i.e. five bits on the right) represent the binary number of the winning neuron's row.

This approach is an efficient method of coordinate representation that provides the network with a 2-D view of spatial context. It is an improvement over an initial approach, which represented the previous winning neuron using only a binary representation of its number within the SOM. Such a representation prevented the network from seeing similarities between neighboring neurons in adjacent columns. For example, neuron 20 and neuron 40 are neighbors on the SOM shown above and will therefore be representative of similar patterns. However, the binary representation of the numbers 20 (i.e. 010100) and 40 (i.e. 101000) are dissimilar. Thus similar input patterns may result in dissimilar context causing similar sequences to be clustered to significantly different regions of the SOM. It is envisaged that this would reduce the network's ability to generalize.

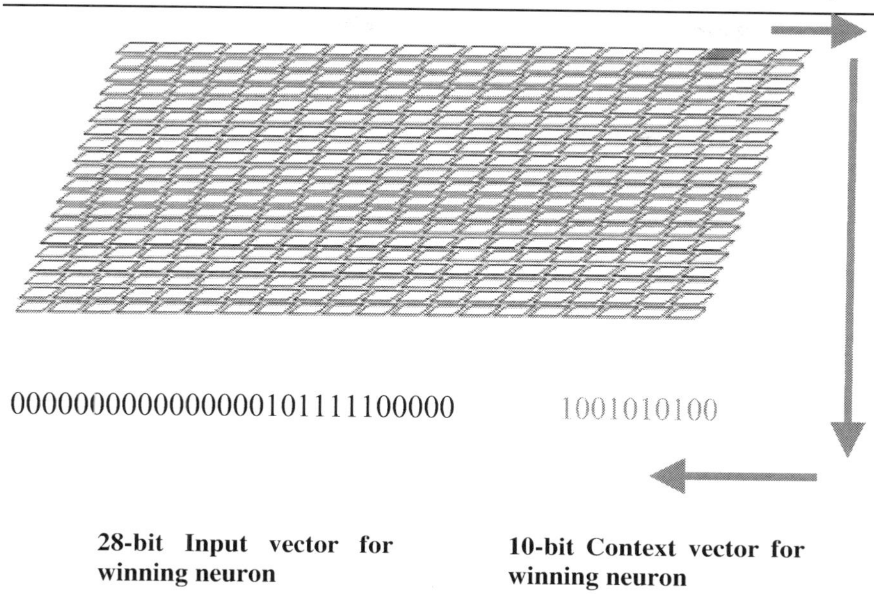

Fig. 1 Network showing recurrent feedback

The coordinate system of context representation solves this problem by effectively providing the network with a 2-D view of winning neurons. In the example given above, neuron 20 would be represented as 0000110100, while neuron 40 would be represented as 0001010100. (Note that only two bits are different in this example as opposed to four bits in the example above).

As with the standard SOM, the recurrent SOM presented here uses a neighborhood function to update the weights of neurons in a region around the winning neuron. Both the weight vector and the context vector of neighboring neurons are moved towards those of the respective input and context vectors. The network uses a Gaussian neighborhood function to calculate the learning rate that will be applied to these neurons. This function allows immediately neighboring neurons to experience similar weight changes to those of the winning neuron, while distant neurons experience minimal weight changes. However, in order to improve computational efficiency, the neighborhood function uses a cut-off value, beyond which neurons do not take part in weight updates at all.

3 Experiments

Initially, the new network is being applied to a corpus-based natural language task (Fig. 2) using the Lancaster Parsed Corpus (LPC) [7]. At present, the main objective of the research is to identify coarse phrase boundaries (e.g. noun phrases or verb phrases with little or no embedding) that may emerge on the topological map from exposure to linear sequences of words (sentences) that have been pre-tagged with symbols denoting the word's part-of-speech (e.g. noun, adjective, verb etc) [2].

A network with an output layer of 20 × 20 neurons was trained in two phases, following Kohonen's research on training SOMs [3]. The first convergence phase consisted of 1000 epochs, in which the learning rate was linearly reduced from an initial value of 0.1, but was not allowed to fall below 0.01. This was followed by a second fine-tuning phase in which a learning rate of 0.01 was applied for 2500 epochs. While the number of epochs in the first phase conforms with Kohonen's research [3], the number of epochs in phase two is considerably smaller than the number suggested. At this initial stage in the research, this reduction is necessary due to time and computational constraints. However, experimental analysis has not shown a significant reduction in the quality of results when training times in phase two are reduced.

A sample of 654 sentences from the LPC [7] were presented to the network. Presentation occurred in random order to improve training efficiency and to prevent the weights from becoming stuck during the low neighborhood value in phase two. The context vector is set to zero between each sentence to prevent contextual information from previous sentences interfering with subsequent sentences.

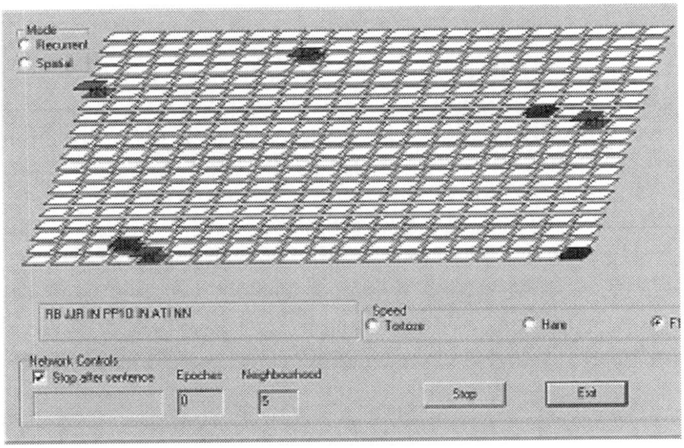

Fig. 2 – Screenshot from the current network. The raised, coloured polygons represent winning neurons for the sentence of tags presented to the network.

4 Results

The preliminary results are encouraging, as they show that word tags are being clustered in locations consistent with their context. The results in Figs. 3–5 show three simple artificially constructed sentences of varying tense. Despite these variations in tense, each exhibits a similar trace pattern over the map. We refer to these traces as signatures.

Fig. 6 shows two simple noun phrases with and without a preposition. While both sentences show similar signatures for the noun phrase, the effect of the preposition can clearly be seen to alter the signature of the second phrase.

It is hoped that further analysis will reveal the extent to which the network can exploit the context and show what kind of temporal syntactic patterns the network can find in input sequences. A major benefit of finding such patterns in an unsupervised manner is that, unlike supervised techniques, there is no dependency on manually annotated corpora, which are not widely available due to the high costs associated with manually annotating raw language data. In fact it is envisaged that, should the unsupervised system prove successful in extracting syntactic structure, it would serve as an automatic syntactic annotation system thus reducing the need and cost of manual annotation.

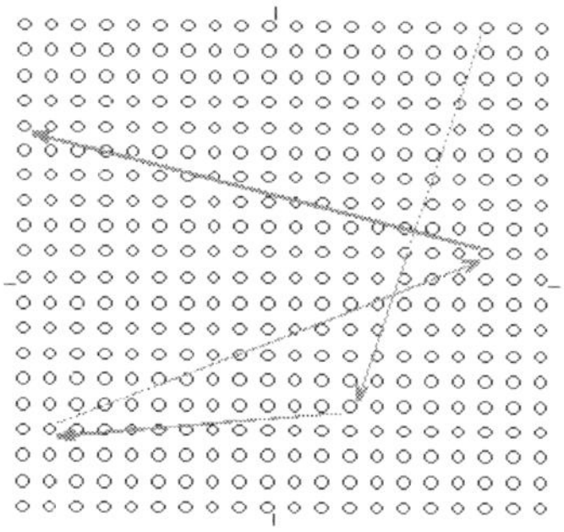

Fig. 3 – *Signature for sentence: "she goes down the stairs"*

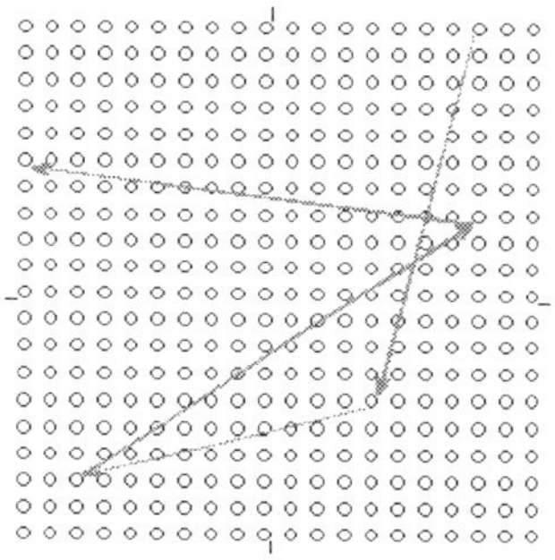

Fig. 4 – *Signature for sentence: "she went down the stairs"*

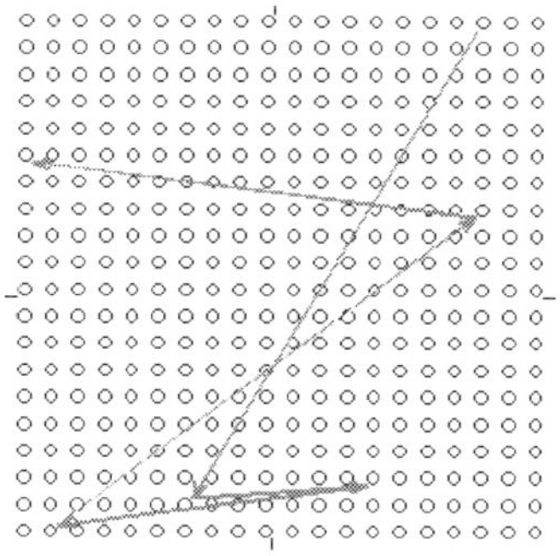

Fig. 5 – *Signature for sentence: "she is going down the stairs"*

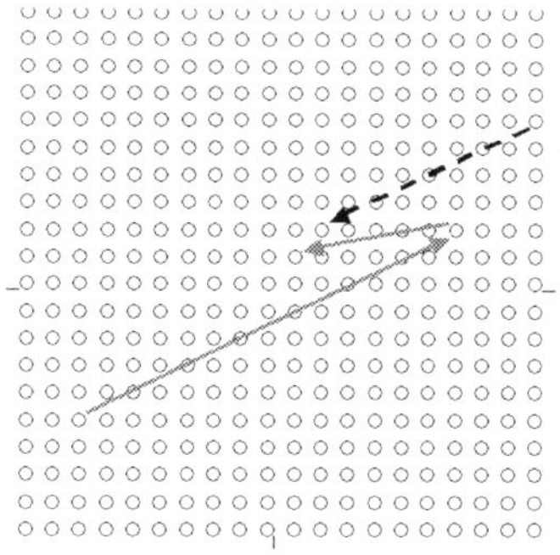

Fig. 6 – *Noun phrase with and without preposition*

The home

In the home

5 Conclusions and future work

We have presented a novel recurrent SOM and applied it to the problem of position-variant recognition. We have shown that the network forms signatures in response to temporal sequences present in the inputs.

In addition to the natural language task, research is also being conducted into enhancing the recurrent SOM using lateral connections and a temporal Hebbian learning [4] mechanism. The purpose of such a mechanism is to attempt to control the recurrency, allowing feedback to occur only when the winning neurons, whose representations are to be fed-back, are stable. This temporal Hebbian learning mechanism has been used in a previous experimental neural network and it is hoped that it will reduce the SOM's training time.

In the next phase of this investigation, hierarchical clustering methods based on temporal SOMs will be developed to obtain finer-grained syntactic groupings. Future work will focus on the context representation that is fed back. The representation may be enlarged to give more emphasis to the context vector than the input vector, and it may also be optimized using genetic algorithms. Further experiments will be performed in the domain of natural language processing; specifically the network will be used to attempt to detect phrase boundaries. Additionally, if the network proves successful, it may also be used in a number of other areas including computer virus detection, speech recognition and image analysis.

On a wider scale, the recurrent SOM could be used as the core of a temporal neural processing system. For example, the recurrent SOM clusters patterns based on input featural similarities whilst a supervised neural network uses these reduced representations to perform a mapping to a corresponding set of desired outputs.

References

[1] Barreto G and Arajo A (2001) Time in self-organizing maps: An overview of models. Int. J of Computer Research, 10(2):139-179

[2] Garside R, Leech G and Varadi T (1987) Manual of information to accompany the Lancaster Parsed Corpus. Department of English, University of Oslo

[3] Haykin S (1999) Neural Networks: A Comprehensive Foundation, Prentice Hall

[4] Hebb D (1949) The Organization of behaviour, John Wiley

[5] Kohonen T (1984) Self-Organization and Associative Memory, Springer-Verlag

[6] Mozer M (1994) Neural net architectures for temporal sequence processing, in.Weigend A and Gershenfeld N (eds), Time Series Prediction, pp 243–264

[7] Tepper J, Powell H and Palmer-Brown D (2002) A corpus-based connectionist architecture for large-scale natural language parsing. Connection Science, 14 (2)

[8] Schmidhuber J (1991) Adaptive history compression for learning to divide and conquer, Int. Joint Conf. on Neural Networks, Vol 2, pp 1130–1135

[9] Varsta M and Heikkonen J (1997) Context learning with the self-organizing map, Proc. Workshop on Self-Organizing Maps, pp 197–202

Optimisation of Neural Network Topology and Input Parameters for Prosody Modelling of Synthetic Speech[1]

Václav Šebesta[*] and Jana Tuckova[**]

[*] Institute of Computer Science, Academy of Sciences of the Czech Republic, and Faculty of Transportation, Czech Technical University, e-mail: vasek@uivt.cas.cz
[**] Faculty of Electrical Engineering, Czech Technical University,
e-mail: tuckova@feld.cvut.cz

Abstract. Variability and the probabilistic nature of speech are motivations to use a statistical approach, for example artificial neural network (ANN) in text-to-speech processing, especially for improving the quality of synthetic speech prosody. We try to investigate methods for extracting knowledge from existing continuous speech databases for optimisation of neural network topology to improve the generalization ability of ANN.

The ANN for the modelling of fundamental frequency and duration of a speech unit for a Text-to-Speech (TTS) synthesis are trained by natural speech. The principle of synthesizer is based on the concatenation of speech units. We also use speech unit segmentation of a text for prosody modelling. The phonemes are the basic units in our neural network approach.

GUHA method (General Unary Hypotheses Automaton) and a neural topology pruning process are applied for the choice of the most important input parameters.

Keywords. Feature selection, Text-to-Speech processing, Prosody modelling, Speech synthesis, Neural network utilization.

1 Introduction

The Text-to-Speech synthesis is one of the possible methods which can be used for speech syntheses. This method is characterized by the ability to synthesize an arbitrary sentences or longer texts from all thematic areas. It will be necessary, e.g. for real dialogue between the computer and the operator. An important characteristic of the synthetic speech quality is not intelligibility only but naturalness as well. We will try to contribute to achieving this aim. The grammatical rules that are often

[1] This research was supported by grant GA CR No 102/02/0124.

used for prosody modelling in the TTS synthesis do not follow exactly the speech naturalness. We have applied a neural network approach for prosody modelling.

The Czech language prosody is dependent on a lot of characteristic properties from which we can use only such a part that can be automatically extracted from the text. We cannot use parameters containing information that can be extracted only from the speech signal. We will try to find relevant features (speech parameters) having the basic influence on the fundamental frequency F_0 and duration D of speech units (phonemes) by the GUHA method.

2 Input data creation for prosody modelling from the database

The influence of a phonemes context is very important from the point of view of prosody modelling and therefore we will use a "moving window" over several phonemes in the training process. In our first attempt we will take into account the context of only three phonemes. The prosodic features are manifested especially in bigger speech units, i.e. in phrases or sentences. This fact was confirmed by our previous experience with prosody modelling by ANN [Šebesta and Tučková (1999), Tučková and Šebesta (2000)].

We will use a text and its speech signal for the training process of ANN, but only the text will be available for real prosody modelling. We cannot use completely all information extracted from a natural speech signal in automatic input data creation. For example, the so-called prominence demonstrating of different weights of a stress in a sentence, cannot be differentiated. Some experiments [Šebesta and Tučková (1999)] documented considerable improvement of the naturalness of synthetic speech but this approach required completion of input features values by hand. It needs a lot of time in the case of big files. We will try to improve the prosody by an other approach.

Changes of F_0 during a sentence create the melody of the sentence (the intonation). The intonation is also connected with the meaning of the sentence and with an emotional timbre. The types of sentences (i.e. two types of questions) can be distinguished by the intonation. It is not possible to change the word meaning by the intonation in the Czech language. It would be necessary to make a complicated and very difficult syntactic and semantic analysis of the text for a correct melody contour.

We are going to determine only such input properties of the Czech language, which can have an important influence on the fundamental frequency F_0 and duration D. We have selected ten basic input parameters for each phoneme according to our previous experience. These input parameters are used three times in the moving window (i.e. for a left-hand phoneme, a current phoneme and a right-hand phoneme). These parameters are described in Table 1. We must ensure a suitable relation among the different categories of parameters by their values. Similar properties of a single parameter are represented by the neighbouring numerical values. The knowledge about these relations was obtained from phonetic and phonologic studies.

Table 1. The characteristic properties of the Czech language, which can be used as inputs for prosody modelling by the ANN. The fundamental frequency and duration of phonemes are outputs of ANN.

Position of speech unit in relation with current unit			Property
Left	Current	Right	
P1	P11	P21	Silent pause identification
P2	P12	P22	Stress unit identification
P3	P13	P23	Syllable nucleus identification
P4	P14	P24	Punctuation mark identification
P5	P15	P25	Phoneme identification
P6	P16	P26	Height of vowel
P7	P17	P27	Length of vowel
P8	P18	P28	Voice of consonant
P9	P19	P29	Creation mode for consonant
P10	P20	P30	Number of phonemes in the current word

The first parameter (the silent pause identification) determines the type of pause. The pause duration influences the speech continuity and speech naturalness. We will differ three types of speech pauses: start/end sentence pauses, start/end word or stress-unit pauses[2] and pauses after punctuation marks in the intra-sentence. The rhythm of a sentence is also created by the distribution of the sentence to stress units. By the second parameter we will differentiate two classes: start of the stress unit and continuation of stress units. Rhythm naturalness relates to syllable in Czech. An exact rule for determination of syllables does not exist. We will identify this fact by vowel, diphthong or consonants r, l, j occurrence in the third parameter. The fourth parameter identifies punctuation marks. The type of a sentence and the type of intonation relate to each other. In some cases we can differentiate sentences described by the same text but with a different F_0 contour. We need to distinguish punctuation marks inside the sentence and at the end of the sentence.

Phoneme identification, which is related to the fundamental frequency is represented by the fifth parameter. Two groups are distinguished - vowels and consonants.

The sixth parameter will serve to distinguish the so-called articulate triangle for identification the height of a vowel. We will distinguish three modes of vowels, namely low, middle and height vowel. The length of a vowel is related to the duration of the phonemes. Two categories, short and long, are identified. The voice of a consonant is important from the point of view of the fundamental frequency

[2]The stress unit is created by one or more polysyllable words. Monosyllable words (e.g. prepositions) are added to the next word.

proportion. The basic groups are voiced consonants, unvoiced consonants and sonar.

Creation mode for consonants is a very important information from the point of view of phoneme (consonant) duration. We will distinguish two groups, one is plosives or affricates, and other is voiced and unvoiced fricatives. The last parameter is the number of phonemes in the word. Usually, the syllables length and also the phonemes length are reduced in the pronunciation of longer words.

3 Data Mining and ANN pruning

An ANN with one hidden layer was applied for the determination of prosody parameters. Important input parameters for the ANN training were determined by the GUHA method. The entire process of input parameters reduction (features selection) is described in [Šebesta and Tučková (1999)]. The processed data form a rectangular matrix where the rows correspond to the phonemes and the columns correspond to the attributes of the phonemes, i.e. input parameters P_1 to P_{30} (see Table 1) and output parameters (fundamental frequency F_0 and duration of phonemes). These attributes are split into antecedents and succedents. The GUHA program generates and evaluates the hypotheses between the conjunctions of the antecedents and the succedents. According to the number of the accepted hypotheses we can determine the importance of individual parameters.

Table 2 contains the numbers of the accepted hypotheses about the existing relations between P_I and fundamental frequency and duration. If we take 300 hypotheses as the limit, we can decide to eliminate input parameters P_1, P_4, P_{22}, P_{23}, P_{25} P_{26}, P_{27}, $P_{28,}$ P_{29} and P_{30} . The amount of input parameters was reduced from 30 to 20. This reduction must be verified by the mathematical model of verification (see Fig. 1 and 2) and also by listening. Both these tests confirmed the correctness of this reduction.

Besides the input parameters, which have only a small influence on the output parameters, also further input parameters having the greatest mutual correlations can be omitted. Their influence on the output parameter can be very similar. We were looking for such parameters in the next step. All 20 parameters P_I were taken as both the antecedent and the succedent. The number of conjunctions in the antecedent was 2 and in the succedent only 1. This is the reason why the values in Table 3 are not symmetrical. We can extract from Table 3 the highest mutual dependency between P_{11} and P_{14}. Therefore we eliminated P_{11} and repeated the entire operation. In the next step we found mutual dependency between P_{13} and P_{18}. Parameter P_{18} was extracted and the total number of input parameters was reduced to 18. After the elimination of P_{19} we have only 17 inputs. According to the mathematical and listening verification we recognized that several testing sentences are better, but in some cases the quality of sentence prosody is worse. It means that the optimal number of input parameters is somewhere among 17 and 19.

Table 2. Numbers of hypotheses about the relations between input parameters and prosody parameters

PARAMET.	NUMBER OF HYPOTHESES ABOUT F_0	NUMBER OF HYPOTHESES ABOUT D	TOTAL NUMBE OF HYPOTHESES	PARAMET.	NUMBER OF HYPOTHESES ABOUT F_0	NUMBER OF HYPOTHESES ABOUT D	TOTAL NUMBER OF HYPOTHESES
P_1	124	106	178	P_{16}	442	430	650
P_2	223	220	332	P_{17}	429	414	631
P_3	253	231	361	P_{18}	451	407	640
P_4	158	133	235	P_{19}	428	400	613
P_5	217	198	311	P_{20}	322	351	504
P_6	300	253	412	P_{21}	408	368	569
P_7	258	237	367	P_{22}	114	96	156
P_8	293	272	419	P_{23}	160	126	205
P_9	301	272	427	P_{24}	450	412	633
P_{10}	209	205	309	P_{25}	194	165	263
P_{11}	222	198	333	P_{26}	201	175	276
P_{12}	339	348	517	P_{27}	140	121	193
P_{13}	467	418	664	P_{28}	72	57	94
P_{14}	265	242	400	P_{29}	73	47	87
P_{15}	375	354	549	P_{30}	68	54	88

4 Experimental Results

The graphs in Fig. 1 show the results of the data mining approach for the prosody modelling by the ANN for a selected test sentence with 21 phonemes. The solid bold line represents the target values (duration in ms and fundamental frequency in Hz), and all the others show the results from the ANN before pruning (GUHA-30) and after several steps of pruning. The results of other sentences are very similar. These results have been also confirmed by listening tests.

In the TTS syntheses process one must take into account not only the graphical part of the utterance and the speech signal, but also the acoustical form of the text. The acoustical form is extremely context dependent. This dependency is called co-articulation. The contexts of all 3 phonemes combinations (CCC, VVC, VCV, VV, VCC, CCV, CVC)[3] were analysed for the determination of opportunities for a next improvement of prosody. The results of the error improvement in the case of co-articulation CVC are shown in Fig. 2. All vowels, a,e,i,o,u,ou were analysed separately in entire training and testing sets, "average" means sum of errors, "MSE" means sum of squares, "max" means maximal error, "0" means errors before the feature selection for all 30 input parameters and "opt" means errors after pruning of input and hidden layers for 18 input parameters. In spite of the reduc-

[3] C is the consonant, V is the vowel

tion of the input parameters one can see that in nearly all cases (with only 3 exceptions) total errors decreased.

Table 3. Numbers of hypotheses about the mutual correlation between the input parameters.

	P_7	P_8	P_9	P_{10}	P_{11}	P_{12}	P_{13}	P_{14}	P_{15}	P_{16}	P_{17}	P_{18}	P_{19}	Σ
P_2	68	75	77	56	34	44	30	36	22	28	29	32	32	166
P_3	139	167	169	86	51	32	28	54	21	20	22	30	30	327
P_5	96	112	116	55	75	29	34	80	25	15	17	37	37	286
P_6	113	113	118	82	134	31	69	139	44	33	40	86	86	397
P_7	77	132	135	74	46	30	25	49	16	19	19	30	30	258
P_8	124	131	152	81	36	62	52	39	46	68	56	38	38	331
P_9	128	152	131	81	44	62	58	47	51	68	57	43	43	358
P_{10}	60	58	58	45	55	22	1	55	0	0	0	1	1	111
P_{11}	47	53	56	45	209	37	122	253	120	107	87	165	163	507
P_{12}	27	33	35	23	105	60	118	111	100	97	97	123	125	297
P_{13}	20	33	34	21	138	100	132	137	155	180	162	200	199	371
P_{14}	67	87	90	51	242	42	133	257	130	118	94	172	171	612
P_{15}	23	26	26	19	106	71	128	113	78	140	117	137	135	279
P_{16}	44	80	80	25	95	91	164	98	145	130	138	130	131	348
P_{17}	23	53	53	7	83	88	160	84	127	149	107	157	155	319
P_{18}	6	0	4	1	166	112	207	164	159	161	168	177	195	357
P_{19}	6	0	4	1	177	112	209	175	169	163	170	197	177	370
P_{20}	0	0	0	21	30	67	87	28	74	77	80	79	79	114
P_{21}	99	95	95	86	48	41	27	48	25	27	23	24	25	183
P_{24}	107	111	111	89	76	46	44	78	45	53	43	49	50	284
Σ	192	259	268	123	345	145	272	242	265	227	284	287	152	1216

5 Conclusion

The comparison of the listening tests approved that the length of speech units is very important for the quality of speech prosody. More co-articulation effects can be involved by the modification of the synthesizer from diphones to triphones. The synthetic speech quality depends also on the method of syntheses (a parametric LPC model, TD-PSOLA type system, etc.), on the prosody modelling and on the size and quality of the speech corpus.

Our work is a continuation of the research concentrated at the Faculty of Electro-engineering CTU in Prague, at the Institute of Radio-engineering and Electronics, Academy of Sciences of the Czech Republic, at the Institute of Phonetics of Faculty of Arts, Charles University in Prague and at the Faculty of Applied Sciences, UWB in Pilsen.

An original software NNLab in MATLAB, V5.2 and NN-Toolbox for automation of the data base creation, neural network training and graphical results processing is under construction.

The listening tests are used for verification of our hypothesis by the MODPROZ programme, based on the diphones concatenation and developed on IRE [Tučková and Horák (1997)] or on the new software ARTIC, based on the triphones and developed at University of West Bohemia in Pilsen [Matoušek (2000)].

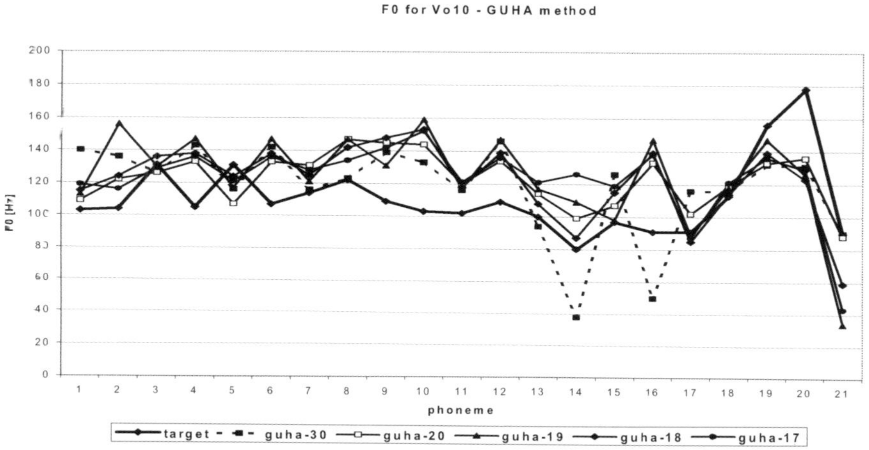

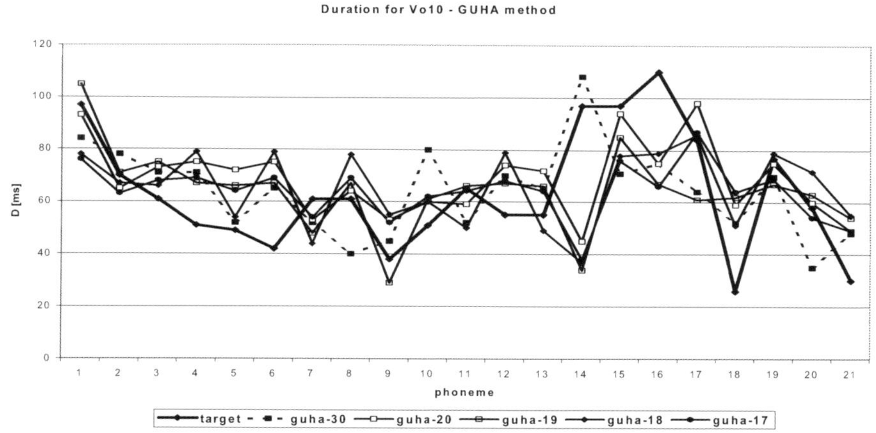

Fig. 1. Fundamental frequencies and durations of 21 phonemes in a test sentence for different numbers of input parameters

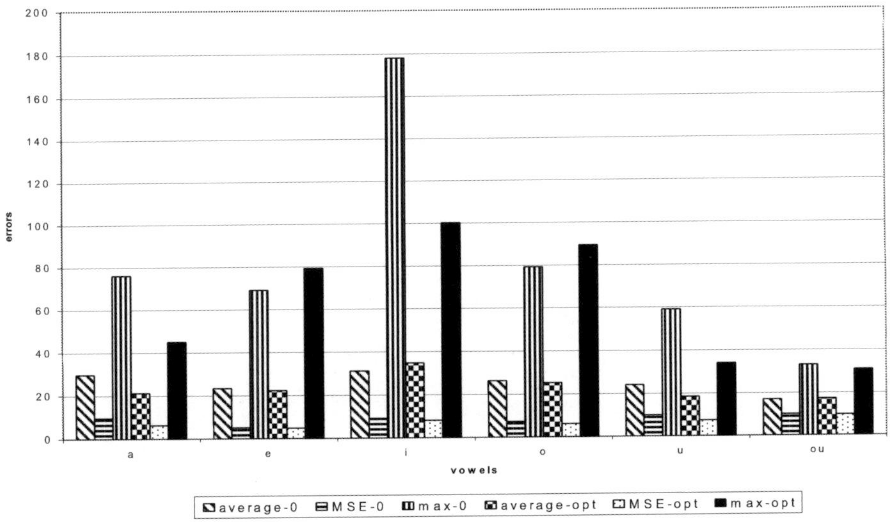

Fig. 2. Three types of errors for all CVC triphones co-articulation combinations

References

1. Vích R (1995) Pitch Synchronous Linear Predictive Czech and Slovak Text-to-Speech Synthesis. Proc. of the 15th Internat. Congress on Acoustics, ICA'95, Trondheim, Norway
2. Matoušek J, Psutka J (2000) ARTIC: A New Czech Text-to-Speech System Using Statistical Approach to Speech Segment Database Construction. Proceedings of ICSLP'2000, vol IV, Beijing, pp 612-615
3. Hájek P, Sochorová A, Zvárová J (1995) GUHA for personal computers., Computational Statistics and Data Analysis, vol 19, North Holland, pp 149-153
4. Šebesta V, Tučková J (1999) Selection of Important Input Parameters for a Text-to-Speech Synthesis by Neural Networks. 1999 International Joint Conference on Neural Networks IJCNN'99 (Proc. On CD-R), Washington, DC, USA
5. Tučková J, Šebesta V (2000) Prosody Modeling for a Text-to-Speech System by Artificial Neural Networks. Proc. IASTED Int. Conf. "Signal and Image Processing 2000", IASTED/ACTA Press, Las Vegas, USA, pp 312 - 317
6. Tučková J, Horák P (1997) Fundamental Frequency Control in Czech Text-to-Speech Synthesis. Third Workshop on ECMS'97, Toulouse, France
7. Matoušek J (2000) Text-to-Speech System Using Statistical Approach to Speech Segment Database Construction. PhD dissertation, UWB in Plzeň, Czech Republic, (in Czech)
8. Traber CH (1995) SVOX: The implementation of the Text-to-Speech System for German. PhD thesis, ETH Zurich, Switzerland

Using ART1 Neural Networks to Determine Clustering Tendency

Louis Massey

Royal Military College, Kingston, Ontario, Canada, K7K7B4

1 Introduction

Clustering is an unsupervised, data driven learning paradigm that aims at discovering natural groups in data [8, 9]. This type of learning has found many useful applications in domains with large amount of data where labeling of a training set for supervised learning is cost prohibitive or where autonomy is essential [1, 10, 11, 12]. However, clustering algorithms generally rely on some prior knowledge of the structure present in a data set. For instance, one needs to know whether or not clusters actually exist in data prior to applying a clustering procedure. Indeed, clustering applied to a data set with no naturally occurring clusters would merely *impose* meaningless structure. The procedure that consists in examining a data set to determine if structure is actually present and thus determine if clustering is a worthwhile operation is a poorly investigated problem known as *cluster tendency* determination [8].

Research in the area of cluster tendency has mainly focussed on the somewhat related problem of establishing the true number of clusters present in the data [6], often as part of *cluster validity*, the evaluation of clustering output quality [8]. Of course, should it be ascertained that the best clustering contains only one group, then null tendency must be concluded. The main problem with these approaches is that they either rely on yet other optimization procedures and similarity metrics (just as the clustering procedure itself), or depend on some parameter estimation. We show how to avoid these problems by using Adaptive Resonance Theory (ART) neural networks [3, 7] to determine clustering tendency of binary data. The binary version of ART (ART1) is used.

2 Adaptive Resonance Theory

ART neural networks are known for their ability to perform plastic yet stable on-line [14] clustering of dynamic data sets [4]. ART detects similarities among data objects, typically data points in an N-dimensional metric space. When novelty is detected, ART adaptively and autonomously creates a new category. Another advantageous and distinguishing feature of ART is its ability to discover concepts in data at various levels of abstraction [12, 16]. This is achieved with the *vigilance parameter* $\rho \in (0,1]$. First, a similarity measure S (Eq. 1) is computed to determine if an existing cluster prototype T_j appropriately represents the critical features of an input pattern X_k.

$$S = ||X_k \wedge T_j|| / ||X_k|| \quad (1)$$

Then, the *vigilance test* compares S with the vigilance parameter (Eq. 2):

$$S \geq \rho \quad (2)$$

Eq. 2 determines whether the current input pattern X_k will be recognized as a known concept or as a novel one. Indeed, if the vigilance test fails for all existing prototypes during the network search phase, X_k is deemed to be novel and new concept formation is triggered. At high vigilance ($\rho \to 1$), a large number of specific (low generality) clusters will be detected in the data. Conversely, at low vigilance ($\rho \to 0$), objects will be assigned to fewer, more general categories. Given a data set $X = \{X_k \mid k = 1, 2, \ldots, R\}$, one then intuitively expects a function relating the number of clusters M to the vigilance ρ. This function is expected to have a minimum value of M=1 for $\rho \to 0$ and a maximum of M=R for $\rho=1$.

3 Minimal and Maximal Vigilance

Of interest is S_{min}, the minimal non-zero value for S. The minimum non-zero value for the numerator of Eq. 1 is 1, that is one common bit between the prototype and the input data. The theoretical[1] maximal value for the denominator is N. Hence, we obtain:

$$S_{min} = 1/N. \quad (3)$$

Similarly, we develop a non-unit maximal value for S:

$$S_{max} = (N-1)/N. \quad (4)$$

Based on Eqs. 2, 3 and 4, one can establish the corresponding *minimal and maximal useful values* for vigilance. This is illustrated as follows. Lets suppose M_0 clusters are obtained with ρ_0. However, the application requires $M_1 < M_0$ clusters, so the input set is re-submitted to the ART1 neural network with $\rho_1 < \rho_0$. The expectation is that with a lower vigilance, fewer and more general groups will be formed: this is normal ART1 behavior. Under certain conditions, this will not occur. Indeed, if $\rho_1 < S_{min}$, then any further reduction of the vigilance will not result

[1] In practice $||X_k||$ is expected to be a small fraction of N, so the actual S_{min} (Eq. 3) will always be larger or equal to the theorical value, but this is inconsequent for the use we make of it.

in less clusters[2]. The last vigilance value for which a reduction in the number of clusters was achieved is the minimal useful vigilance, ρ_{min}. A similar reasoning applies for the maximal useful vigilance ρ_{max} in which case a maximum number of clusters $M \leq R$ is reached. Hence, in reality, the intuition that varying the vigilance within its whole range of permissible value can give rise to an arbitrary number of clusters $M \in [1,R]$ is incorrect.

Minimal and maximal useful vigilance may be problematic for applications that require concept granularities beyond or below possible limits. However, they are also useful structure detector. Indeed, the phenomenon just described can be interpreted as an inherent ability of ART1 to detect natural similarities in the data. For instance, *less clusters cannot be discovered in the data below minimal vigilance simply because the data is naturally not susceptible to assemble into less groups*.

4 Tendency Determination

By setting vigilance to a value below minimum useful vigilance, one obtains an indication of the natural tendency of the data to cluster. This is deduced from the number of clusters formed. The following situations can occur for $\rho<\rho_{min}$:

1. If M>1: it can be concluded that the data has a natural tendency to cluster.
2. If M=1: from the most general point of view[3], the data does not have a tendency to form clusters.
3. If M→R: this means weak tendency, with the extreme case of M=R clusters (trivial clustering) corresponding to no tendency.

For situation 2, one must progressively increment the vigilance parameter from its minimal useful value to determine if eventually M>1 clusters form. Hence, one must consider the notion of *generality* when computing clustering tendency. This can be illustrated by considering fauna classification: at the highest level of the hierarchy (the more general cluster), all animals are clustered into category "animal" (i.e. M=1). However, this does not mean that there is no structure in the data. By lowering the generality, M>1 potentially useful clusters may be found (for example, corresponding to animal classes or families).

We have described a method to establish if natural groups occur in the data. The residual question is whether those groups are the result of mere coincidence. Indeed, it can easily be demonstrated [13] both analytically and empirically that clusters do occur in a random data set. Evidently, such clusters are meaningless and clustering tendency of such origin must be appropriately detected. We now show how maximal vigilance is used for that purpose.

Increasing vigilance means that the ART network will form more and smaller clusters since it is being more demanding about features matching, as per the vigilance test (Eq. 2). Random data should therefore have a natural predisposition to

[2] Under certain circumstances, the number of clusters may decrease slightly [15].

[3] Recall that ρ_{min} implies the more general clustering possible.

split into many small clusters more rapidly than data that contains actual structure because it is less likely to have the required number of bit matching to pass the more stringent vigilance test. Maximal vigilance for random data, which we denote by ρ^r_{max}, will thus be smaller than for non-random. This idea can be used to establish non-random tendency as shown by the following inequality:

$$\rho^r_{max} < \rho_{max} \tag{5}$$

Maximal vigilance can be approached incrementally for a given data set and the relation between M and ρ plotted. Observing how fast M tends towards R compared to a baseline random data set allows for the detection of clustering tendency of a random origin. Such a graphic and qualitative approach may not be ideal, but it suits our current objective by giving an idea of whether or not clusters may be due to mere chance. Elements of a quantitative approach are given in [13].

5 Empirical Validation

Patterns are bit strings of length N. In the first experiment, we consider the case where tendency is determined by failing to reach M=1 at $\rho < \rho_{min}$. The data set with R=10 patterns and N=50 (Fig. 1a) is submitted to ART at $\rho < \rho_{min}$ ($\rho = 0.01$), then re-submitted at progressively incremented vigilance. The effect of minimal vigilance is visible in Fig. 1b. M=3 clusters are detected at below minimum vigilance, which allows one to conclude that there is clustering tendency for the data set. Visual inspection of figure 1a confirms this finding. Applying a clustering procedure to the data is therefore a meaningful operation.

In the second experiment, any data set in which not a single feature overlap exists can be used. This kind of data has no inherent clustering tendency due to an absolute lack of similarity between objects. By clustering this data at below minimum vigilance, we obtain M=R, hence confirming null tendency. Applying a clustering procedure to this data set would impose artificial structure and would therefore be a meaningless operation.

In the third experiment, we verify that the rate at which M grows when ρ is progressively incremented to ρ_{max} detects tendency caused by chance as per Eq. 5. Three random[4] data sets are processed with ART at vigilance varying from below

```
10111110110000000001000000000000000000000000000000
11011011010000000000000000000000000000000000000000
11101101110000000000000000000000000000000000000000
11110111100000000000000000000000000000000000000000
00000000000000000000000000000000000000001101011110
00000000000000000000000000000000000000001110011101
00000000000000000000000000000000000000001111011011
00000000000000000000000000000000000000001011010111
00000000000000000111010000000000000000000000000000
00000000000000000111100000000000000000000000000000
```

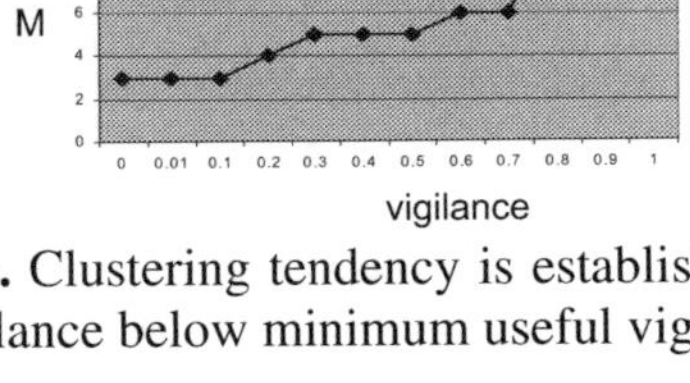

Fig. 1a. Dataset for experiment 1. **Fig. 1b.** Clustering tendency is established by observing that 3 clusters are formed at vigilance below minimum useful vigilance.

[4] The random bit patterns are actually pseudo-random data generated with java.util.Random.

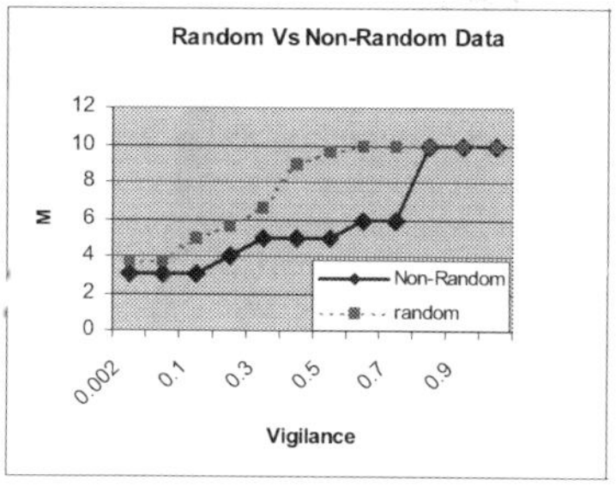

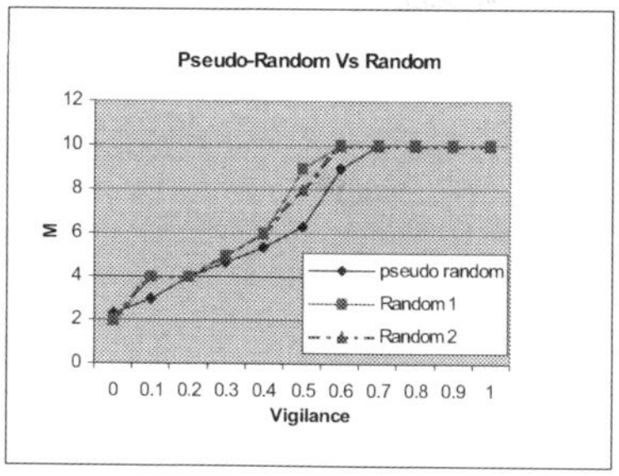

Fig. 2a. Baseline pseudo-random data reaches maximal vigilance faster than non-random data. This indicates that clustering tendency is not caused by chance clustering
Fig. 2b. True random data reach maximal vigilance faster than the baseline, which is an indication that their clustering tendency is caused by mere chance.

ρ_{min} to 1. The number of clusters formed is averaged for each vigilance value. Other data sets will be compared with this baseline to determine if their clustering tendency is caused by chance. The random data must possess the same characteristics (N, R and P(1), the bit set probability) as the dataset to be compared against. Here we used N=50, R=10 and P(1)=0.2. In Fig. 2a, the number of clusters is plotted for the random baseline and for the data used in experiment 1. It can be observed that ρ^r_{max} for the random baseline is smaller than ρ_{max} for the tested data. As per Eq. 5, this is an indication that cluster tendency is not caused by random structure in the data. In Fig. 2b, two other data sets are compared to the random baseline. For these data sets, $\rho^r_{max} > \rho_{max}$, which means that clustering tendency is caused by mere chance clustering. Indeed, these two data sets were obtained from known random sources (random 1 comes from radio atmospheric noise and was obtained at random.org. Random 2 comes from radioactive source decays and was obtained from HotBits (http://www.fourmilab.ch/hotbits/)). Other similar experiments have been conducted with several real-life or benchmark data sets. One of these experiments is documented in [12].

6 Conclusion and Future Work

We have shown how the vigilance parameter of a binary ART neural network can be used to determine the clustering tendency of a data set. The idea is based on the fact that at the highest level of generality, that is for vigilance set below its minimal theoretical level, ART should collapse all clusters into a single group. However, if the data possesses inherent structure, it will not. It was furthermore argued that clustering tendency can be achieved at various levels of generality. We also described the use of maximal vigilance to detect cluster tendency caused by chance. Hence, tendency is determined in two simple steps: first, verify that the data does not contain a trivial number of clusters (M=1 or M=R) at minimal vigilance; and second, verify that non-trivial clustering are not caused by chance by considering the rate at which maximum vigilance is reached compared to baseline random data. The method to determine clustering tendency as described in this paper is applicable to binary data inputs only; investigation of the non-binary ART versions, such as ART2 [2] and fuzzyART [5] on real-valued continuous data would be an interesting area of future research.

References

[1] Bezdek JC (1993) Review of MRI Images Segmentation Techniques Using Pattern Recognition. Medical Physics vol 20 no 4 pp1033-1048

[2] Carpenter GA, Grossberg S (1987) Art 2: Self-organisation of stable category recognition codes for analog input patterns. Applied Optics Vol 26 pp 4919-4930

[3] Carpenter GA, Grossberg S (1987) Invariant pattern recognition and recall by an attentive self-organizing ART architecture in a nonstationary world. In: Proceedings of the IEEE First International Conference on Neural Networks pages II-737-745

[4] Carpenter GA, Grossberg S (1995) Adaptive Resonance Theory (ART). In: Handbook of Brain Theory and Neural Networks Ed: Arbib MA, MIT Press

[5] Carpenter GA, Grossberg S, Rosen DB (1991) Fuzzy art: Fast stable learning and categorization of analog patterns by an adaptive resonance system. Neural Networks Vol 4 pp 759-771.

[6] Dubes RC (1987) How Many Clusters are Best? – An Experiment. Pattern Recognition Vol 20 No 6 pp 645-663.

[7] Grossberg S (1976) Adaptive pattern classification and universal recording. Biological Cybernetics Vol 23 pp 121-134.

[8] Jain AK, Murty MN, Flynn PJ (1999) Data Clustering: A Review. ACM Computing Surveys Vol 31 No 3.

[9] Kaufman L, Rousseeuw PJ (1990) Finding Groups in Data: An Introduction to Cluster Analysis. Wiley-Interscience.

[10] Kohonen T, Lagus K, Salojärvi J, Honkela J, Paatero V (2000) Self Organization of a Document Collection. IEEE Transactions On Neural Networks Vol 11 No 3.

[11] Li C, Biswas G (1995) Knowledge-based scientific discovery in geological databases. In: Proceedings of the First International Conference on Knowledge Discovery and Data Mining, Montreal, Canada, pp 204–209.

[12] Massey L (2002) Structure Discovery in Text Collections. In: Proc of KES'2002 Sixth International Conference on Knowledge-Based Intelligent Information & Engineering Systems, Podere d'Ombriano, Crema, Italy, pp161-165.

[13] Massey L (2002) Determination of Clustering Tendency With ART Neural Networks. In: Proceedings of 4th Intl. Conf. on Recent Advances in Soft Computing, Nottingham, U.K., 12 & 13 December 2002.

[14] Moore B (1988) ART and Pattern Clustering. Proceedings of the 1988 Connectionist Models Summer School pp174-183.

[15] Sadananda R, Sudhakara Rao GRM. (1995) ART1: model algorithm characterization and alternative similarity metric for the novelty detector. In: Proceedings IEEE International Conference on Neural Networks Vol 5 pp2421 –2425.

[16] Vlajic N, Card H-C (1998) Categorizing Web Pages using modified ART. In: Proc. of IEEE 1998 Conf. on Electrical and Computer Engineering Waterloo, Vol1 pp313-316.

High Performance Associative Memories and Structured Weight Dilution

S.P Turvey, S.P.Hunt, N.Davey and R.J.Frank

Department of Computer Science, University of Hertfordshire, College Lane, Hatfield, AL10 9AB. United Kingdom
email: {s.p.turvey, s.p.hunt, n.davey, r.j.frank}@herts.ac.uk

Abstract. The consequences of two techniques for symmetrically diluting the weights of the standard Hopfield architecture associative memory model, trained using a non-Hebbian learning rule, are examined. This paper reports experimental investigations into the effect of dilution on factors such as: pattern stability and attractor performance. It is concluded that these networks maintain a reasonable level of performance at fairly high dilution rates.

Keywords. Associative Memory, Hopfield Networks, Weight Dilution, Capacity, Basins of Attraction, Perceptron Learning.

1 Introduction

The associative memories examined here are based on the standard Hopfield architecture [10]. It has been known for some time [1] that networks with superior performance to that of the original model can be built. Performance may be improved by using an alternative learning rule: either one that finds an approximation to the projection weight matrix, or one that implements perceptron-style learning. (See [6,7,14] for a comparison of performance of different models).

Weight dilution is a technique for reducing the degree of connectivity within a network. Connections are removed after training has taken place (*post-training dilution*). For one-shot Hebbian learning, as employed in the 'standard' Hopfield model, it is known [13] that capacity drops linearly with the fraction of connections removed. It has even been suggested that an associative memory may be trained by starting with a fully connected network with random fixed weights and systematically removing a fraction of the connections [12].

2 Models Examined

In each experiment we train a network of N units with a set *bipolar* (+1/-1) training vectors, $\{\xi^p\}$. The resulting weight matrix is denoted by $\mathbf{W}$. During recall the net input, or *local field*, h, of a unit and its *next* state, S', are calculated as follows:

$$h_i = \sum_{j \neq i} w_{ij} S_j \qquad S'_i = \begin{cases} 1 & \text{if } h_i > \theta_i \\ -1 & \text{if } h_i < \theta_i \\ S_i & \text{if } h_i = \theta_i \end{cases}$$

where w_{ij} is the weight on the connection from unit j to unit i, S_j is the *current* state of unit j, and θ_i is the *threshold* on unit i (zero in all of our experiments).

Unit states are updated asynchronously in random order. This and a symmetric weight matrix guarantee simple point attractors in the network's state space, each of which is a stable state of the network.

A training vector, ξ, will be a stable state of the network if the *aligned local fields*, $h_i\xi_i$ are non-negative for all i (assuming all θ_i are zero). Each training vector that is a stable state is known as a *fundamental memory* of the trained network. The *capacity* of a network is the maximum number of fundamental memories it can store. The *loading*, α, on a network is calculated by dividing the number of vectors in the training set by the number of units in the network, N.

Learning Rules

Two learning rules have been employed in this work. The first, described by Blatt & Vergini [3], approximates the projection matrix generated using the pseudo-inverse rule (see [8] for details). The second is Gardner's perceptron-like symmetric local learning rule [6,9].

Blatt & Vergini's Rule (BV)

Blatt & Vergini's [3] rule is an iterative method for approximating the projection weight matrix. The algorithm is guaranteed to find an appropriate weight matrix within a finite number of presentations of each pattern if such a matrix exists.
The minimum number of presentations of the training set to perform, P, is calculated as being the smallest integer conforming to:

$$P \geq \log_k \left(\frac{N}{(1-T)^2} \right)$$

where k and T are real valued constants such that $1 < k \leq 4$ and $0 \leq T < 1$. k is referred to as the *memory coefficient* of the network; the larger it is, the fewer steps are required to train the network. In this work, k=4 and T=0.5 for all networks trained by this rule.

The algorithm is as follows:

BEGINNING WITH A ZERO WEIGHT MATRIX
FOR EACH PATTERN IN TURN
APPLY THE PATTERN ONTO THE NETWORK
FOR $m := 1$ TO P
FOR EACH PROCESSING ELEMENT IN TURN
UPDATE INCOMING WEIGHTS ... $\Delta w_{ij} = \left(\frac{k^{m-1}}{N}\right)\left(\xi_i^p - h_i\right)\left(\xi_j^p - h_j\right)$
REMOVE ALL SELF-CONNECTIONS

Note: patterns are added *incrementally*.

Symmetric Local Learning (SLL)

Gardner [9] pointed out that an iterative perceptron-like training rule could be made to produce symmetric weights by simply updating both w_{ij} and w_{ji} when either changes. Gardner also showed that such algorithms would find a symmetric weight matrix, if one existed, for a particular training set. The SLL algorithm is:

BEGIN WITH A ZERO WEIGHT MATRIX
REPEAT UNTIL ALL LOCAL FIELDS ARE CORRECT
SET THE STATE OF NETWORK TO ONE OF THE ξ^p
FOR EACH UNIT, i, IN TURN
IF $h_i^p \xi_i^p$ IS LESS THAN T THEN
UPDATE WEIGHTS ON CONNECTIONS INTO AND OUT OF UNIT i :

$$\Delta w_{ij} = \Delta w_{ji} = \frac{\xi_i^p \xi_j^p}{N}$$

OTHERWISE DO NOTHING

This is a symmetric version of perceptron learning, with a fixed margin, T, and learning rate $1/N$. We call T the *learning threshold* for the network. Since a set of patterns is stable when the aligned local fields of those patterns are all non-negative, we could set T to zero. However, based on previous results [6], we choose T=10 in order to achieve a better attractor performance for the networks.

Training Sets

Throughout this work we employ training sets made up of pseudo-random training vectors. An uncorrelated training set is one in which the patterns are *completely* random. Correlation can be increased by varying the probability that a given bit in a training pattern is +1 (or –1). We refer to the probability of any bit being +1 in each training vector as the *bias*, b, on the training set, so $\forall i,p \bullet prob\ (\xi_i^p = +1) = b$. Thus, a bias of 0.5 corresponds to an uncorrelated training set and a bias of 1 corresponds to a completely correlated one (as does a bias of 0).

Weight Dilution

We present two approaches to weight dilution. In the first, pairs of units are chosen at random and, if the units are connected, *both* connections between them are removed, until the correct proportion of connections has been removed from the network. By removing both connections between units we ensure that the weight matrix remains symmetrical.

The second approach involves using an heuristic to select the connections to be removed, in an attempt to ensure that the most efficacious connections are retained [2,4,5]. In this case, the connection pair with the weight of least magnitude is identified, and both connections in the pair are removed from the network. This is repeated until the required number of connections have been eliminated.

3 Analysing Performance

Two aspects of network performance were measured: *pattern stability* and *attractor performance*. A series of experiments were carried out on networks of size N=100 using training sets of bias 0.5 and 0.9 and at a fixed loading of α=0.50. Networks were trained using either the BV rule, or the SLL rule. The connections in the networks were then diluted according to the methods described above.

Pattern Stability

The proportion of fundamental memories that are stable after dilution provides an indicator of the robustness of a model. Networks were trained to below maximum capacity, so all training patterns were fundamental memories prior to dilution. Figs 1 and 2 show the *proportion* of stable training patterns at various dilutions.

Attractor Performance

For an associative memory to be effective, the training patterns should be both stable states of the network and attractors in its state space.

We use R, the normalized mean radius of the basins of attraction [11], as a measure of attractor performance. It is defined as:

$$R = \left\langle\left\langle \frac{1-m_0}{1-m_1} \right\rangle\right\rangle$$

where m_0 is the minimum overlap an initial state must have with a fundamental memory for the network to converge on that fundamental memory, and m_1 is the largest overlap of the initial state with the rest of the fundamental memories. The angled braces denote an average over sets of training patterns. Details of the algorithm used can be found in [11].

4 Results

Figs 1 and 2 below show the effect of varying the degree of weight dilution within the network on the proportion of trained patterns that are stable, whilst figs 3 and 4 show the effect of varying the degree of weight dilution on attractor performance (N=100, α=0.50 throughout).

In each figure, the left hand graph shows performance of networks trained with uncorrelated patterns (b=0.5), whilst the right hand graph shows performance of networks trained with correlated patterns (b=0.9). The upper line on each of the eight graphs represents the results obtained using informed dilution.

Pattern Stability

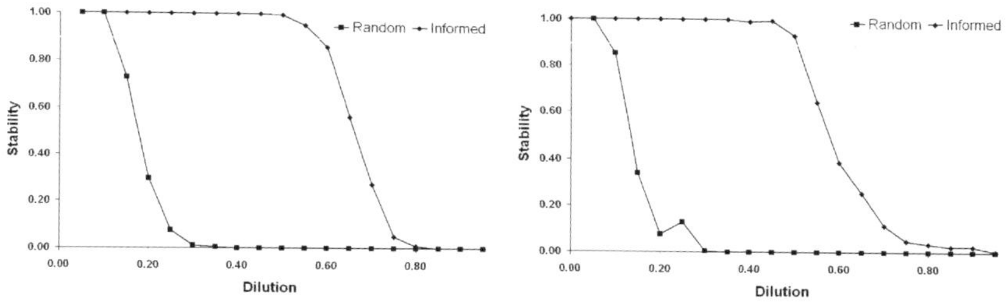

Fig 1. Pattern stability in networks trained with the BV rule.
Left-hand graph: bias=0.5. Right-hand graph: bias = 0.9.

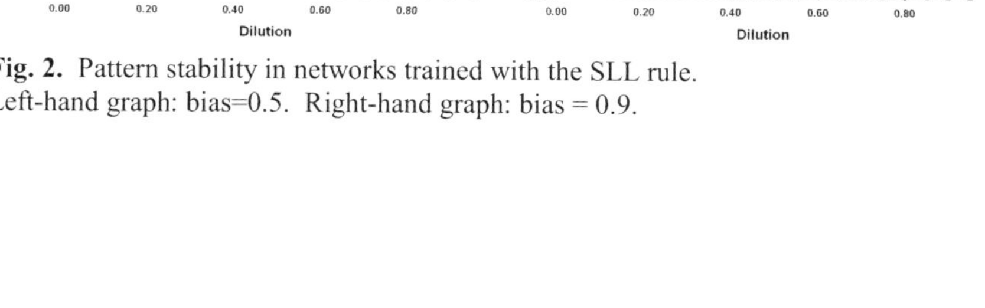

Fig. 2. Pattern stability in networks trained with the SLL rule.
Left-hand graph: bias=0.5. Right-hand graph: bias = 0.9.

Attractor performance

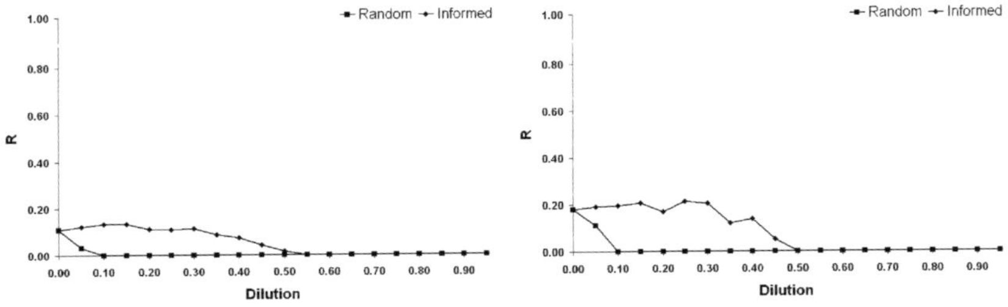

Fig. 3. Attractor performance in networks trained with the BV rule.
Left-hand graph: bias=0.5. Right-hand graph: bias = 0.9.

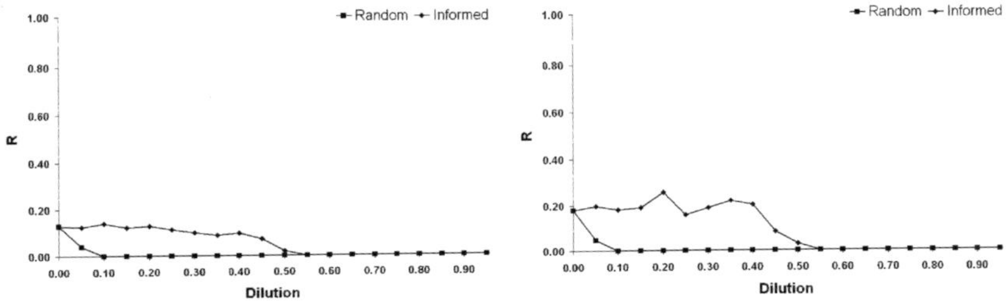

Fig. 4. Attractor performance in networks trained with the SLL rule.
Left-hand graph: bias=0.5. Right-hand graph: bias = 0.9.

5 Discussion

Observations on Pattern Stability

We make four key observations here:

1) Informed dilution gives a significant improvement in pattern stability over simple random dilution. This takes the form of an increase in the level of dilution at which the networks retain memory of all the trained patterns.
2) It is possible to remove around 50-60% of the networks' connections without a serious decline in the stability of the trained patterns.
3) The bias in the training set makes very little difference to the pattern stability.
4) The learning rule used appears to make little difference to the effect of dilution on pattern stability.

Observations on Attractor Performance

The pattern here is similar to that for pattern stability. Specifically:

1) Informed dilution performs significantly better than simple random dilution.
2) It is possible to remove up to approximately 40% of the networks' connectivity without serious damage to the attractor performance of the network.
3) The bias in the training set makes little difference to the attractor performance.
4) The learning rule used appears to make little difference to the effect of dilution on attractor performance.

Conclusions

This paper reports two important results:
1) Informed dilution is markedly better than random dilution.
2) Informed dilution demonstrates that a large number of connections are redundant in networks of this type and at these loadings.

As the loading of these networks is α=0.5 they are below their maximum storage capacity; it may be of interest to repeat these experiments at higher loadings where the networks may be under greater stress.

Failure, when it occurs, proceeds with great rapidity. There is a sharp decrease in both proportion of stable patterns and attractor performance once the networks begin to lose their stability and ability to act as attractors. In this respect, our results differ from those of Sompolinksy, whose work on randomly diluting the traditional Hopfield network [13] showed a linear decline in pattern stability.

The system of informed dilution we have presented is very simple; no retraining of the network is required. It is possible that in biological systems complex strategies may be similarly unnecessary. Chechik *et al* [5] have noted that during brain maturation there is a reduction in connectivity that is expensive to maintain from an energy perspective. It is interesting that we have also found redundancy in connectivity, albeit in a much simpler system.

Informed dilution, as implemented in this work, is functionally equivalent to the system of *annealed dilution* proposed by Bouten *et al.* [4] in which the dilution is performed as part of the learning process. Our results concur with their prediction that 60% dilution is the approximate limit beyond which network capacity is compromised.

The work presented here concentrates on the dilution of *fully connected* networks, whereas our current work focuses on networks that have been created as sparsely-connected *tabula rasa*. Training these networks has presented new challenges and performance characteristics. We expect to be able to present these new findings in the near future.

References

[1] Abbott, L.F (1990) Learning in neural network memories. *Network: Computational Neural Systems*, **1**, 105-122

[2] Barbato, D.M.L. and O.Kinouchi (2000) Optimal pruning in neural networks. *Physical Review E* **62**(6), 8387-8394

[3] Blatt, M.G. and E.G.Vergini (1991) Neural networks: a local learning prescription for arbitrary correlated patterns. *Physical Review Letters* **66**(13), 1793-1797

[4] Bouten, M., A.Engel, A.Komoda, and R.Serneels (1990) Quenched versus annealed dilution in neural networks. *Journal of Physics A*, **23**, 4643-4657

[5] Chechik, G., I.Meilijson and E.Ruppin (1998) Synaptic Pruning in Development: A Computational Account, *Neural Computation* **10**, 1759-1777

[6] Davey, N., R.G.Adams, and S.P.Hunt. High Performance Associative Memory Models and Symmetric Connections (2000) *Proceedings of the International ICSC Congress on Intelligent Systems and Applications (ISA'2000): Symposium on Computational Intelligence (CI'2000)*, **2**, 326-331

[7] Davey,N. and S.P.Hunt (2000) A Comparative Analysis of High Performance Associative Memory Models. *Proceedings of 2nd International ICSC Symposium on Neural Computation (NC 2000)* 55-61

[8] Diederich,S. and M.Opper (1987) Learning of Correlated Patterns in Spin-Glass Networks by Local Learning Rules. *Physical Review Letters*, **58**, 949-952

[9] Gardner,E., H.Gutfreund and I.Yekutieli (1989) The Phase Space of Interactions in Neural Networks with definite Symmetry, *Journal of Physics A*, **22**, 1995-2008

[10] Hopfield, J.J (1982) Neural networks and physical systems with emergent collective computational abilities. *Proceedings of the National Academy of Sciences (USA)*, **79**, 2554-2558

[11] Kanter, I. and H. Sompolinsky (1987) Associative Recall of Memory Without Errors. *Physical Review A*, **35**(1), 380-392

[12] López,B. and W. Kinzel (1997) Learning by dilution in a neural network. *Journal of Physics A*, **30** 7753-7764

[13] Sompolinsky, H. (1986) Neural Networks with nonlinear synapses and a static noise, *Physics Review* A **34**, L519-L523

[14] Turvey, S.P., S.P.Hunt, N.Davey and R.J.Frank (2001) An experimental assessment of the performance of several associative memory models. *Proceedings of the 5th International Conference on Artificial Neural Networks and Genetic Algorithms (ICANNGA 2001)*, 70-73

Neural Network Analysis of Estimation of Data

Endang Triastuti[*], Robin Crockett [*], Phil Picton[*], and Alasdair Crockett[**]

[*] School of Technology & Design, University College Northampton, Northampton, NN2 6JD, UK
[**] UK Data Archive, University of Essex, Colchester, CO4 3SQ, UK

Abstract. A set of data may be 'coarsened' as a result of enumerators' or compilers' efforts to estimate (or falsify) observations. This type of coarsening typically results in excesses of 'convenient' numbers in the data sets, such as multiples of 5 or 10 in decimal number systems, apparent as patterns of periodic unit-width spikes in the frequency distributions.
We report on the development of novel Radial Basis Function neural-network techniques for detecting numerical data coarsened by rounding/estimation (or falsification) and quantifying that rounding/estimation. The objective is to provide an alternative to conventional statistical approaches based on missing-data techniques: data coarsened thus are not actually missing, solely shifted in size. The results show that the neural networks can successfully detect and classify the coarsening in data-sets and, hence, yield insights into the ways in which people count when performing enumeration or other numerical data-compilation exercises.

Keywords. Data quality, data coarsening, missing data, neural networks, radial basis functions.

1 Introduction

This research is being carried out at the University College Northampton in collaboration with the University of Essex. It is a development of an ongoing research programme initiated in response to problems encountered in the analysis of data from the Religious Census of 1851 for England and Wales [2,3]. The initial work resulted in a new model for analysing rounding and estimation present in the data-sets but, whilst sufficient for the original purpose, the constraints of this analysis indicated that new techniques would be advantageous.

The general nature of a data-set coarsened by rounding is apparent from the frequency distribution (of congregation-sizes in this instance) shown in Figure 1. Instead of the expected reasonably smooth curve, indicating a 'predictable' distribution of observations, this frequency distribution shows excesses of observations at distinct, well-defined sizes which are periodic with intervals (in this instance) of 5, 10, 20, 50 and 100. The periodic nature of the coarsening can be confirmed by autocorrelation techniques.

This particular type of periodic structure arises from estimation: certain of the original enumerators did not count exactly but estimated the congregation-sizes. When estimating thus, the returned value is a 'round' number, i.e. a convenient multiple of a base-unit, with the closeness of the returned value to real observed value being dependent on a variety of factors including the observation size, as well as the ability and diligence of individual enumerators. The base-units are generally convenient numbers within the number system being used, such as 5 (counting 'by fives') and 10 (counting 'by tens') in decimal number systems, or 6 (half-dozens) and 12 (dozens) in duodecimal systems.

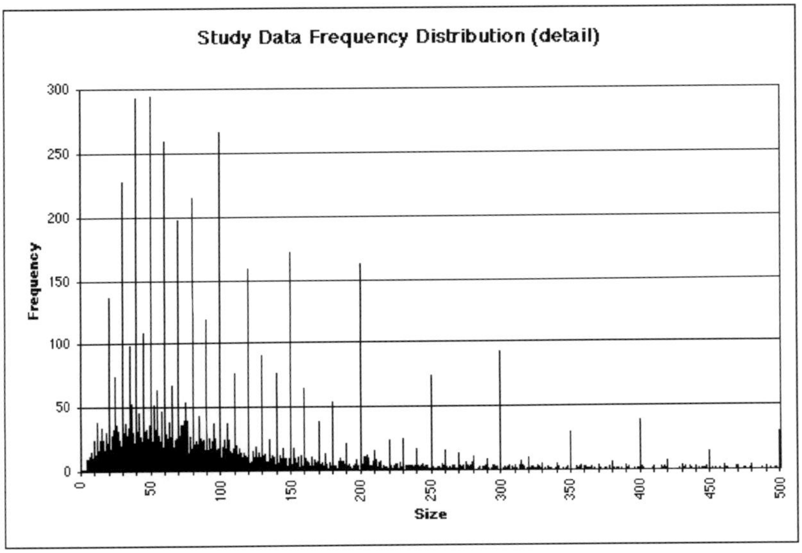

Fig. 1. Study Data Frequency Distribution

Conventionally, analysis of such coarsening utilises statistical methods developed originally for missing-data problems [4,5]. Data coarsened by estimation are not missing, solely shifted in size as a result of the rounding inherent in the estimation process. Thus, application of missing-data techniques, which seek to replace missing data according to probability distributions of sizes, are liable to distort coarsened data by effectively duplicating the shifted observations at probabilistically determined 'real' sizes. Therefore, techniques which analyse the shifts in size, such as those under development by the authors, potentially avoid introducing the distortions to which missing-data oriented techniques can be susceptible, as well as providing information regarding the counting behaviour.

Neural networks have been shown to be universal function approximators in that they can approximate any function to an arbitrary degree of accuracy; thus, in theory at least, should be capable of fulfilling the user's hope on finding the underlying periodic structure(s) in the data [6]. This property of neural networks, particularly Radial Basis Function (RBF) networks potentially makes them unbiased modelling tools capable of detecting the presence of rounded data [8,9]. In essence, the neural networks recognise the periodic 'patterns' which such coarsening creates in the frequency distributions.

2 The Current Model: The Benchmark

Without other information, it is not possible to distinguish an exact 'round' observation from an estimated one: for example, an observation (count) of size 65 could be exact or could represent underlying observations of size 63 or 68 counted 'by fives'; or one of size 70 could be exact or could represent underlying observations of size 68 or 73 counted 'by tens' or 'by fives'. However, it is possible to estimate the excesses of 'round' observations and the corresponding deficits of 'non-round' observations resulting from estimation processes and, within this, determine the relative importance of, for example, base-5 and base-10 estimation processes.

Our current model of such data-sets assumes that the enumerators can be divided into groups according to the base-unit used for estimation: to continue with the foregoing example, the base-5 and base-10 subsets comprise all observations counted 'by fives' and 'by tens' respectively. Thus, for the data summarised in Figure 1, there are base-5, base-10, base-20, base-50 and base-100 estimation subsets and a base-1 exact-count subset (1 being the effective base-unit of exact counts). Whilst it is not possible to distinguish individual observations within the whole data-set as being exact or rounded/estimated to particular base-units, we can determine probabilistically how many observations are contained within each subset [1]. For the study data, the subsets are: base-1, 52%; base-5, 12%; base-10, 20%; base-20, 6%; base-50, 7%; base-100, 3%.

The base-units are determined by 'modulo-testing' the frequency distribution, a procedure entailing investigation of the numbers of observations exactly divisible by given base-numbers (integers). The base-number results for the data summarised in Figure 1 are shown in Table I. These results are then analysed statistically both in relation to the expectation and for (sub-) multiple effects in order to determine the actual estimation base-units: these being the base-numbers whose multiples appear statistically significantly more frequently in the data than expected, independently of (sub-) multiple effects.

The accuracy of the model is dependent upon the accuracy to which the probability distributions of the various estimation behaviours can be determined. The details are data-set dependent but, in general, the subset probability distributions for the small base-units, such as base-1 (exact counts), base-5 and base-10, closely follow the underlying frequency distribution. The probability distributions of larger base-units, however, show more variation due to variation of counting/estimation behaviour with size of observation.

3 Radial Basis Function Neural Network Model

A RBF neural network consists of input, hidden and output layers, each layer containing nodes which are connected to nodes in adjacent layers and each node being a processing unit containing weight and summation functions. The input and hidden layers use radial basis and linear transfer functions respectively.

A RBF network is a special case of regularisation network. The method solves an interpolation problem by constructing a set of linear equations of the basis function. This method constructs a linear function space which depends on the positions of the known data points according to an arbitrary distance, *d*:

$$d = \sum_{i=1}^{n} \left| x_i - w_{ji} \right| b \tag{1}$$

where, d is the distance between input nodes and weights.
x_i are the input nodes.
w_{ji} are the weights in layer j, transpose of x_i in this case.
b is the bias, equal to 1 in unbiased layers.

The method uses the Gaussian basis function (normal distribution) [1]:

$$f(d) = e^{-d^2/2\sigma^2} \tag{2}$$

where, $f(d)$ is the (radial) basis function.
σ is a parameter specifying the width of the basis function.

The hidden layer computes a linear transfer function and then feed into a competitive function which selects the largest y_i-value to yield the classification (the most significant base-number).

$$y_j = \sum_{i=1}^{n} x_i w_{ji} \tag{3}$$

where; y_j are the output nodes in layer j (hidden layer).

4 RBF Neural Network Diagnosis of Coarsening

In these early stages of the research, the neural network modelling has been performed using Matlab and the Neural Network Toolbox. Although both function approximation and classification have been used in the research [7], the model now solely uses classification. As the work proceeds and the model is refined, a custom neural network, optimised for this type of analysis is being progressively developed.

A pair of training-data input and classification output is required for each base-number (i.e. potential estimation subset base-unit) or combination of base-numbers being investigated. The basic training data-sets are 'dummy' frequency distributions comprising appropriate patterns of 0s and 1s (e.g. 1-1-1-1... for base-1, 0-1-0-1... for base-2, 0-0-1-0-0-1... for base-3, etc.) shaped and scaled according to the underlying probability distribution of the data-under-investigation. The target outputs are classes corresponding to the base-numbers under investigation.

The classification process is iterative, the iterations progressively identifying (potential) estimation base-units.

- First iteration. Each potential estimation base-unit has an individual training data-set corresponding to it. A ranking of base-numbers is produced: the pattern associated with the top-ranked base-number is being recognised as being the most significant pattern in the frequency distribution at this stage.
- Second iteration. The highest ranked base-number (from the previous iteration) is considered in combination with each of the other base-numbers: each training data-set comprises a combination of two training data-sets from the first iteration. A refined ranking of base-numbers is produced: the pattern associated with the top-ranked pair of base-numbers is being recognised as being the most significant pattern in the frequency distribution at this stage.
- Third iteration. The highest ranked pair of base-numbers (from the previous iteration) is considered in combination with each of the other base-numbers: each training data-set comprises a combination of three training data-sets from the first iteration. A further refined ranking of base-numbers is produced: the pattern associated with the top-ranked triplet of base-numbers is being recognised as being the most significant pattern in the frequency distribution at this stage.
- Fourth and higher iterations. The process is then repeated, adding one base-number to the combination per iteration, until the probability of the top-ranked combination ceases to increase from one iteration to the next.

The hypothesis is that at each iteration, the combination of base-units which best fits the real frequency distribution is selected and that, whilst the probability of the top-ranked combination continues to increase with each successive iteration, the fit is improving. Once this probability stops increasing, the goodness of fit stops improving. Thus, the combination which yields the highest probability should contain all the estimation base-units plus, possibly, some base-numbers which are (sub-) multiples or multiplicative combinations of them. Further analysis is then required to determine which of the base-numbers thus identified are 'real' subset base-units and which are merely (sub-) multiples of them: resolving these (sub-) multiple effects is one of the objectives of the ongoing research.

5 Results

The probability assigned to each base-number or combination thereof by the neural network — and by which these are classified and ranked — is the probability that the pattern associated with that base-number (or combination) would be selected as being the frequency distribution of the target data. Thus, this probability is a measure of significance of a base-number (or combination) within the data. In the first iteration (with individual base-numbers) all base-numbers from 1 to 100 are used but for subsequent iterations it has proved sufficient to use those base-numbers ranking higher than (the higher of) base-11 or base-13 in the first iteration. This criterion is empirical: unless there is other information or it is clearly apparent from the first iteration to the contrary, it is highly improbable that numbers as 'inconvenient' as base-11 or base-13, the first two prime numbers greater than ten, would form the basis of estimation processes (baker's dozens notwithstanding) and so it is even less probable base-numbers ranking below these would form the basis of estimation processes.

The results, for the data summarised in Figure 1, are presented in Tables 1 and 2. Table 1 shows the results of the first iteration and also shows the results yielded by the modulo test (the basis of the current model) as a *de facto* benchmark. The results are restricted to the highest ranked twenty-six in accordance with the criterion described above; base-13 ranks 27 and base-11 ranks 34 in this case. There are some differences between the ranking produced by neural network that produced by the modulo test. However, it is clear that the neural network produces rankings similar to that of the modulo test, indicating that the neural network can both 'see' and identify the patterns which result from coarsening. As described previously [7], when there are no patterns (i.e. no periodic structure) in the data set, as is the case with random data, the neural network produces the ranking which would be expected, e.g. more multiples of base-1 than of base-2, more multiples of base-2 than of base-3, etc.

Table 2 shows the top-ranked combinations of base-numbers and their probabilities. The increasing trend in probability breaks after the ninth iteration, yielding the best-fit combination (in descending order) of base-1, base-10, base-5, base-20, base-50, base-100, base-30, base-40 and base-70. Up to the sixth iteration (base-100), the neural network recognises the estimation base-units determined by modulo-testing with no spurious results. The probability, however, continues to increase for a further three iterations, additionally indicating base-30, base-40 and base-70, results which differ from the modulo-test benchmark. These reasons for this are not fully understood at present but are due, at least in part, to interactions between base-numbers and how these affect the patterns actually being 'seen' by the neural network.

6 Conclusions

The techniques developed here seek to complement conventional statistical approaches by detecting and quantifying the levels of estimation (and/or falsification) present in numerical data-sets. The results obtained so far indicate that suitably configured RBF neural networks can detect the presence of estimation in data-sets via recognising the coarsening patterns it produces. This is an important step towards being able to use neural networks to identify and model estimation (and/or falsification) behaviour present in the compilation of numerical data-sets.

The results also show that the neural networks can identify and rank base-numbers, i.e. potential subset base-units, according to their significance producing results very similar to the modulo-test benchmark.

The neural network, however, indicates further base-units in comparison to the benchmark and work is ongoing to determine the reasons for this and, hence, possible modifications to the method to eliminate these 'false positives'. It should be noted, however, that if the subset model is applied to all nine base-units indicated by the neural network, then the three spurious base-units (base-30, base-40 and base-70) would yield subsets of zero size and so their elimination would be possible at this stage.

Table 1. Neural Network and Modulo Test Rankings of Base-Numbers

Rank	Modulo Test		Neural Network	
	base	prob.	base	prob.
1	1	1.0000	1	0.5576
2	2	0.7018	5	0.5475
3	5	0.5833	2	0.5324
4	10	0.4734	10	0.5189
5	4	0.3868	4	0.4652
6	3	0.3452	20	0.4590
7	20	0.2710	3	0.4446
8	6	0.2427	15	0.4410
9	15	0.2022	6	0.4371
10	8	0.1934	30	0.4323
11	25	0.1905	25	0.4292
12	50	0.1667	8	0.4288
13	30	0.1622	40	0.4266

Rank	Modulo Test		Neural Network	
	base	prob.	base	prob.
14	40	0.1366	50	0.4245
15	7	0.1284	12	0.4149
16	12	0.1269	60	0.4119
17	9	0.0993	7	0.4103
18	100	0.0847	35	0.4093
19	60	0.0833	9	0.4067
20	14	0.0786	14	0.4050
21	16	0.0767	100	0.4044
22	35	0.0684	70	0.4042
23	11	0.0588	16	0.4041
24	13	0.0584	45	0.4038
25	18	0.0572	80	0.4036
26	75	0.0493	18	0.4009

Table 2. Neural Network Identification of Base-Units

Iter.	Bases (least significant to right)	Prob.
1	b1	0.5576
2	b1, b10	0.6847
3	b1, b10, b5	0.7713
4	b1, b10, b5, b20	0.8080
5	b1, b10, b5, b20, b50	0.8349
6	b1, b10, b5, b20, b50, b100	0.8439
7	b1, b10, b5, b20, b50, b100, b30	0.8529
8	b1, b10, b5, b20, b50, b100, b30, b40	0.8596
9	b1, b10, b5, b20, b50, b100, b30, b40, b70	0.8618

References

[1] Chen S, Cowan CFN, Grant PM (1991) Orthogonal Least Squares Learning Algorithm for Radial Basis Function Networks, *IEEE Transactions on Neural Networks*, 2, 2, March 1991, pp 302-309.

[2] Crockett RGM, Crockett AC (1999) Historical Sources: How People Counted. A Method for Estimating the Rounding of Numbers, *History and Computing*, 9, 1/2/3, 1999, pp 43-57.

[3] Crockett RGM, Crockett AC (2000) People and counting: how people count in enumeration exercises, *3rd European Social Science and History Conference*, Vrije Universiteit, Amsterdam, April 2000.

[4] Heitjan DF, Rubin DB (1990) Inference from Coarse Data via Multiple Imputation with Application to Age Heaping, *Journal of the American Statistical Association*, 85, 410, 1990, pp 304-314.

[5] Heitjan DF, Rubin DB (1991) Ignorability and Coarse Data, *Annals of Statistics*, 19, 4, 1991, pp 105-117.

[6] Polhill JG, Weir MK (2001) An Approach to Guaranteeing Generalisation On Neural Networks, *Pergamon, Neural Networks* 14 (2001), pp 1035-1048.

[7] Triastuti E, Crockett RGM, Picton PD, Crockett AC (2002) Neural Network Analysis of Estimated Data, *Proc. Eunite 2002*, Albufeira, Portugal. 19-21 September 2002; pp 161-166.

[8] Turner SJ, Crockett RGM, Picton PD, Triastuti E (2001) Genetic Algorithms for Simulating Counting Behaviour, *19th Biennial Conference on Numerical Analysis*, University of Dundee, 26-29 June 2001.

[9] Turner SJ, Triastuti E, Crockett RGM, Picton PD, Crockett AC (2002) Intelligent Techniques for Detecting Estimated and Falsified Data, *Proc Sixth Multi-Conference on Systemics, Cybernetics and Informatics*. Orlando, Florida, USA. 14-18 July 2002, pp 445-450.

A Compiler and Simulator for Partial Recursive Functions over Neural Networks

João Pedro Neto [*], José Félix Costa [**], Paulo Carreira [***] and Miguel Rosa [+]

[*] Dept. de Informática, Faculdade de Ciências,Lisboa, PORTUGAL
email: jpn@di.fc.ul.pt
[**] Dept. de Matemática, Instituto Superior Técnico, Lisboa, PORTUGAL
email: fgc@math.ist.utl.pt
[***] Oblog Consulting S.A. Rua da Barruncheira, Carnaxide, PORTUGAL
email: paulo.carreira@oblog.pt
[+] email: mar@fccn.pt

1 Introduction

The field of Artificial Recurrent Neural Networks (ARNNs), mainly in the last two decades, was able to solve engineering problems while keeping the simplicity of the underlying principles that allow them to mimic their biological counterparts. All this attracts people from many different fields such as Neurophysiology and Computer Science. We introduce our subject from a Computer Science perspective: the ARNN is seen as a computing mechanism able to perform computation based on a program coded as a specific arrangement of neurons and synapses. This work implements a compiler and a simulator based on [4]. In [3,7,5] similar ideas are presented but they are based on higher-level languages. We start by presenting the underlying theoretical context on which this work is based. In section 2 we give a brief review of the concept of partial recursive function. In section 3 we present our approach for building neural networks from partial recursive functions. The explanation of how we adapted the design of [4] into a compiler is given in section 4. Section 5 refers to the simulator and usage examples and section 6 concludes this paper. The simulator is freely available at *www.di.fc.ul.pt/~jpn/ netdef/nwb.html.*

2 Partial Recursive Functions

The language that we will use to express computation is the one of partial recursive functions (PRF). Although primitive when compared to modern computer languages, it is simple and powerful enough to program any mechanism with the

same computing power as a Turing machine. Surely, building complex programs with this language would be very difficult, and more appropriate languages exist. For our purposes however, this language is well suited. The PRF theory identifies the set of computable functions with the set of partial recursive functions. We shall use $a(x_1,\ldots,x_n) \equiv b(x_1, \ldots, x_n)$ to denote that for all $x_1,\ldots,x_n$, $a(x_1,\ldots,x_n)$ and $b(x_1,\ldots,x_n)$ are both defined and coincide or are both undefined.
The axioms – also called primitive functions – are:

- **W** that denotes the *zero-ary constant* 0;
- **S** that denotes the *unary successor function* $S(x)=x+1$;
- **U(i,n)** that for i and n fixed, $1 \le i \le n$, denotes the *projection function* $U_{i,n}(x_1, \ldots, x_n) = x_i$.

The construction rules are:

- **C**, denoting *composition*. If $f_1, \ldots, f_k$ are n-ary PRFs, and g is a k-ary PRF, then the function h defined by composition, $h(x_1, \ldots, x_n) \equiv g(f_1(x_1, \ldots, x_n), \ldots, f_k(x_1, \ldots, x_n))$, is a PRF;
- **R,** denoting *recursion*. If f is a n-ary PRF and g is a (n+2)-ary PRF, then the unique (n+1)-ary function h, defined by

  1) $h(x_1,\ldots, x_n, 0) \equiv f(x_1,\ldots,x_n)$ and
  2) $h(x_1,\ldots, x_n, y+1) \equiv g(x_1,\ldots,x_n, y, h(x_1,\ldots, x_n, y))$;

  is a PRF;
- **M,** denoting *minimization*. If f is a (n+1)-ary PRF, then $h(x_1,\ldots, x_n) \equiv \mu_y(f(x_1,\ldots, x_n, y)=0)$ is also a PRF, where $\mu_y(f(x_1,\ldots, x_n, y)=0) =$

Error!

For instance, $f(x,y)=x+1$ is a PRF and is described by the expression `C(U(1,2),S)`. The function $f(x,y)=x+y$ is also a PRF described by the expression `R(U(1,1), C(U(3,3),S)`. In fact, it can be shown that every Turing computable function is a PRF. Details on PRF theory are found in [1,2].

3 Coding PRF into ARNNs

Finding a systematic way of generating ARNNs from given descriptions of PRF's greatly simplifies the task of producing neural nets to perform certain specific tasks. It also provides a proof that neural nets can effectively compute all Turing computable functions (cf. [4]). In this section we briefly describe the computing rule of each processing element, i.e., each neuron. Furthermore, we present the coding strategy of natural numbers that is used to load the network. Finally, we will see how to code a PRF into an ARNN.

3.1 How do the processing elements work?

We make use of σ-processors, just as in [4]. In each instant t each neuron j updates its activity x_j in the following non-linear way:

$$x_j(t+1) = \sigma\ (\ \textbf{Error!} + \textbf{Error!} + c_j\ ) \qquad \sigma(x) = \textbf{Error!}$$

where a_{ji}, b_{jk}, and c_j are rational weights; N is the number of neurons, M the number of input streams u_k; and σ is the continuous function defined above.

3.2 How do we represent the numbers?

We use an unary representation where each natural number is represented as a rational number by means of a mapping α where, for each n, α(n) is given by

$$\textbf{Error!}10^{-i-1}$$

3.3 How to build the ARNN?

The following three net schemata were implemented to compute the corresponding three axioms of recursive function theory. Changes were made with respect to [4]. First, the W axiom is provided with two additional neurons. Second, each U(i,n) axiom is constructed with only five neurons making it more efficient.

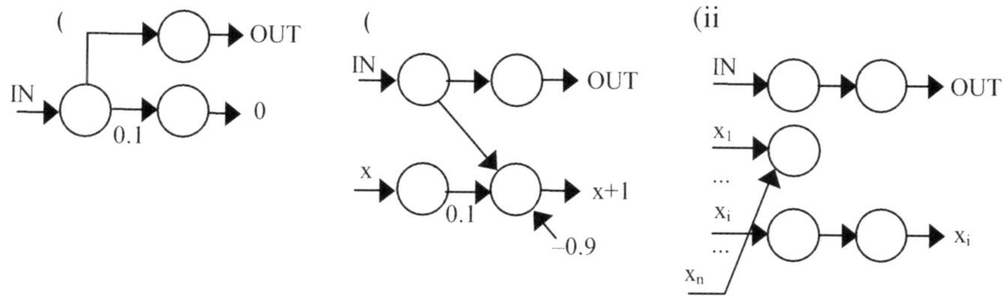

Fig. 1. Axioms (i) W; (ii) S; (iii) U(i,n).

The net schemata of figures 3, 4 and 5 (see appendix) illustrate these rules, where gray colored circles represent repeated neurons. The Sg box is the subnet that finds whether a number is positive or zero.

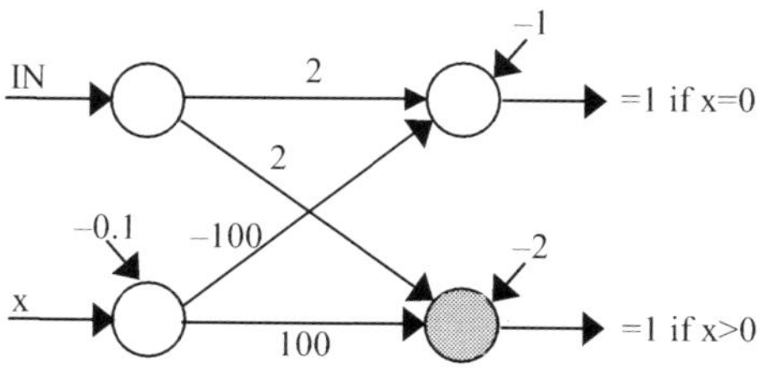

Fig. 2. The signal (Sg) network.

For each PRF expression an ARNN is generated using the structure of the expression. The construction of the ARNNs is made in a top-down fashion, beginning with the outermost ARNN and then continuously instantiating ARNNs until reaching the axioms. For an expression `R(U(1,1),C(S,U(3,3))` we would first build the network for the recursion, then instantiate it with the projection axiom network and with the composition rule network, that in turn would accommodate the successor and projection axiom networks. This instantiation mechanism for the network schemata consists of replacing the boxes by compatible networks. A box is said to be compatible with a network schemata if both have the same number of inputs and outputs. The substitution operation of a box by network schemata consists of connecting the box inputs (outputs) to the network inputs (outputs).

4 Putting the pieces together

The tool we want to present to the user should be capable of handling more complex functions than the simple examples used in the previous sections. Specifying a complex function implies writing a complex expression. This motivates a modular and yet simple language to increase readability.

A PRF description is a sequence of statements where each statement is a pair that contains an identifier and an expression. Each identifier labels an expression. Each expression is written in the language described in section 2 and may refer to a previous statement. For example, when specifying the product PRF we can initially write the statement for the sum PRF and follow it by a shortened product PRF expression in which the sum expression is substituted by the initial `sum` identifier as shown below (so, expressions make use of those defined previously).

```
sum R(U(1,1), C(U(3,3), S))
product R(Z, C(U(3,3), U(1,3), sum))
```

5 Using the simulator

An application was built to provide the user not only with a compiler of PRF into ARNNs but also with a simulator that, among other things, allows step-by-step simulation and inspection of groups of neurons. In our application there are two ways of building ARNNs: by compiling the PRFs or by reading the dynamic equations system that define the neural network. The simulator takes the already built ARNN and configures it to read the input. Then it requests each neuron to process its input data. This process is iterated until a signal is seen in the OUT line of the outmost network. During the ARNN computation, the program allows the user to inspect the computations that are taking place inside the net. The user can also see the corresponding equation of each neuron and define a limit for the iteration number. Each neuron has a name given by the compiler with some special meaning. The name of each neuron expresses information about itself:

- *Func* is one of W, S, U(i,j), R, C, M.

• *Type* is one or more of the following: *In*, stands for the IN line supporting neurons; *Out*, stands for the OUT line supporting neurons; *Data*, stands for neurons that receive parameters from external networks or that send result of computation to other networks; and *Res* stands for neurons that handle results of functions; *Hid*, stands for neurons not in a border layer.
• *Num* is the number of the neuron inside its network schemata.
• *Depth* is the same as the depth in the corresponding PRF parsing tree.
• *Id* is used to ensure an unique identifier for each neuron.

Some examples of function descriptions are given in the table below:

Table 1. Some PRF descriptions

0Function	Expression
f=0	`W`
f(x)=x+1	`S`
f(x,y,z)=y	`U(2,3)`
f(x)=0	`R(W, U(2,2))` → *`also called Z`*
f(x,y,z)=z+1	`C(U(3,3),S)`
f(x,y)=x+y	`R(U(1,1),z+1)`
f(x,y)=x*y	`R(Z,C(U(3,3),U(1,3),x+y))`
f(x)=sg(x)	`R(C(W,S),C(W,U(1,2)))`
f(x,y)=x-1	`R(W,U(1,2))`
f(x,y)=x-y	`R(U(1,1),C(U(3,3),x-1))`
f(x,y)=|x-y|	`C(x-y,C(U(2,2),U(1,2),x-y),x+y)`

As an example, let us consider the following description of the *sum* function:

```
proj/1  U(1,1)
proj/2  U(3,3)
comp    C(proj/2,S)
sum     R(proj/1,comp)
```

Notice that this function is not limited by any integer value (there is no MAXINT), since the rational precision is not limited (it is, however, finite). After the compilation of these four statements we have an ARNN with 39 neurons and 70 synapses. Below we present the first lines of the dynamic equation system:

XRin_1_0(i+1) = σ(Ein(i))
XRhid2_1_1(i+1) = σ(XRin_1_0(i))
XRhid3_1_2(i+1) = σ(0.1*XRin_1_0(i) + XRhid3_1_2(i) + XRhid4_1_3(i) - XRhid5_1_4(i) - XREout_1_15(i))
XRhid4_1_3(i+1) = σ(0.1*XRhid3_1_2(i) + XRhid5_1_4(i) - 0.9)
XRhid5_1_4(i+1) = σ(XRhid6_1_5(i))
XRhid6_1_5(i+1) = σ(XSout_4_7(i))
XRhid7_1_6(i+1) = σ(XRhid3_1_2(i) + XRhid15_1_14(i) - 1)
. . .

The simulator is freely available at *www.di.fc.ul.pt/~jpn/netdef/nwb.html.*

6 Conclusion

We hope to have taken one step further in understanding the relationship between programming languages and ARNNs. We started presenting a theoretical framework of a very simple programming language. Next, we described a systematic way of obtaining ARNNs from programs written in the PRF language and concluded with the presentation of a compiling tool. Although the ideas here presented may seem oversimplified when facing real-world applications, the problem itself (can some type of ARNNs perform exact symbolic computation?) has an affirmative answer. Moreover, there is an automatic translation from the mathematical description of any Turing computable function into a Neural Network able to compute that function. This establishes a computational equivalence of this simple model of ARNNs and Partial Recursive Functions (itself an Universal model of computation), which is modular-oriented (making complexity management of large programs easier to deal with). Adapting them to other more elaborate frameworks like those of high-level languages is straightforward, as in [5]. Finally, we also defend that the search for a more suited language to the architecture of ARNNs can give us fresh views on efficient approaches for systematic construction of ARNNs.

Acknowledgements

We would like to thank the technical and financial support given by LabMAg and the Portuguese R&D Agency Fundação de Ciência e Tecnologia. Also, special thanks to Hava Siegelmann for all the scientific collaboration within this project.

References

[1] Boolos, G. and Jeffrey, R., *Computability and Logic* (2nd ed.), Cambridge Press, 1980.
[2] Cutland, N., *Computability – An Introduction to Recursive Function Theory*, Cambridge University Press, 1980.
[3] Gruau, F., Ratajszcza J., and Wiber, G., *Fundamental study – A neural compiler*, Theoretical Computer Science, Elsevier, 141, 1995, 1-52.
[4] Neto, J., Siegelmann, H., Costa, J., and Araújo, C., *Turing universality of neural nets (revisited)*, Proceedings of Computer Aided Systems Theory – EUROCAST'97, 1997.
[5] Neto, J., Siegelmann, H., and Costa, J., *Symbolic processing in neural networks*, to be published at the Journal of Brasilian Computer Society, 2001.
[6] Siegelmann, H. and Sontag, E., *On the computational power of neural nets,* J. of Computer and System Sciences, Academic Press, [50] 1, 1995, 132-150.
[7] Siegelmann, H., *On NIL: the software constructor of neural networks*, Parallel Processing Letters, [6] 4, World Scientific Publ. Company, 1996, 575-582.
[8] Siegelmann, H., *Neural Networks and Analog Computation, Beyond the Turing Limit*, Birkhäuser, 1999.

Appendix

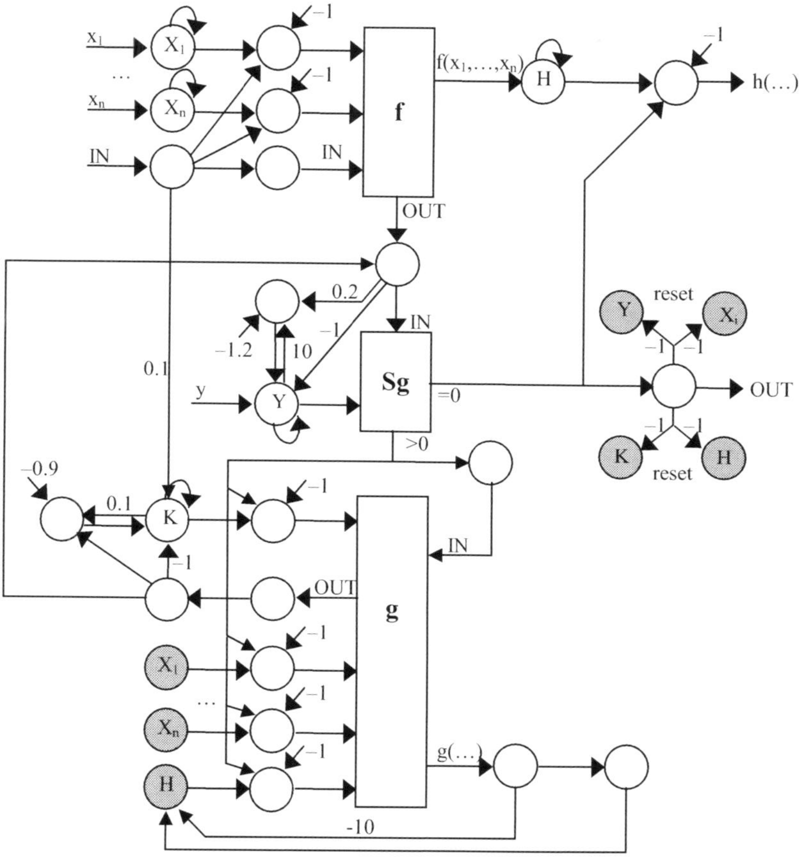

Fig. 3. Recursion.

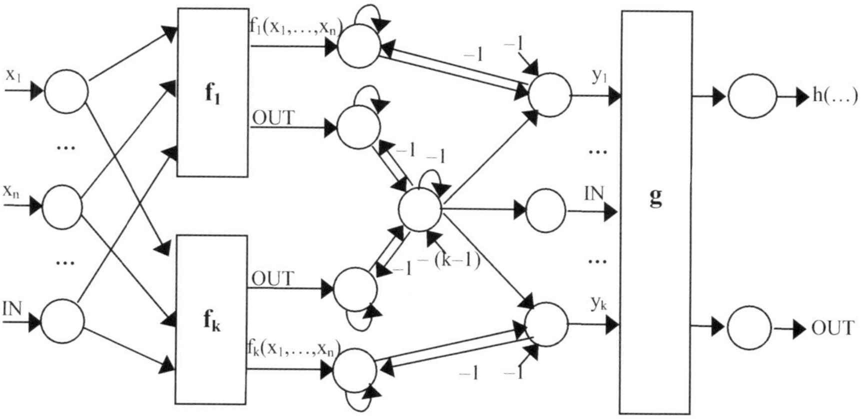

Fig. 4. Composition.

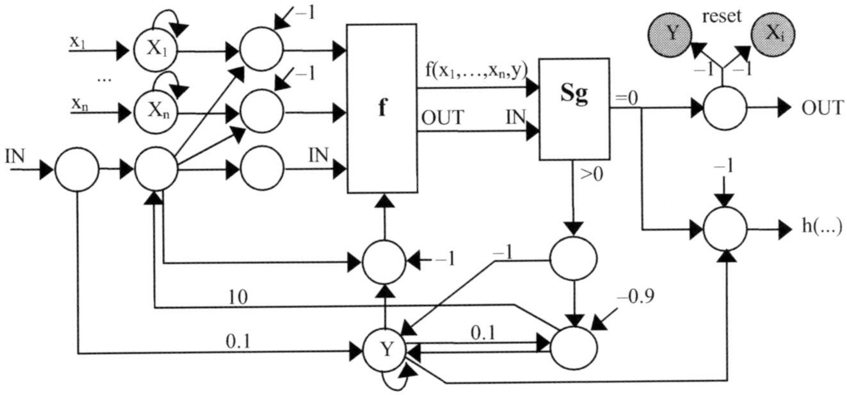

Fig. 5. Minimization.

Multistage Neural Networks: Adaptive Combination of Ensemble Results

Shuang Yang*, Antony Browne*, and Philip Picton**

* School of Computing, London Metropolitan University, London EC3N 1JY, UK, e-mail: (syang@lgu.ac.uk)
** School of Technology and Design, University College Northampton, Northampton NN2, 6JD, UK

Abstract. In the past decade, more and more research has shown that ensembles of neural networks (some times referred to as committee machines or classifier ensembles) can be superior to single neural network models, in terms of the generalization performance they can achieve on the same data sets. In this paper, we propose a novel trainable neural network ensemble combination schema: multistage neural network ensembles. Two stages of neural network models are constructed. In the first stage, neural networks are used to generate the ensemble candidates. The second stage neural network model approximates a combination function based on the results generated by the ensemble members from the first stage. A sample of the data sets from UCI Machine Learning Depository are modeled using multistage neural networks and a comparison of the performance between multistage neural networks and a majority voting scheme is conducted.

Keywords. neural network ensemble, combination strategy, multistage ensembles.

1 Introduction

It is well known that combining a set of neural network classifiers whose error distributions are diverse can lead to the generation of superior results than those achieved by any single classifier. The most common combination strategies used to combine the results of individual ensemble members are simple averaging [9,11], weighted averaging [5], majority voting [3] and ranking [2,4]. These are catalogued as fixed combination schemes, which require no prior training. One deficiency of such schemes feature is that weightings for the importance of the output of each ensemble member must be pre-chosen and then applied to produce the combination result.

There is a trend to explore trainable combination methods which aim to find the optimal weights to be combined [6,8,13,14]. In this paper, we propose a new model for ensemble combination, based on another neural network layer, called multistage neural network ensembles. Unlike the conventional combination schema, the weights assigned to ensemble members will be generated automatically by another neural network layer. The new model outlined here inherits some ideas from stacking [12] and develops them further.

2 Multistage Neural Networks

It appears attractive to make the combination process adaptive, so that no a-priori (and possibly incorrect) combination weightings need to be chosen. Therefore, a model is proposed where the procedure of combining ensemble classifiers is turned into the training of another neural network model. The structure of the multistage neural network ensembles is shown in Figure 1.

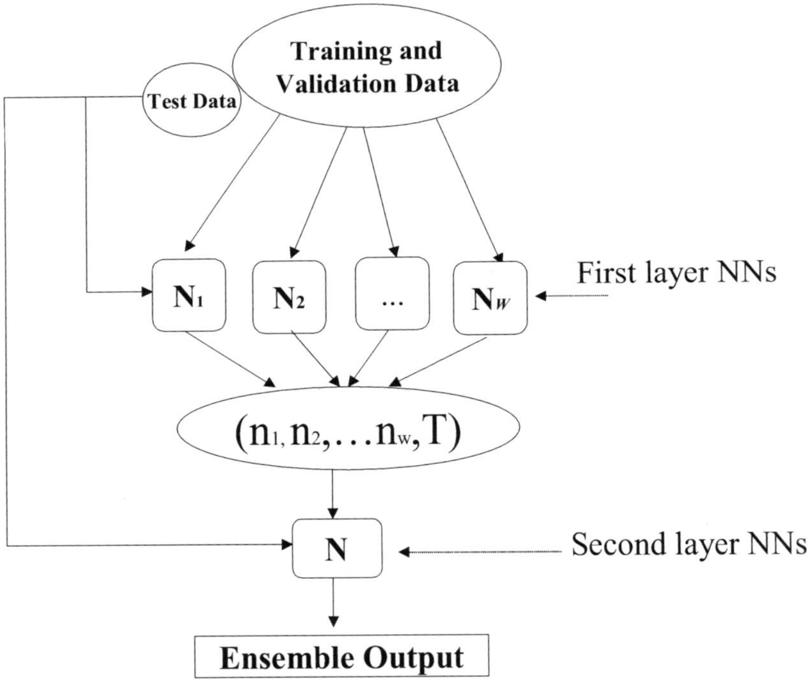

Fig 1. Illustration of multistage ensemble neural networks

Referring to figure 1, suppose there is a source data set S{ $s_{1,}s_2,...s_n$ } with its corresponding target data set T{ $t_1, t_2...t_m$ } are partitioned into three parts: test data, training data and validation data. The training data usually are pre-processed by various methods in order to generate diverse results before they are applied to the first layer's neural network models: $N_1, N_2 ... N_W$. The pre-processing methods on training data set include distributing sample presentation sequences randomly, noise injection, bagging, boosting or other methods. After training, the test data set are applied to these ensemble candidates to access their performance. Afterwards the whole training data is applied and each first layer neural networks' corresponding results ($n_1, n_2 ... n_w$) are used as the second layer neural network model's inputs. The second layer neural networks is trained by using the first layers generated results on the whole training data as inputs combined with their target data set.

The advantages of multistage ensemble neural networks are:

- For each ensemble candidate, the normal data preparation methods can be applied and ensemble candidates will be trained separately by using various different neural network models and algorithms. The generalization of these first layer neural networks are expected to be as diverse as possible in order to get the best ensemble result. Because of this, the choices for the first layer neural network training are very flexible and allow a wide range of selections.
- The second layer of a single neural network is used as an ensembles' combiner to adaptively apply different weights to those first layer ensemble members in order to approximate a best combination function. There is a strong reason to believe that it can more accurately adjusting the weights which are assigned to ensemble members than manual methods.
- Multistage ensemble neural networks can be applied to both classification and regression problem, unlike some other ensemble combination techniques.

3 Experimental Results

Five classification data sets taken from UCI Machine Learning Depository (http://www.ics.uci.edu/mlearn/MLRepository.html) were used for these experiments. The details of these data sets are listed in Table 1.

Table 1. Summary of the data sets

Data	Number of Cases	Number of Input features	Number of Output features
Breast-cancer –w	682	9	2
Bupa-Liver	345	6	2
Ionosphere	351	34	2
Iris*	150	4	3
Pima-Diabetes	768	8	2

Notations: * - multi-class data set.

For each ensemble candidates' training, each data set was randomly partitioned into three parts: training, validation and test data. The sequences of training data were randomly distributed before they were applied to the neural networks model (i.e., steps were taken to prepare the most diverse data among ensemble members for training). There were no overlapping data instances inside the training data sets. For most of the data sets listed in Table 1, this approach was effective in generating diverse training data sets. The exception to this was Iris data set, where the bagging [1] method was applied, due to the source data's small size.

Nine single hidden layer neural networks trained by the backpropagation algorithm were generated as ensemble candidates for each ensemble. These nine candidates were constructed with different sequences of training data, different neural network structures (numbers of hidden neurons) and different initialization. Each neural network model was trained 20 times with random initialization of starting weights. The number of hidden neurons was changed from two to the number of inputs of each data set (time considerations prevented the exploration of networks with hidden layers larger than this). During training, the validation data set was used to prevent overfitting. As the experiments in this paper was concentrated on comparing the performance of two ensemble combination methods, to make the things simpler, the test data set was applied to the ensemble candidates after their training. Those ensemble candidates with the best generalization performance were kept. Majority voting and multistage neural networks were then applied to these same ensemble members to generate the combination results. These nine single neural network's average performances of each data set on the test data, which were generated from ten-fold cross validation approach are illustrated in table 2.

Table 2. Ensemble candidates performance

Data	Single Neural Networks Performance (%)								
	1	2	3	4	5	6	7	8	9
Breast –cancer-w	97.21	97.50	97.21	96.62	96.76	97.06	97.06	97.35	97.50
Bupa-Liver	72.06	72.65	71.76	73.82	72.06	72.94	71.76	72.35	72.65
Ionosphere	95.43	94.86	95.14	93.14	94.29	93.71	94.86	90.57	92.57
Iris *	97.60	96.80	98.4	96.00	96.00	98.00	98.40	99.20	97.60
Pima-Diabetes	73.03	74.08	72.50	74.21	71.32	69.87	68.68	70.00	69.74

After the training of the first layer's ensemble candidates, the whole source data was randomly distributed again. Ten-fold cross-validation [7] was applied to the second layer's neural network training in order to estimate the average performance. Hence it was not necessary to use any cross-validation approach. The procedure of the second layer of neural network model training is: First, the training data was injected into the ensemble members and their outputs were concatenated together with the corresponding target values as the input of the second layer's neural network. Second, The parameters of the second layer neural network model were chosen to be fixed: learning rate was 0.1 and the number of hidden neurons was two. These parameter settings were based on the empirical observation of a large number of training runs. It was noticed that the changing of parameters setting didn't effect the final combination result much, based on the experiments using the above datasets. In most of cases the best parameters combination was:

learning rate of 0.1 and two hidden neurons. However the initialization does contribute strongly to the final result. Therefore, the second layer neural network model was trained 1000 times with random initialization of starting weights.

The result of the majority voting applied to three, five and nine ensemble members were then compared with the performance of multistage neural networks by averaging over ten-fold cross-validation. In Table 4, all the results shown for the voting and multistage neural network are based on the same ensemble members for each group. The details for each combination group can be seen in Table 3. The figures present the average performance of the two ensemble combination methods were shown in Table 4.

Table 3. The selection of ensemble members

Data	Th actual ensemble members for each groups of ensemble models		
	3	5	9
Breast-cancer –w	1,2,5	1,2,3,4,5	1,2,3,4,5,6,7,8,9
Bupa-Liver	1,2,3	1,2,3,4,5	1,2,3,4,5,6,7,8,9
Ionosphere	2.3.4	1,2,3,4,5	1,2,3,4,5,6,7,8,9
Iris *	1,2,3	1,2,3,4,5	1,2,3,4,5,6,7,8,9
Pima- Diabetes	2,3,4	1,2,3,4,5	1,2,3,4,5,6,7,8,9

Table 4. Ensemble performance comparison between multistage neural networks and majority voting

Data	Voting Accuracy (%)			Multistage Neural Networks Accuracy ((%)		
	3	5	9	3	5	9
Breast-cancer -w	97.21	97.35	97.65	97.65	97.65	97.8
Bupa-Liver	72.65	73.53	73.82	73.82	74.12	74.41
Ionosphere	96.0	96.0	95.72	96.0	96.29	96.57
Iris *	98.8	97.6	98.8	98.8	98.8	99.2
Pima- Diabetes	74.47	73.95	73.42	75.13	75.53	76.58

From these results, it can be seen that multistage neural networks always perform better than majority voting based on the same ensemble members, regardless the number of ensemble members used in combination was three, five or nine. The results have demonstrated that, on a wide range of data sets (including simple categorization and multiple categorization), multistage neural network ensembles offer improved performance when compared with majority voting as an ensemble combination method.

The intention of using different number of ensemble members being combined was to find whether the ensemble numbers will effect the performance of multistage neural networks. The current experimental results show that the multistage approach generalization improved steadily while the number of ensemble numbers was increased.

4 Conclusion

This paper has demonstrated that multistage ensembles, where the adaptive properties of a second layer network are used to combine the outputs of the individual ensemble members, offer enhanced performance over a simple voting based combination method. Currently, experiments are being carried out to investigate the exact theoretical reason for this performance improvement. The reason probably lies within differences between the kinds of decision surfaces a second layer network can model when compared to those decision surfaces that can be produced by majority voting. However, a clearer analysis and description of this area needs to be developed. In future, the effect of the choice of training, validation and test data sets, and choice of neural network architecture for the second layer combiner will be investigated. It may be that these factors interact, and will allow this research to push the performance of such ensembles even further.

References

1. L. Breiman (1996). Bagging predictors, *Machine learning*, vol. 26(2), pages 123-140.
2. K. Ghoneim and B.V.K. Vijaya (1995). Learning ranks with neural networks, *Applications and Sci. of Artificial Neural Networks: Proc. of the SPIE*, vol. 2492, 446-464.
3. L. Hansen and P. Salamon (1990). Neural network ensembles. *IEEE Trans. on Pattern Analysis and Machine Intelligence,* vol. 12, pages 993-1001.
4. T.K. Ho, J.J. Hull and S. N. Srihari (1994). Decision combination in multiple classifier systems. *IEEE Trans. Pattern Analysis and Machine Intelligence*, vol. 16(1), 66-76.
5. R.A. Jacobs (1995). Methods for combining experts' probability assessments. *Neural Computation*, vol. 7, pages 867-888.
6. J. Kittler (1998). Combining Classifiers: A Theoretical Framework. *Pattern Analysis* & Application, vol. 1, pages 18-27.
7. A. Krogh and J. Vedelsby (1995). Neural Network Ensembles, Cross Validation, and Active Learning. *Advances in Neural Information Processing Systems*, MIT press, G. Tesauro, D.S. Touretzky and T.K. Lee (eds.), vol. 7, pages 231-238.
8. D.S. Lee and S.N. Srihari (1993). Handprinted Digit Recognition: A Comparison of Algorithms. *Pre-Proc. 3RD International Workshop On Frontiers In Handwriting Recognition,* Buffalo, USA. pages 153-162.
9. W. Lincoln and J. Skrzypek (1990). Synergy of clustering multiple back propagation networks. *Adv. in Neural Inf. Proc. Sys.-2,* M Kaufmann, D. Touretzky (ed.), 650-657.
10. D. Partridge and N. Griffith (1995). Strategies for Improving Neural Net Generalisation. *Neural Computing and Applications,* vol. 3, pages 27-37.
11. K. Tumer and J. Ghosh (1995). Order statistics combiners of neural classifiers. *Proc. of the World Congress on Neural Networks* , Washington D.C.:INNS press, 31-34.
12. D.H. Wolpert (1992). Stacked generalization. *Neural Networks,* vol. 5, 241-259.
13. S. Yang, et-al (2002) Multistage neural network ensembles. *Proc. of the Third Int. Workshop on Multiple Classifier Systems*, Springer., vol. 2364, Pages 91-97.
14. X. Zeng and T.R. Martinez (2000). Using a Neural Network to Approximate an Ensemble of Classifiers. *Neural Processing Letters,* vol. 12, pages 225-237.

On the Fuzzy Spatio-temporal Specification of Multimedia Synchronisation Scenarios

André L.V. Coelho and Ivan L.M. Ricarte

DCA-FEEC-Unicamp, Campinas, Brazil
{coelho,ricarte}@dca.fee.unicamp.br

Abstract. In this paper, we present an authoring language for the high-level specification of temporal and spatial interrelations between multimedia document objects. This work follows a synchronisation methodology based upon concepts of fuzzy sets and software components. As result, the designer is endowed with a more expressive and flexible modelling tool for creating multimedia scenarios, which is done by means of hierarchical fuzzy descriptive plans. As well, such approach gives support to the representation of inaccurate relationships and unpredictable situations. After discussing the semantics behind the language directives, an illustrative scenario is presented.

1 Introduction

It is many times very arduous to completely describe the activities underlying a multimedia document (MMDoc) scenario as the synchronisation relations between media elements may not be properly represented by accurate specifications of well-defined, crisp spatio-temporal interdependencies [2]. Modelling imprecision, therefore, needs to be properly captured by novel authoring languages in order to convey to designers the necessary mechanisms for the flexible (and better aesthetic) characterisation of the scenes, dismissing peculiarities such as exact presentation times or precise spatial arrangements.

Towards this goal, we have developed a methodology (MuSCoF) approaching the high-level, designer-oriented, descriptive specification of intermedia temporal/spatial associations. MuSCoF has features borrowed from most of the multimedia synchronisation approaches, such as hierarchical, constraint, and causality-based ones [5]. Basically, it subsumes a three-step process and is based on two complementary models, namely, *Fuzzy Synchronisation Model* (FSM) and *Multimedia Component Model* (MCM). The steps encompass: *(i)* the user input layout of the multimedia presentation scenes by means of a fuzzy script plan (fuzzy synchronisation specification); *(ii)* the parsing of such sketch for recognising the multimedia entities that compose the scenes and for the consistency checking of the fuzzy spatio-temporal relationships among them; and *(iii)* the generation of temporal, spatial, and action managers (synchronisation mechanisms) associated to each multimedia entity.

In this work, we focus mostly on the characterisation of the MuSCoF fuzzy authoring language, whose temporal and spatial directives (to be used for the construction of the aforementioned fuzzy scripts) were produced according to the FSM guidelines. We advocate that, by employing fuzzy sets theory, interdependencies in multimedia specifications can assume a more manageable perspective, improving both the understandability and feasibility of the whole presentation scheduling process. As well, this gives space to the representation and handling of inaccurate/uncertain relationships [6] as well as of non-predictable situations. Such aspects are very appealing when one wishes to take advantage of the soft character behind most multimedia synchronisation requirements–as for MMDoc scenarios–, or when it is desirable to give support to user interaction.

Following the guidelines from Gibbs' work [1], in MuSCoF we assume a layout model wherein every MMDoc entity is modelled after a particular type of multimedia component (MMComp), and these components should be logically arranged through composite constructs and communicate one to the other through the interchange of signal events. The idea is to embed into the same container all structural and logical information related to a particular media object. This is a means to decentralise the co-ordination activities necessary for the elements' display, and also to provide graceful performance degradation in case of component failure. Media contents represent pieces of raw (binary) information classified into pairs of MIME types/subtypes and may be associated to one or more media presentation objects that contain rendering and layout information for distinct environments or circumstances. The decision of which presentation object (each possibly with a distinct duration interval) to deploy at runtime is due to the designer's choice, something very attractive for interactive purposes. Such flexibility, however, has great impact on the synchronisation service to be deployed, as by now synchronisation points between media objects are forced to assume a relative basis, therefore justifying the high-level, fuzzy synchronisation model proposed herein.

The *temporal (spatial) manager* is responsible for maintaining, interpreting and executing all temporal (spatial) directives associated with the MMComp it represents. For such, it should be configured by the parsing step of MuSCoF with a base of event rules containing all the information necessary for the triggering of synchronisation events it shall produce along its existence. Such information typically indicates *when* and to *whom* a synchronisation message should be forwarded. Another kind of management mechanism relates to the MMComp state maintenance. In MuSCoF, there are two orthogonal means of classifying a MMComp state. In the first categorisation, a MMComp may be in one of two states: *passive* or *active*. In the second, a MMComp may be labelled either as *alive*, or *expired*, or *out-of-function*. A "passive" entity has no capability of influencing the behaviour/execution of a peer, but may be, otherwise, affected by other "active" components. The latter may have their presentation duration fuzzified and assigned with fuzzy synchronisation points. A MMComp is considered "alive" when, even inert in the sense of rendering, it may still affect the state/behaviour of other elements. It may be switched between "active" and "passive" along its execution time. MMComps are termed "expired" when they do not have any more power to influence other entities, and are said "out-of-function" when its rendering has been hampered by some malfunction behaviour. Finally, *action managers* cope with all activities performed by a particular MMComp in respect to the behaviour perceived by the other scene elements. Some actions have well-known semantics, such as VCR operations (start, stop, pause, resume), and may be already supported by the multimedia environment, whereas others may be conceived by the designer for specific purposes.

2 MuSCoF Fuzzy Multimedia Synchronisation Model and Language

According to this modelling framework, the multimedia components' interrelations are portrayed via a set of directives where fuzzy causal rules and constraints behave as basic synchronisation mechanisms. First, we focus on temporal aspects; afterwards, on spatial issues.

2.1 Fuzzy Temporal Synchronisation Model

In active MMDocs, *temporal objects* are the model elements upon which relations are specified. Pazandak and Srivastava [5] enumerate three classes of temporal objects: *media objects* (here encapsulated as MMComps), *events* (referring to the endpoints of media objects as well as user- or application-defined signals) and *timepoints* (instants with no duration). Events may also be relative to temporal *intervals*, which in turn explicitly represent media duration. The last three elements have their semantics fuzzified in our model, giving rise to what we call *fuzzy temporal objects*. By this means, MuSCoF provides support both to *sync-point models* and *interval-based models* [2, 3]. In the first, synchronisation is made in a point-to-point basis, whereas, in the second, the media relationships are given by logically arranging their time intervals (typically through the intervals' endpoints). Thus, the basic FSM directives are *(i)* fTPoint(F_T, f_n), where F_T is a fuzzy timepoint (possibly a singleton) whose fuzzy number is f_n; *(ii)* fEvent(e, M_O, a, F_T, t), where e is an endpoint event associated to timepoint F_T and related to the action a of M_O (t indicates whether it is a begin or end event); and *(iii)* fTInterval(F_I, F_{Tb}, F_{Tf}) where F_I is the fuzzy interval delimited by two normal fuzzy timepoints, namely F_{Tb} and F_{Tf}, which must be defined in such a way that if $\forall x \in \text{Core}(F_{Tb})$ and $\forall y \in \text{Core}(F_{Tf})$ then $x < y$ (we refer to this as "fuzzy ranking by the core").

We define two types of fuzzy media duration associated to a MMComp: one for its presentation time Δ_p (since a *beginning event* at instant t_b until an *end event* at instant t_f) and the other for its time of liveness Δ_l (temporal length it may affect the presentation of other components). t_b and t_f should be two fuzzy timepoints related to a fuzzy temporal interval. By this means, a designer can capture temporal uncertainties through the use of fuzzy numbers, such as $t_b =$ *about 5min*, and vague relations between temporal objects may be represented through the employment of both types of fuzzy duration. The corresponding MuSCoF language directives are fPresDur(M_O, p_s, F_I) and fLiveDur(M_O, F_I), where M_O has a presentation object p_s (or liveness property) with duration defined by fuzzy interval F_I.

It is also possible to apply fuzzy arithmetic operations between fuzzy temporal objects in a way such as to express inexact or flexible restrictive relations between them. By this means, in MuSCoF, we define a *fuzzy temporal constraint* as a constraint in which at least one of its elements is quantified by a fuzzy value. This may be described as $e_1 \theta e_2 \phi d$, where $\theta \in \{<, =, >\}$, $\phi \in \{\oplus, \ominus, \otimes\}$ (fuzzy addition, subtraction, and multiplication, respectively) and e_i and d represent fuzzy events and fuzzy timepoints, respectively. To represent all possible fuzzy constraint relations of this kind we have devised a generic directive, namely fTConstraint(e_1, e_2, d, o), which indicates that the fuzzy timepoint associated to fuzzy event e_1 is calculated by applying operator o over the fuzzy timepoint of e_2 and the fuzzy (or crisp) delay d. In this context, it would be feasible to capture that "the media object O_1 should startup approximately two seconds after the end of media object O_2 presentation": "startup" and "end" translate to fuzzy events and "approximately two seconds" is a fuzzy delay.

Interval-based models are more flexible (and "high-level") than instant-based models as they relieve the designer from having to update the synchronisation specification each time the presentation time of an object has to be changed. Typically, interval-based media orderings follow Allen's set of logical relations, namely *before*, *meets*, *overlaps*, *finishes*, *during*, *starts*, and *equals* [2, 3]. In MuSCoF, these seven temporal relations are made less stringent and more flexible in order to represent vague temporal constraints among MMComps. In this regard, we assume that each media time interval is defined via the fTInterval($\cdot$) primitive, i.e. they should have fuzzy timepoints for initial and final instants. The fuzzified set of MuSCoF directives (given in Fig. 1) is fBefore(M_{O1}, M_{O2}), fTMeets(M_{O1}, M_{O2}), fOverlaps(M_{O1}, M_{O2}), fFinishes(M_{O1}, M_{O2}), fDuring(M_{O1}, M_{O2}), fStarts(M_{O1}, M_{O2}), and fTEquals(M_{O1}, M_{O2}). In order to resolve such fuzzy constraints while maintaining their well-defined semantics (and thus not violating Allen's proofs), the MuSCoF parsing phase (or even the runtime presentation scheduler, in case of dynamically-generated scenarios) should resort to a fuzzy ranking method to be applied over the intervals' endpoints. For instance, by using the above "fuzzy ranking by the core" criterion for fBefore($\cdot$), the parser should detect whether the condition $\{x = \sup(\mathrm{Core}(T_{f1})), y = \inf(\mathrm{Core}(T_{b2})) \mid x < y\}$ is satisfied or not (here, T_{f1} and T_{b2} are the final and beginning fuzzy timepoints for intervals F_{I1} and F_{I2}, respectively).

To give more flexibility to the specification of temporal interdependencies in a multimedia scenario, the duration Δ_p may, otherwise, be represented as a linguistic variable defined over a temporal scale. In MuSCoF, this was taken into account in order to represent distinct presentation phases (e.g., *beginning*, *middle*, and *finish*). The designer is the one who stipulates the amount, the meaning, and the format of the membership functions associated with the linguistic terms (phases) ranging over each Δ_p partition (so, the Δ_p of each media object can be customised). For so, he/she has to make use of fTVar($M_O, u, \{p\}$) directive, where u refers to the universe of discourse and $\{p\}$ is the set of terms, representing possible coarse-grained,

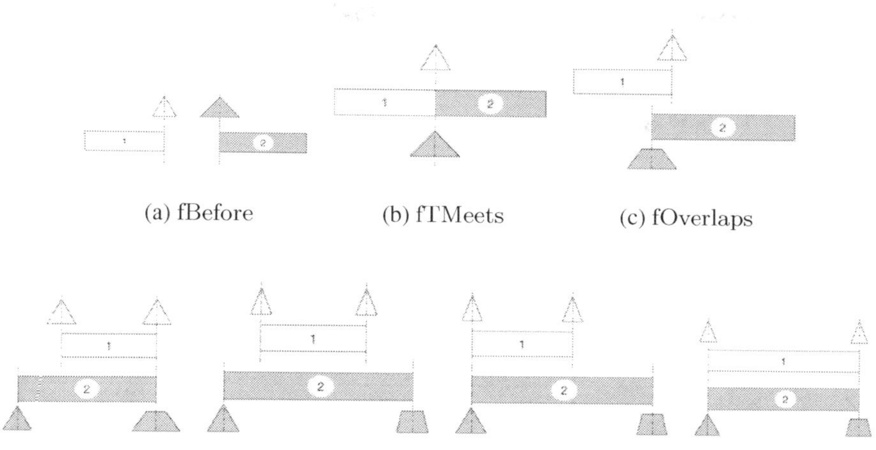

(a) fBefore (b) fTMeets (c) fOverlaps

(d) fFinishes (e) fDuring (f) fStarts (g) fTEquals

Fig. 1. Consistency rules for the fuzzified Allen relations in MMComp synchronisation

fuzzy synchronisation points (FSPs) for relating two media presentations. Each FSP, say P_i, represents a subset (interval) of values delimited by endpoints P_{ib} and P_{if}, whose values, assuming the above partition, are given by $P_{ib} = \inf_{t \in \Delta_p} \varphi_\tau$ and $P_{if} = \sup_{t \in \Delta_p} \varphi_\tau$, where the phase $\varphi \in \{\textit{beginning, middle, finish}\}$ and $\varphi_\tau = \{t \mid \varphi(t) \geq \tau\}$ is the α-cut in relation to the universe of discourse defined by Δ_p and to the *threshold* level defined by τ. The role of the latter is to explicitly indicate the minimal and maximal values delimiting the FSP for assuming that "M_{O1} has fired a fuzzy synchronisation event in relation to M_{O2}." This strategy introduces uncertainty in the occurrence of a synchronisation event and allows for the "relative firing" between different pairs of MMComps. Therefore, $\langle\langle M_{O1}, \textit{finishing}, \textit{almost}, \tau_1\rangle \rightarrow \langle M_{O2}, \textit{start}, \textit{state}_2\rangle\rangle$ (M_{O2} will commence when M_{O1} is "almost finishing") and $\langle\langle M_{O1}, \textit{beginning}, \textit{very}, \tau_{1'}\rangle \rightarrow \langle M_{O3}, \textit{start}, \textit{state}_{2'}\rangle\rangle$ (M_{O3} will commence after the "very beginning" of M_{O1}) should incur distinct synchronisation points/events for the same media object M_{O1}. By this means, it is possible to represent fuzzy causality between the activities realised by media objects. Pazandak and Srivastava's temporal model [5] also supports the notion of causality, albeit they do not envisage situations where the "cause-effect" may be fuzzified, as proposed here. For this purpose, we have defined a general directive, namely fTCausal($\langle M_{O1}, p, [H], \tau\rangle \rightarrow \langle M_{O2}, a, s\rangle$), which should be interpreted as "If M_{O1} reaches the sync point delimited by the linguistic term p, the set of zero or more modifiers H and threshold τ, so perform action a of M_{O2} and changes its state to s."

2.2 Fuzzy Spatial Synchronisation Model

In this subsection, we deal with topological arrangements of MMComps. The purpose is to model both the spatial position of the media objects as well as their spatial interdependencies. We kept quite the same reference model designed for the FSM temporal modelling, although hereafter considering only static 2D information. The basic primitives include *(i)* $\text{fSPoint}(F_S, f_n)$, where F_S is a fuzzy one-dimensional spatial co-ordinate (maybe a singleton) with f_n as its fuzzy number; and *(ii)* $\text{fSInterval}(F_I, F_{Sa}, F_{Sb})$, where F_I is the 1D fuzzy interval defined by two normal fuzzy spatial values, namely F_{Sa} and F_{Sb}, which comply with the above fuzzy ranking criterion. Here, we have followed the guidelines of recent work on topological modelling for spatial databases [2, 4] and made use of the concept of *minimum bounding box* (MBB). The idea is to approximate the visual area of a given MMComp with the rectangle having the minimum area among those containing the displayed media and whose sides are parallel to the system reference axes. An MBB is completely described by its height, width, and the co-ordinates of its centre. In MuSCoF, we extend such conceptual entity by "fuzzifying" its spatial parameters, giving birth to $\text{fMBB}(M_O, \langle F_{Sx}, F_{Sy}\rangle, F_{Ih}, F_{Iw})$, whose interpretation is "M_O has a fuzzy MBB with fuzzy center co-ordinates given by F_{Sx} and F_{Sy} and with height and width values given by fuzzy spatial intervals F_{Sh} and F_{Sw}, respectively."

For introducing "causal-effect" relations concerning the spatial handling of MMComps by the user, we have followed the same idea as before and made use of α-cuts and linguistic variables as supportive concepts. However, instead of using the temporal presentation duration Δ_p, we have considered here the MBB presentation area A_{MBB} as a linguistic variable whose linguistic terms (to be partitioned and characterised by the designer) represent distinct manipulation regions by the user. In this regard, we have devised the directive $\text{fSVar}(M_O, \langle u_h, u_w\rangle, \langle\{p_h\}, \{q_w\}\rangle)$, where the M_O's area is given by $u_h \times u_w$ (the space delimited by the Cartesian product of the universes of discourse relative to M_O's height and width, resp.) and the possible *fuzzy synchronisation regions* (FSRs) are defined by the Cartesian product of the sets of linguistic terms $\{p_h\} \times \{q_w\}$ over each range. After defining the area of each manipulative M_O with $\text{fSVar}(\cdot)$, it is possible to model interactive situations like $\langle\langle M_{O1}, \textit{mouse\_click}, \langle \textit{middle}, \textit{middle}\rangle, \textit{near}, \tau\rangle \rightarrow \langle M_{O2}, \textit{start}, s_2\rangle\rangle$ (to be interpreted as "M_{O2} will commence its presentation and move to state s_2 whenever the user clicks with the pointer on a 'near-middle' region of M_{O1}") by means of the directive $\text{fSCausal}(\langle M_{O1}, \textit{user\_action}, \langle p, q\rangle, [H], \tau\rangle \rightarrow \langle M_{O2}, a, s\rangle)$.

Finally, for dealing with 2D topological constraint-based relations, we have extended the binary spatial relations defined by Papadias *et al.* [4] and also employed by Kwon *et al.* [2]. These relations have their semantics easily grabbed with the pictorial representations given in Fig. 2, and may be checked out for consistency by following the aforementioned "fuzzy ranking by the core" criterion applied over the MBB's parameters. For brevity of space, we omit such consistency rules here.

3 Specification Example

To demonstrate how to make use of the MuSCoF fuzzy authoring language, a high-level specification plan of a possible scenario is presented in Fig. 3. In this figure, $\text{script}(\cdot)$ and $\text{embed}(\cdot)$ are layout primitives for creating hierarchical specifications

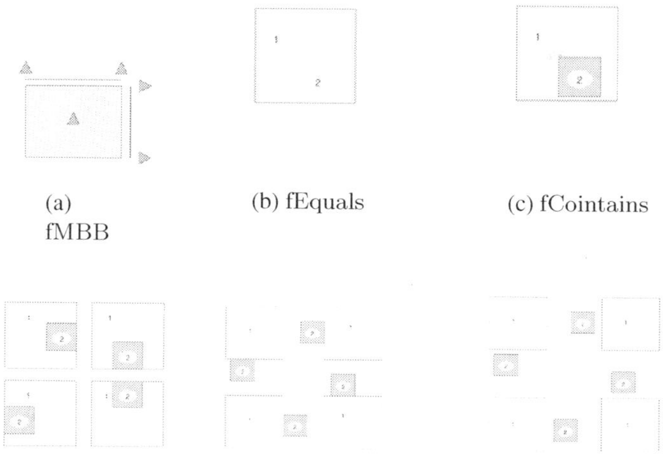

(a) fMBB (b) fEquals (c) fCointains

(d) fCovers (e) fSMeets (f) fDisjoint

Fig. 2. Fuzzified MBB spatial relations for MMComp synchronisation

(by nesting subscripts into scripts) and for encapsulating MMComps into others, respectively. The idea is just to give the reader a flavour of some of the functionalities brought forth by our proposal, namely, *high expressiveness*, *simplicity*, *flexibility*, and *support to uncertainty*. A textual representation of this specification could be generated by a MuSCoF-based authoring environment.

4 Final Remarks

In this chapter, we have presented a fuzzy authoring language for the spatio-temporal synchronisation of MMDoc components. The use of fuzzy systems theory in the field of multimedia synchronisation is yet barely explored, despite its potential to offer flexible specifications for MMDoc display. In our model, the use of fuzzy causal rules and constraints allow the creation of high-level descriptive plans for the design of multimedia presentations. Among the advantages of this model, we may highlight the flexible support to human interaction and the enhancement of the specification's understandability (more closely related to human reasoning). As future work, we should deal more closely with the consistency issues encompassing the second step of MuSCoF. The idea is to provide some automatic tool to assist the designer in the validation of (dynamically) created scenarios. Fuzzy Petri Nets seem to be a good alternative in such regard, and may also be employed as synchronisation mechanisms.

$\mathrm{script}(M_O)\{\mathrm{mmcomp}(v); \mathrm{embed}(M_O, v); \mathrm{fTVar}(v, [0, 100], \{\textit{beginning, middle, finish}\});$

$\mathrm{mmcomp}(a); \mathrm{embed}(M_O, a); \mathrm{fTPoint}(t_1, \triangle(10, 8, 11)); \mathrm{fTPoint}(t_2, \triangle(20, 18, 21));$

$\mathrm{fTInterval}(i_1, t_1, t_2); \mathrm{fPresDur}(a, \text{".wav"}, i_1);$

$\mathrm{fSPoint}(s_1, \triangle(122, 120, 144)); \mathrm{fSPoint}(s_2, \triangle(190, 185, 195)); \mathrm{fSInterval}(j_1, s_1, s_2);$

$\mathrm{fSPoint}(s_3, \triangle(155, 150, 157.5)); \mathrm{fSPoint}(s_4, \triangle(190, 185, 195)); \mathrm{fSInterval}(j_2, s_3, s_4);$

$\mathrm{fSPoint}(s_5, \triangle(158, 154, 170)); \mathrm{fSPoint}(s_6, \triangle(180, 170, 185)); \mathrm{fMBB}(a, \langle s_5, s_6\rangle, j_1, j_2);$

$\mathrm{mmcomp}(l); \mathrm{embed}(M_O, l); \mathrm{fTPoint}(t_3, \square(7, 9, 10, 12)); \mathrm{fTPoint}(t_4, \triangle(20, 17, 20.5));$

$\mathrm{fTInterval}(i_2, t_3, t_4); \mathrm{fPresDur}(l, \text{".html"}, i_2);$

$\mathrm{fSPoint}(s_7, \triangle(140, 130, 148)); \mathrm{fSPoint}(s_8, \triangle(175, 170, 180)); \mathrm{fSInterval}(j_3, s_7, s_8);$

$\mathrm{fSPoint}(s_9, \triangle(125, 120, 128)); \mathrm{fSPoint}(s_{10}, \square(145, 150, 155, 160)); \mathrm{fSInterval}(j_4, s_9, s_{10});$

$\mathrm{fSPoint}(s_{11}, \triangle(158, 164, 170)); \mathrm{fSPoint}(s_{12}, \triangle(140, 135, 145)); \mathrm{fMBB}(l, \langle s_{11}, s_{12}\rangle, j_3, j_4);$

$\mathrm{fTCausal}(\langle v, \textit{beginning}, \text{"quite"}, 0.7\rangle \rightarrow \langle a, \text{start, active}\rangle); \mathrm{fTEquals}(a, l); \mathrm{fSMeets}(a, l); \ldots\}$

Fig. 3. A possible high-level specification plan for a presentation M_O including a video v, an audio track a and a legend l. $\triangle(\cdot)$ and $\square(\cdot)$ represent the fuzzy triangular and trapezoidal membership functions, respectively.

Acknowledgement. CNPq has sponsored the first author via Ph.D. scholarship #140719/99-7

References

1. G. Gibbs. Composite multimedia and active objects. In *Procs. of OOPSLA*, pages 97–112, 1991.
2. Y-M. Kwon, E. Ferrari, and E. Bertino. Modeling spatio-temporal constraints for multimedia objects. *Data & Knowledge Engineering*, 30:217–238, 1999.
3. T. Little and A. Ghafoor. Synchronization and storage models for multimedia objects. *IEEE Journal of Selected Areas in Communications*, 8:413–427, 1990.
4. D. Papadias, Y. Theodoridis, T. Sellis, , and M. Egenhofer. Topological relations in the world of minimum bounding rectangles: A study with r-trees. In *Procs. ACM SIGMOD Conference*, pages 92–103, San Jose, 1995.
5. P. Pazandak and J. Srivastava. A multimedia temporal specification model and language. Technical Report 94-33, Dept. Comp. Science, Univ. Minnesota, 1994.
6. L. Sabry-Ismail, N. Layaïda, and C. Roisin. Dealing with uncertain durations in synchronized multimedia presentations. *Multimedia Tools and Applications Journal*, 18:213–231, 2002.

Part II

Evolutionary Computing

An Approach to On-line Design of Fuzzy Controllers with Evolving Structure[1]

P. P. Angelov

Department of Communications Systems, Lancaster University
Bailrigg, Lancaster, LA1 4YR, UK; e-mail: P.Angelov@Lancaster.ac.uk

1 Introduction

Fuzzy logic controllers of the Takagi-Sugeno (TS) type are an attractive alternative for effective control of complex non-linear systems. Their main advantages are their computational efficiency, transparency and flexibility [13, 14]. This is due to their dual, quasi-linear nature: they are non-linear and therefore suitable for complex systems, but they could also be treated as linear in respect to the consequent parameters. This fact makes possible the application of efficient recursive techniques. TS models are bridging the classical control theory [8] and the fuzzy set theory [12]. They generalize the well-established concepts of gain scheduling and local linear modeling and control [4].

The off-line design of fuzzy rule-base is studied extensively during the last decade [9-13]. The performance of a controller or a model build using such off-line approaches is not guaranteed for the real data, which differ from the validation data used [2]. The reason can be the change in the object of control or the environment [3]. The object can change due to due to a number of reasons like aging, wearing, change in the mode of operation or development of a fault. The environment usually changes seasonally or due to unexpected situation [3]. A tool for solving such problems exists in the conventional control theory - this is the adaptation. For the linear case it is well-developed [8]. It assumes fixed model structure and applies to parameters only [8]. For the more general and more realistic non-linear case, however, there is no existing general solution [2, 3, 4, 5, 6].

The approach presented in this contribution exploits the specific quasi-linear and dual (semi-fuzzy, semi-crisp) nature of the TS rule-based for *on-line* adaptation of its structure and parameters [2-6]. It combines unsupervised *on-line* recursive clustering with a version of recursive weighted least squared estimation of the consequent parameters of the model [4, 5]. The approach is applied in a combination with the indirect learning technique, proposed in [15] and developed in [1] to

[1] The author acknowledges use of the data courtesy of ASHRAE; generated from the ASHRAE funded research project RP1020

the *on-line* design of non-linear fuzzy controllers. The neural network, which is used in [1] and [15] and is trained *off-line* a newly developed *evolving* rule-based model [4, 5, 6] is used. In this way the controller structure is designed *on-line* by adding more informative rules than those, that previously formed the model. The existing rules are replaced with new ones based on the informative potential of the data samples, measured by their spatial proximity to the potential rules centers. In a driven evolution of the controller structure the rule-base gradually evolves inheriting and updating the existing rules and their parameters using the most informative new data.

This controller is flexible due to its evolving structure. It is highly transparent due to the use of linguistic expressions. It has a compact form due to the conservative mechanism of rule-base formation. It is computational efficient due to the use of the quasi linear TS model and combination of the unsupervised clustering with the recursive least squares estimation. Having in mind these qualities this controller is a promising candidate for effective *on-line* control.

The new evolving fuzzy controller is demonstrated on the example of indoor temperature control.

2 Evolving fuzzy controller

The proposed control scheme blends two basic paradigms:

- indirect learning [15];
- evolving fuzzy rule-based models [2, 6].

The first one approximates the inverse dynamics of a plant. It assumes a model-free concept. It feeds back the integrated (or delayed one-step back) output signal instead of feeding back the error between the plant output and the reference signal.

Fig. 1 represents the basic control scheme for the indirect learning approach. The switch has two positions: 1 for control and 2 for learning. During the learning phase, the actual output signal (y_{k+1}) at the time-step ($k+1$) is fed back and the switch is in position 2. The other signal supplied to the input of the controller comes from the integrator (or a delay block). In discrete form it can be realised by memorising the past output signals (y_k). The output is the control signal (u_k). During the control phase the switch is at the position 2. Then the reference signal (y_r) is supplied as input. It replaces the predicted next step output (y_{k+1}) in this phase of the training.

The already trained in the previous learning phases controller produces as an output such a control signal (u_k), which brings the output of the plant at the next time step (y_{k+1}) as closer as possible to the reference signal (y_r).

The indirect learning-based control realises the controller as a neural network trained *off-line* [1, 15] using the set of data (y_k, y_{k+1}, u_k; $k=1,2,\dots,N$; where N is the number of the training data samples). The structure of the neural network used in [1] and [15] to represent the controller is trained to by the inverted plant model. It, however, has a fixed structure (the number of hidden neurons is determined *a pri-*

ori) and the training is *off-line*, because of its iterative nature [1, 15]. Additionally, an important disadvantage of neural networks in comparison to the fuzzy rule-based models is their lack of interpretability and transparency [11, 12].

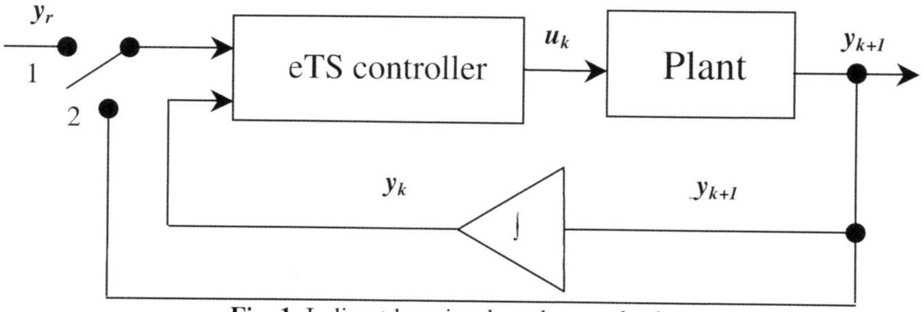

Fig. 1. Indirect learning-based control scheme

At the same time the indirect learning control scheme is inherently convenient for recursive, *on-line* implementation. On the other hand, the fuzzy rule-based models are proven universal approximators [16] and the recently developed evolving Takagi-Sugeno (eTS) models [4] combine the recursive on-line implementation with the high transparency and approximation power [5, [6]. We have applied eTS instead of the neural network in the indirect learning control scheme, which learns and evolves its structure (fuzzy rules and fuzzy sets) and its parameters (focal points of the rules and local model parameters) *on-line* in the process of control itself.

The eTS model used in the indirect learning control scheme (Fig.1) can be represented as follows:

$$\mathfrak{R}_i\text{: } \textbf{\textit{IF}}\ (y_{k-1}\ \text{is}\ Y_i)\ \textbf{\textit{AND}}\ (y_r\ \text{is}\ R_i)\ \textbf{\textit{THEN}}\ (u_{ki}=c_{i0}+c_{i1}y_{k-1}+c_{i2}\,y_r) \tag{1}$$

where $\mathfrak{R}_i$ denotes the ith fuzzy rule; $i\in\{1, R\}$; R is the number of fuzzy rules; Y_I and R_i denote fuzzy sets of the antecedent part of the controller; u_{ki} denotes output (control signal) of the ith local controller at time instance $\boldsymbol{k}=2,3,\ldots$; c_{il} are its parameters, $l\in\{0,1,2\}$.

The level of contribution of corresponding linear sub-model to the overall output of the controller is determined as [2, 3, 4, 5, 6] assuming Gaussian antecedent fuzzy sets:

$$\mu_{ij} = e^{-\alpha\left\|x_{ij}-x_i^*\right\|};\ i=1,2,\ldots,R;\ j=1,2 \tag{2}$$

where α is a positive constant, which defines the zone of influence of the ith model; $x=[\,y_k,\ y_r]^T$ is the inputs vector; x_i^* is the focal point of the ith rule antecedent.

The level of activation of the rules are defined by conjunction or Cartesian product of the respective fuzzy sets for this rule:

$$\tau_i = \mu_{i1}(y_k)\times\mu_{i2}(r)\times\mu_{i3}(F_k) = \bigcap_{j=1}^{2}\mu_{ij}(x_j) \qquad i=1,2,\ldots,R \tag{3}$$

The control signal is calculated by weighted averaging of individual rules' contributions:

$$u_k = \sum_{i=1}^{R} \lambda_i u_{ki} = \sum_{i=1}^{R} \lambda_i x_e^T \pi_i \; ; \; \lambda_i = \frac{\tau_i}{\sum_{j=1}^{R} \tau_j} \tag{4}$$

where λ_i is the normalized firing level of the ith rule; u_{ki} represents the output of the ith linear model; $\pi_i = [c_{i0}\, c_{i1}\, c_{i2}]^T$, $i=1,2, \ldots,R$, is the vector of parameters of the ith linear model; $x_e = [1\; x^T\;]^T$ is the expanded data vector.

3 Experimental results with the evolving fuzzy controller: indoor temperature control

The problem of indoor temperature control is considered as an illustration of the approach presented in this contribution. A subsystem of a real air-conditioning system is considered. Data generated from the ASHRAE funded research project RP1020 has been used. The cooling coil cools the warm air flowing onto the coil prior its use for maintaining comfortable conditions in the occupied space [2]. The main load on the coil is generated due to the supply of ambient air required to maintain a minimum standard of indoor air quality.

The outlet temperature from the coil (current and delayed values) has been used to train the ($\boldsymbol{T^{out}}$) are used as inputs to the controller, including its structure. Volumetric flow rate of air (m_a=1.0,kg/s), moisture content (g=0.008,kg/kg), and the temperature of the water (T_w=10,°C) are supposed to be constant, and the sampling interval is one minute. Results of simulation with the evolving fuzzy controller are depicted in Fig. 2.

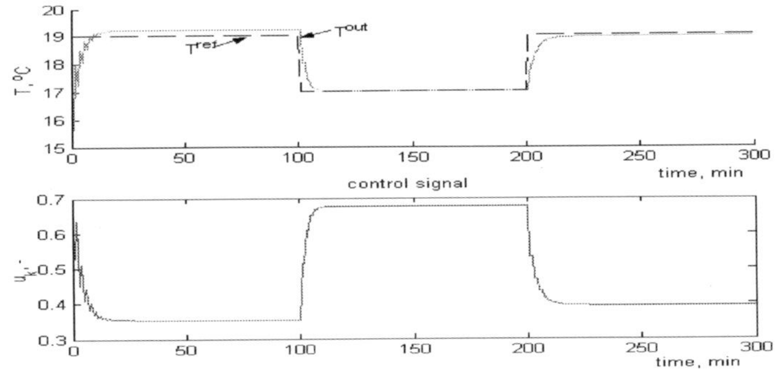

Fig. 2. Control of the indoor temperature by eTS controller

The initial structure of the controller, which consists of three fuzzy rules, has been generated *off-line* using the first 100 data points:

R_1: **IF** (T_{k-1}^{out} is H) **AND** (T^{ref} is H) **THEN** ($u_{k1}=c_{10}+c_{11}\,\mathbf{y_{k-1}}+c_{12}\,\mathbf{y_r}$) (5)

R_2: **IF** (T_{k-1}^{out} is M) **AND** (T^{ref} is M) **THEN** ($u_{k2}=c_{20}+c_{21}\mathbf{y_{k-1}}+ c_{22}\mathbf{y_r}$)

R_3: **IF** (T_{k-1}^{out} is H) **AND** (T^{ref} is H) **THEN** ($u_{k3}=c_{30}+c_{31}\mathbf{y_{k-1}}+ c_{32}\,\mathbf{y_r}$)

where the linguistic labels (*H* for *High, L* for *Low, VL* for *Very Low, M* for *Medium*) are assigned to membership functions representing the respective fuzzy rule.

In the *on-line* phase, the incremental learning and control are sequentially performed at each time step in respect to each new data point. The new couple of data (T_k^{out} , u_k) has been recorded in *real time* and the value of T_k^{out} has been memorised. At the next time step the triplet (T_{k-1}^{out} , T_k^{out} , u_{k-1}) is formed, which includes the integrated (memorised) value of the output (T_{k-1}^{out}). It includes both the inputs and the output of the controller and is used its training. Both the structure and the parameters of the eTS controller are updated by a *recursive* procedure as

detailed in [4]. Since the procedure [4] is non-iterative, the calculations take several seconds making possible real-time application. The structure of the eTS controller has been stable and the only change was made at time instant *k=103min*. A new fuzzy rule has been added to the rule-base, because this new data point has high enough informative potential:

R_4: **IF** (T_{k-1}^{out} is L) **AND** (T^{ref} is L) **THEN** ($u_{k4}=c_{40}+c_{41}\mathbf{y_{k-1}}+c_{42}\,\mathbf{y}_r$) (6)

There was also one data point with informative potential high enough to modify the rule-base although it was not enough to be added as a new rule. This data point was close to the centre of the rule R_3. Based on this, the fuzzy rule R_3 has been replaced by the more informative rule R_5 in a way described in more details in [2]-[6]:

R_5: **IF** (T_{k-1}^{out} is M) **AND** (T^{ref} is RH) **THEN** ($u_{k5}=c_{50}+c_{51}\mathbf{y_{k-1}}+ c_{52}\mathbf{y_r}$) (7)

4 Conclusion

The problem of on-line design of Takagi-Sugeno fuzzy controllers, which have evolving structure, is treated in the paper. An approach for generating the fuzzy rules, representing the structure of the controller based on the data, which are collected during the process of control, is proposed. It is based on the recently introduced technique for on-line identification of Takagi-Sugeno type of fuzzy models. The controlled plant's output and the control signal are used to train on-line in a non-iterative, recursive way the Takagi-Sugeno fuzzy controller. The control rule-base structure is gradually evolving taking into account the new data and exploiting the quasi-linear nature of Takagi-Sugeno models. This approach is combined with the indirect adaptive control approach. A practical engineering example of an air-conditioning system control is given with illustrative purposes.

References

1. H.C. Andersen, F.C. Teng, A. C. Tsoi (1994). Single Net Indirect Learning Architecture, In *IEEE Transactions on Neural Networks*, vol. 5, No 6, pp.1003-1005.
2. P.P. Angelov (2002). *Evolving Rule-based Models: A Tool for Design of Flexible Adaptive Systems*, Heidelberg: Springer, Physica-Verlag.
3. P.P. Angelov, R. A. Buswell (2002). Identification of Evolving Fuzzy Rule-based Models, In *IEEE Transactions on Fuzzy Systems*, vol.10, No.5, 667-677.
4. P.P. Angelov, D. P. Filev (2003). An Approach to On-line Identification of Takagi-Sugeno Fuzzy Models, In *IEEE Transactions on System, Man and Cybernetics, part B*, to appear.
5. P.P. Angelov, D. P. Filev (2002). Flexible Models with Evolving Structure, In *Proc. IEEE Symposium on Intelligent Systems* IS'02, Varna, Bulgaria, 10-12 September 2002, vol.2, pp.28-33.
6. P.P. Angelov, R. A. Buswell (2001). Evolving Rule-based Models - A Tool for Intelligent Adaptation, In *Proc. of the joint 9th IFSA World Congress and 20th NAFIPS Annual Conference*, Vancouver, BC, Canada, 25-28 July, 2001, pp.1062-1067, invited paper.
7. P.P. Angelov, V.I. Hanby, R. A. Buswell, J.A. Wright (2001). Automatic Generation of Fuzzy Rule-based Models from Data by Genetic Algorithms, In *Developments in Soft Computing* (R. John and R. Birkenhead Eds.): Physica Verlag, pp.31-40.
8. K.J. Astroem, B. Wittenmark (1984). *Computer Controlled Systems: Theory and Design*, Englewood Cliffs, NJ, USA: Prentice Hall.
9. B. Carse, T.C. Fogarty, A. Munro (1996). Evolving Fuzzy Rule-based Controllers using GA, In *Fuzzy Sets and Systems*, vol. 80, pp.273-294.
10. C. K. Chiang, H.-Y. Chung, J.J. Lin (1996). A Self-Learning Fuzzy Logic Controller using Genetic Algorithms with Reinforcements, In *IEEE Trans. on Fuzzy Systems*, vol. 5, pp.460-467.
11. M.G. Cooper, J. J. Vidal (1996). Genetic Design of Fuzzy Controllers, In *Genetic Algorithms and Pattern Recognition*, S. K. Pal, P. P. Wang Eds., CRC Press, chapter 13, pp.283-298.
12. D. Driankov, H. Hellendoorn, M. Reinfrank (1993). *An Introduction to Fuzzy Control*, Berlin, Germany: Springer Verlag.
13. W. S. Lin, C.-H. Tsai (2001). Self-organizing Fuzzy Control of Multi-variable Systems using Learning Vector Quantization Network. In *Fuzzy Sets and Systems*, vol.124, pages 197-212.
14. F.-J. Lin, C.-H. Lin, P.-H. Shen (2001). Self-Constructing Fuzzy Neural Network Speed Controller for Permanent-Magnet Synchronous Motor Drive. In *IEEE Transactions on Fuzzy Systems*, vol.9, No.5, pages.751-759.
15. D. Psaltis, A. Sideris, A. A. Yamamura (1988). A Multilayered Neural Network Controller, In *IEEE Transactions on Control Systems Management*, vol. 8, No 2, pp.17-21.
16. Wang L.-X. (1992). Fuzzy Systems are Universal Approximators, In *Proc. IEEE Conference on Fuzzy Systems FUZZ_IEEE*, San Diego, CA, USA, pp.1163-1170.

Can a Parallel Agent Approach to Genetic Algorithms Reduce Search Times

Martin Lefley and Ian D McKew

School of Design, Engineering, and Computing,
Bournemouth University, Poole, DORSET, BH12 5BB. UK.
email: mlefley@bournemouth.ac.uk

Abstract. Genetic algorithms (GAs) are typically applied to complex problem domains where the time taken to find an optimal solution is long, being proportionate to the size and form of search space. This paper describes empirical research using a parallel agent architecture for genetic algorithms to reduce search times. With 8 slaves searching for the same solution the search time for the most difficult search was reduced by a considerable 61%. The main focus, however, related to another aspect, with slaves able to periodically share their best solutions. Here a further 64.3% decrease was attained, an overall saving of 86%, approximately 7 times faster. It is hoped that those pursuing similar paths can follow our strategy and learn from these results.

Keywords. Agents. Aglets. Distributed computing. Genetic Algorithms. Parallel Processing. Search.

1 Introduction

Genetic algorithms are an artificial intelligence technology used to search for optimal solutions in complex problem domains. As complexity and search space size increase, so does time taken to find the optimum. This research investigates the effect of a distributed architecture on search time. A major aim being to explore how often to share solutions. There is also discussion of the effectiveness of the networked parallel architecture used in terms of reliability and visibility.

Holland (1962) and his students laid the foundations for the first genetic applications. Goldberg (1989) defines them as "search algorithms based on the mechanics of natural selection and natural genetics". Solutions to a problem in a given domain are usually represented using binary strings though larger alphabets have been used. There are a number of optimisation strategies although, as Konstam et al. (1992) point out, "reproduction, mutation, and crossover remain the basic strategies used in the exploitation of the GA approach".

2 Genetic Algorithms and Parallel Processing

Talbi and Bessiere (1991) implemented a massively parallel GA, the result being a superlinear speed-up of the *graph-partitioning problem*. Logar et al. (1992) used the 2048-processor MasPar MP-1 data-parallel computer to achieve a relative speed up of 145 times. This chapter focuses on an alternative architecture for parallel GAs, a distributed workstation network. Geyer-Shulz and Kolarik (1992) used a UNIX-based workstation network solving a quadratic optimisation problem to demonstrate performance gains using this approach. They reported a drop in runtime from 322 seconds on one host to 39 seconds on ten hosts. Ochi et al. (1997) applied a parallel GA to a job-sequencing problem and found that running times improved significantly. A concluding remark suggested continuation of experiments to determine optimal migration frequency, a variable scrutinised here. Konstam et al. (1992) used 5 Sun 3 systems; each processor periodically sent a population *slice* (first *n* strings rather than fittest) to a *master* process, which assembled a new starting population forwarded to each processor.

3 Agent Technology

Lange and Oshima (1998) define a software agent from an end-users perspective as "a program that assists people and acts on their behalf by allowing people to delegate work to them." An agent runs independent of the process that created it so it does not need to re-establish communication with this process. An agent operates within an execution environment where it has controlled access to local resources and services. Agents are *goal-driven* as they perform delegated tasks, acting in pursuit of their goals. Agents are *autonomous* as they are able to operate without direct intervention, executing in their own thread. Agents may be *reactive*, they may communicate, move freely or incorporate local intelligence. To reduce execution times for processor-hungry jobs, tasks that can be partitioned can send agents to work concurrently at a number of hosts. However, if the agents are too large, Ghanea-Hercock et al. (1999) warn, "the benefits of solving the problem in parallel may be lost". GAs, are used in complex problem domains where execution time is typically long. Parallel computing offers greater processing speeds but conventional programs are difficult to divide into consistently equal strands. Genetic algorithms offer a paradigm that can utilise certain parallel architectures.

If the intention for the GA application had been for a number of dedicated workstations receiving requests in order to facilitate parallel searching, a different approach, such as client/server may have been used. However a mobile agent environment allows executable code to be moved to the most appropriate hosts rather than to dedicated *servers*. For this reason, agent systems can be more robust and fault-tolerant, persisting despite computer failures. They need not suffer *server*-downtime because agents can relocate to more appropriate hosts. Another advantage is a reduction in network traffic, as continual communication is not required as in the *client/server* approach. Wooldridge and Jennings (1998) state "the non-agent approach should generally be preferred, since it will be better understood by the software engineers involved in the system development, and as a consequence is likely to be more manageable and predictable". Another disadvantage may be the difficulty visualising and debugging multi-agent systems but this is true of all complex software such as GAs and parallel systems.

4 Design and Implementation of the GA Application

The basic requirements were that a GA application be built using agents to enable multiple processors to search for the same solution. A simple agent architecture was used for the GA application, with all *slaves* working on the same problem, rather than a part of it. Manipulation of the main variable under investigation, the reciprocal share frequency, was used to establish how often to share solutions to attain the lowest search times. Knapik and Johnson (1998) provide a useful reference book which was used extensively in developing this system. A *master/slave* pattern was used as a basis for the design. Partitioning was kept simple, with *slaves* all searching concurrently, sharing best solutions and supervised by a *master* processor. It was necessary to use a relatively simple problem, one in which the optimal solution should always be found quickly to enable measurement and comparison. The *Bit Count* problem was chosen; for a given binary string, the number of 1s determines fitness. All bits set to 1 represents optimal solution. By modifying the string/chromosome length, the size of the problem is easily varied.

0 1 1 0 0 0	Fitness = 2.0
1 0 0 0 0 1	Fitness = 2.0
1 1 1 1 1 1	Fitness = 6.0 = Optimal Solution

Fig. 1. Example of optimising the Bit Count problem for three 6-bit binary strings

The genetic system used for our test incorporates roulette wheel sampling with crossover and mutation. As with Ochi et al. (1997) only the fittest string from each host is shared. When selecting strings to replace, Logar et al. (1992) used random substitution because substituting for the worst individuals could lead to "reduced diversity in the population". However, they point out that this "does not preclude replacement of the best strings". A variation of their method was therefore adopted retaining the fittest string from each population (similar to elitism but for entering populations). Research and experimentation led to choosing a crossover rate of 0.7 with a *one-point crossover* strategy. Mutation probability was set at 0.001 and population size set to 1000, relatively large to ensure straightforward successes so as to prevent outliers distorting results. Solution complexity was manipulated using string/chromosome length in multiples of 20, from 40 to 120 bits. Each search was also conducted on 2, 4, and 8 processors. These searches were repeated using reciprocal share frequencies of 30, 20, 15, 10, 5, 3, and 1 (generations) to assess which resulted in the shortest times. Each test was run five times so that an average search time could be calculated. Such procedure is essential for the validity of research involving stochastic elements in populations of solutions.

5 Evaluation of the agent based GA application

For the most difficult search, where the chromosome length was 120 bits a significant improvement in search times was seen each time the processor count was doubled. With two *slaves*, a significant reduction in runtime was achieved by allowing solutions to be shared A considerable improvement of over 2 minutes (a 49.1% reduction) for the best reciprocal share frequency of 1, that is solutions shared every generation. With eight *slaves*, sharing solutions, for the most difficult search, the runtime reduced significantly, from 147.8 seconds to 115.8 seconds, a 22% improvement. The best time of 52.8 seconds was attained when solutions were shared every generation, a 64.3% reduction or 95 seconds.

Table 1. Average execution time for each GA test, in seconds

		CHROMOSOME LENGTH 33333333(IN CHARACTERS / BITS)					**Share Frequency**
		40	**60**	**80**	**100**	**120**	
NUMBER OF PROCESSORS	**1**	7.4	17.5	41.6	87.7	379.9	**Infinite**
	2	7.3	18.9	42.2	86.1	259.1	**Infinite**
		6.8	17.3	38.2	82.8	180.8	**30**
		8.0	18.6	39.9	83.4	177.7	**20**
		7.5	19.9	40.5	77.8	192.0	**15**
		7.6	20.0	41.1	78.5	172.1	**10**
		7.4	18.1	41.3	72.2	165.4	**5**
		7.7	20.0	38.2	71.6	138.7	**3**
		8.6	19.8	46.6	72.6	132.0	**1**
	4	7.4	21.2	40.5	76.0	180.8	**Infinite**
		6.2	17.8	34.0	71.2	165.1	**30**
		7.5	15.6	38.4	68.8	143.1	**20**
		7.1	18.5	37.6	75.2	137.8	**15**
		7.1	16.1	33.9	62.8	122.2	**10**
		8.0	17.2	33.4	57.3	97.5	**5**
		7.7	15.0	29.8	54.0	85.1	**3**
		7.9	18.7	34.8	49.7	76.1	**1**
	8	6.6	16.8	33.8	64.0	147.8	**infinite**
		6.6	15.2	34.0	63.3	115.8	**30**
		6.9	15.0	33.7	64.7	92.3	**20**
		6.9	17.0	33.7	55.4	85.1	**15**
		6.6	17.0	28.8	50.3	87.2	**10**
		6.7	14.3	28.9	46.2	67.6	**5**
		7.9	15.2	28.3	40.5	64.2	**3**
		6.7	15.3	26.4	38.4	52.8	**1**

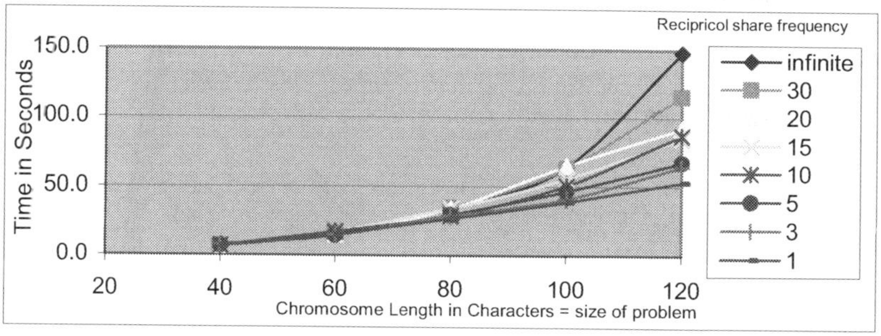

Fig. 2. Search times as chromosome length and share frequency increase

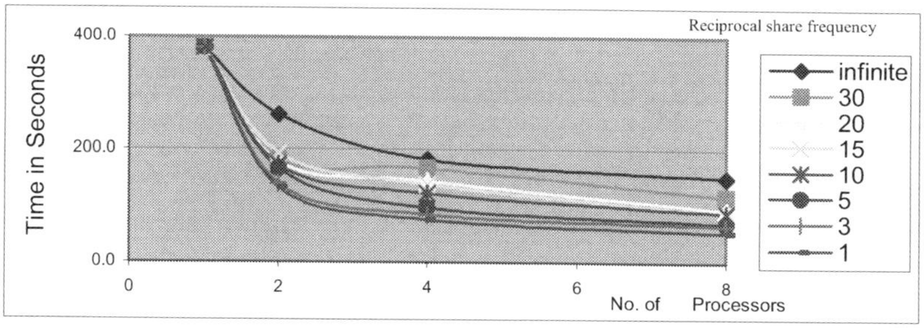

Fig. 3. Comparison of search times as number of processors & share frequency increase, with maximum chromosome length of 120.

Ease of configuration is an important factor when considering an agent architecture but this is something that is true of most, if not all, distributed systems, as the design, implementation, maintenance, and management of these systems is inherently difficult. If execution time is a critical factor then the benefits are likely to outweigh the additional time that is required to configure such systems. The agent approach also proved relatively reliable and robust, as the application was not dependent on particular PCs or intensive communication between them.

6 Conclusions

The evaluation demonstrates the considerable potential agents offer as a means for implementing parallel systems, particularly where execution times are an issue. Considerable benefits were achieved simply by distributing the search, as the most difficult search saw marked performance gains each time the processor count was doubled. Although the example problem was simple, the search system did not 'know' this, so there is no reason to believe that these results would not scale to other, more complex problems. No training data was required for the tests used, real world problems may need to distribute example data. For smaller systems with sufficient local storage this will be a one off distribution, launch is slowed but run times will remain the same, unless redistribution occurs. Internal systems (e.g Ethernet/Intranet) will distribute data more quickly than long distance communication. Where needed, continual reference to large databases over slow networks will slow performance and the agent system may not be ideal.

Further experiments are required to explore solution share size, selection and replacement strategies. Parameter optimisation should be problem dependent, further experiments need be carried out in a range of GA application domains to discover whether consistent patterns emerge.

References

Geyer-Schulz, A. & Kolarik, T., 1992. Distributed computing with APL. *Proceedings of the international conference on APL*, 60-69.

Ghanea-Hercock, R. & Collis, J.C. & Ndumu, D.T., 1999. Co-operating mobile agents for distributed parallel processing. *In: Proceedings of the 3rd annual conference on autonomous agents*, Seattle, WA, USA, 398-399.

Goldberg, D.E., 1989. *Genetic algorithms in search, optimisation, and machine learning*. Reading, Massachusetts, USA: Addison-Wesley Longman.

Holland, J.H., 1962. Outline for a logical theory of adaptive systems. *Journal of the ACM*, 9(3), 297-314.

Knapik, M. & Johnson, J., 1998. *Developing intelligent agents for distributed systems: Exploring architecture, technologies & applications*. NY.: McGraw-Hill

Konstam, A.H., Hartley, J. & Carr, W.L., 1992. Optimisation in a distributed processing environment using genetic algorithms with multivariate crossover. *Proceedings of the 1992 ACM annual conference on communications*, 109-116.

Lange, D.B. & Oshima, M., 1998. *Programming and deploying java mobile agents with aglets*. Reading, Massachusetts, USA: Addison-Wesley Longman.

Logar, A.M., Corwin, E.M. & English, T.M., 1992. Implementation of massively parallel genetic algorithms on the MasPar MP-1. *Proceedings of the 1992 ACM/SIGAPP symposium on applied computing*, 2, 1015-1020.

Ndumu, D.T., Nwana, H.S., Lee, L.C. & Collis, J.C., 1999. Visualising and debugging distributed multi-agent systems. *In: Proceedings of the 3rd annual conference on autonomous agents*, Seattle, WA, USA, 326-333.

Ochi, L.S., Drummond, L.M.A. & Figueiredo, R.M.V., 1997. Design and implementation of a parallel genetic algorithm for the travelling purchaser problem. In: *Proc. of 1997 ACM symposium on applied computing*, SJ, California, 257-262.

Talbi, E.G. & Bessiere, P., 1991. A parallel genetic algorithm for the graph partitioning problem. *Proceedings of the 1991 int.l conf. on supercomputing*, 312-320.

Wooldridge, M. & Jennings, N.R., 1998. Pitfalls of agent-oriented development. In: *Proc. of the 2nd int. conf. on autonomous agents*, Minneapolis, USA, 385-391.

Comparison of Domination Approaches for Diploid Binary Genetic Algorithms

Sima Uyar and A. Emre Harmanci

Istanbul Technical University, Computer Engineering Department
Maslak TR80626 Istanbul – Turkey
e-mail: {etaner, harmanci}@cs.itu.edu.tr

1 Introduction

Most studies in application of genetic algorithms deal with static environments. However there is a class of problems where the environment changes in time. From a genetic algorithm point of view, the change may be in the fitness function, in the constraints or the problem instance itself. In such environments, it becomes important for the solution approach to adapt to this change and follow the new optima. Classical genetic algorithms [2] do not have the necessary mechanisms to address the issues encountered when working in such environments. Thus it becomes necessary either to make modifications to existing algorithms or to incorporate other features. The main issues when dealing with dynamic environments is preserving diversity in the gene pool of the population and being able to converge to a solution on the phenotype level. The modifications or the new features have to take this issue into account and incorporate a balance between preserving diversity and converging to a solution. There are several different approaches in literature which deal with variations of genetic algorithms suitable for working in dynamic environments. Since the choice of a suitable approach depends on analyzing the nature of the change in the environment, it is worthwhile to categorize the change based on specific properties. A good set of criteria for this purpose is given in [1] as the frequency of change, the severity of change, the predictability of change and the cycle length / cycle accuracy. Using diploid representations is one of the approaches for addressing performance and diversity issues in dynamic environments. When using a diploid representation for individuals, the choice of a good domination mechanism is a very important factor in performance. There has been some research done in the area of domination in diploid genetic algorithms and a good survey can be found in [1].
The main aim of this study is to give performance comparisons of some of the existing major domination approaches, show their weaknesses and propose a new, adaptive approach that will perform better in the cases where these fail.

2 Domination Approaches Chosen for Comparisons

There are several diploid approaches in literature and a survey of these may be found in [1]. It has been shown in [3] and [6] that in the case of dynamic environments, some type of a dominance change mechanism is needed. The approaches which do not incorporate this feature do not perform better than a simple, haploid genetic algorithm. Based on these results, the approaches with fixed dominance mechanisms will not be used for comparisons in this study. In [3], the domination approach proposed by Ng-Wong [4] and the additive diploidy approach proposed by Ryan [5] are modified to have a dominance change mechanism as is explained in the following subsections. The results reported in [3] conclude that both approaches are good for environments where the change oscillates between two solutions with fixed intervals but fail to give the same level of performance for the case where the change instances and severities occur randomly.

2.1 Modified Ng-Wong Approach

In the basic Ng-Wong approach [4], there are four alleles, two of which are dominant given by 0 and 1 and the others recessive given by i and o. In the phenotype, the dominant allele is expressed. As given in [3], Ng-Wong allows a dominance change to occur when the fitness of a population member drops by a particular percentage between successive evaluation cycles. Dominance change is achieved by inverting the dominance values of all allele pairs, i.e. 11 becomes ii, 00 becomes oo, 1o becomes i0 and 0i becomes o1.

2.2 Modified Additive Diploidy

In the basic additive diploidy approach [5], alleles are represented by ordered values that are combined using a pseudo-arithmetic to obtain the phenotype. In [5], each value is associated with a number (A:2, B:3, C:7, D:9) and addition is performed. If the resulting sum is above a threshold (chosen as 10), the phenotype becomes 1 and if it is below, it becomes 0. In [3], the above approach is modified to incorporate a dominance change mechanism where alleles in the genotype are demoted or promoted by one grade (demoting a B makes it an A and promoting it makes it a C). As in the Ng-Wong approach, dominance change is allowed to occur when the fitness of a population member drops by a particular percentage between successive evaluation cycles. To achieve the dominance change, for each locus one of the genotypic alleles is chosen at random and if the phenotypic expression for this locus is 1, then the chosen genotypic allele is demoted (unless it is an A) or else if the phenotypic expression for this locus is 0, then the chosen genotypic allele is promoted (unless it is a D).

3 The Proposed Domination Approach

The adaptive dominance approach proposed in this study is loosely based on a natural phenomenon known as the penetrance of a gene [8]. The main factor for this phenomenon in nature is environmental change. The level of penetrance for a gene may be calculated as the proportion of individuals with a given genotype which exhibit a particular phenotype. Under incomplete penetrance, a dominant trait may not be expressed in the phenotype for a given genotype. This implies that the level of penetrance of a gene may change when environmental conditions are altered. For the proposed approach in this study, individuals are represented with three binary strings, i.e. two chromosome strings making up the genotype and one phenotype string. A global domination map is used for genotype to phenotype mapping. The length of the map is the same as the length of the chromosomes. Each location on the map may take on a real value in [0.0,1.0]. This value shows the domination factor of the allele 1 over the allele 0 for the corresponding location on the chromosomes. For example if the value for the 5th location on the domination map is 0.7 and the corresponding 5th locations on the chromosomes of an individual are different, the individual will show the allele 1 in its phenotype for the 5th location with probability 0.7 and the allele 0 with probability 0.3. This domination map is re-evaluated at the end of each generation using Equation 1.

$$dom[i] = \frac{\sum_j (p_{ij} * f_j)}{\sum_j f_j} \quad i = 1,2,.., ChrmLen \;\; and \;\; j = 1,2,., PopSize \tag{1}$$

where p_{ij} shows the ith phenotypic value of the jth individual and f_j shows the fitness of the jth individual. All locations on the domination map are initialized to 0.5, meaning that allele 1 and allele 0 are equally dominant in the beginning. Since the domination map is re-calculated in each generation based on the current fitness values of individuals, the adaptation is automatic and it is not required to explicitly detect the change to make the necessary modifications.

4 Experimentation

In the tests performed for this study, the modified Ng-Wong and additive diploidy approaches are compared against the proposed domination mechanism through a controlled set of changes using a modified 0/1 knapsack problem. In the tests, change is introduced through changing the weight constraint for the knapsack instance. The knapsack is chosen in such a way that for each weight constraint, there is only one possible solution. Two types of change cases are used for comparisons; in case 1 the optima oscillate between two values with fixed intervals and in case 2 the change is random in both occurrence instance and severity. For case 1 the weight constraint oscillates between two fixed values every 500 generations. The change instances for the random change case are given in Table 1. The *generation no.* shows the generation at which the change occurs, the *weight constraint* shows the constraint to be applied, the *optimum value* is the actual optimum solution for the knapsack for the given constraint and the *hamming distance* is the hamming distance between the previous and the current actual optimum solutions.

Table 1. Discussed change instances for the random change case

Generation no.	Weight Constraint	Optimum Value	Hamming Distance
0	22	112648	0
1500	100	131066	9
2100	40	128512	11

All algorithms use binary representation for the genes, two point cross over, roulette wheel selection and non-overlapping populations and are run using the same set of parameters: The population size is 250, the cross-over probability is 0.75 and the mutation probability is 0.005. Controlled tests where the actual best fitnesses for each weight constraint is known is performed. The comparisons are done based on the ability of the dominance approaches in tracking the change in the environment for each different type of change case. The tests are performed for all test cases. The results are given as averaged over 5 runs for each test case. The results obtained for the random change cases for each domination approach is given in Table 2. The *value found* column gives the fitness value obtained by each approach for each change interval and the *generation* column gives the generation number at which this given best fitness is reached by each algorithm for each change interval.The plots for the results obtained are given at the end of the paper. Figures 1a, 1c, 1e show the plots for the case where the change oscillates every 500 generations and Figures 1b, 1d, 1f show the plots for the cases where the change occurs randomly as given in Table 1. In all plots, the x-axis shows the number of generations and the y-axis shows the fitness values found. On all plots, the actual fitnesses are plotted using a dashed line and the fitness values found by the approaches are plotted using solid lines.

Table 2. Found solutions by each method

Ng-Wong		Additive Diploidy		Proposed	
Value Found	Generation	Value Found	Generation	Value Found	Generation
112648	9	112648	25	112648	39
131040	1852	130728	1719	131066	1890
128512	2180	128512	2203	128512	2113

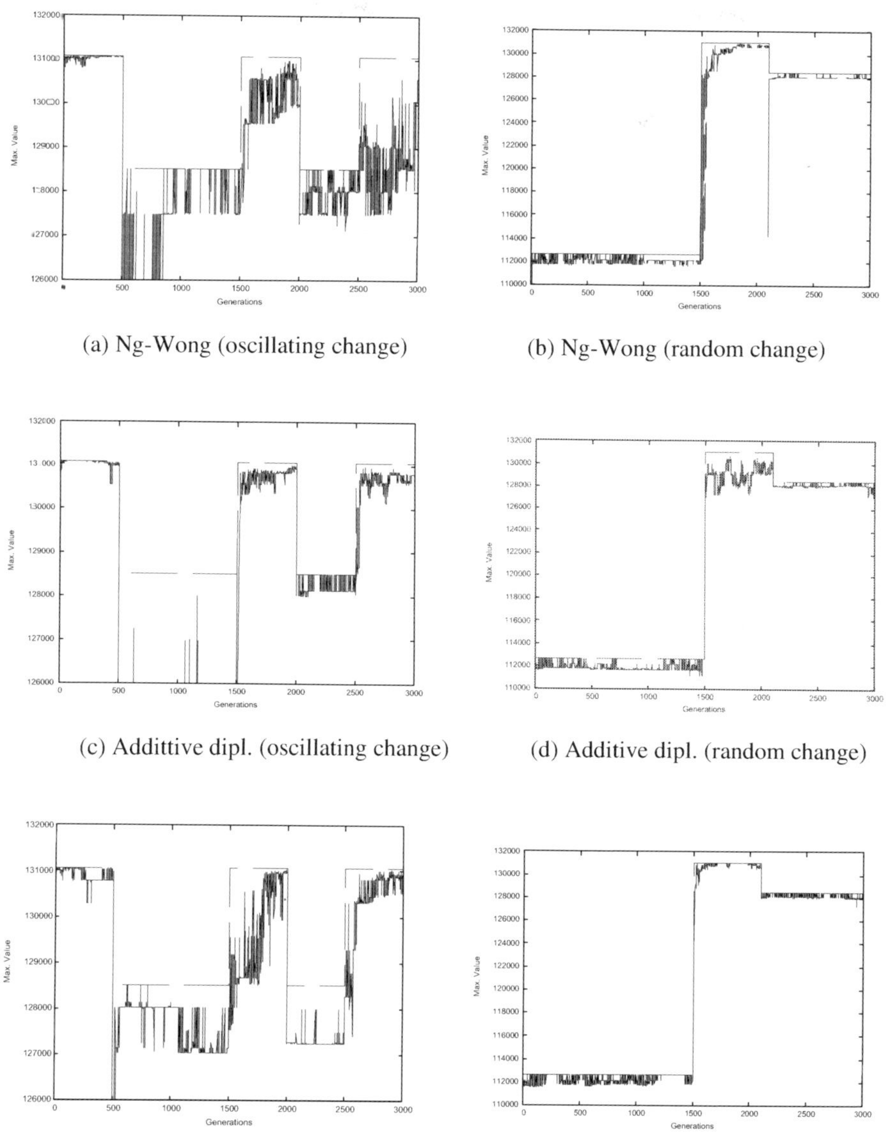

(a) Ng-Wong (oscillating change)

(b) Ng-Wong (random change)

(c) Addittive dipl. (oscillating change)

(d) Additive dipl. (random change)

(e) Proposed appr. (oscillating change)

(f) Proposed appr. (random change)

Fig. 1. Result plots for the tested approaches and the proposed approach for all test cases.

5 Conclusion

The results obtained for the modified Ng-Wong and additive diploidy approaches are in keeping with the ones found in [3]. The proposed approach shows a similar performance to these for the oscillating change case. The re-calculation of the domination map at the end of each generation brings extra computational load, however unlike the other two approaches, for the proposed approach it is not required to detect the change in the environment. Detecting when the change occurs is an important issue in dynamic environments and a good detection scheme affects the performance of the algorithm. In the proposed approach, the adaptation is done automatically based on a feedback obtained from the population. Because of this, there is no need to do calculations to explicitly detect the change. The detection is implicitly achieved by the adaptive property of the domination mechanism. This property makes the proposed approach very effective in dynamic environments. In the cases where the change is random, the two tested algorithms do not give an acceptable performance as noted also in [3]. When the severity of the random change is moderate or low, the proposed approach shows good performance in adapting to the change. When the severity of the change is high, even the proposed approach may not be able to give acceptable results. These types of change instances require very high levels of diversity to be present in the population which diploidy alone may not be able to provide. Additional mechanisms to provide this are needed. These mechanisms are further discussed and explored in [7] by the same authors.

References

[1]Branke J. (2002), Evolutionary Optimization in Dynamic Environments, Kluwer Academic Publishers.

[2]Goldberg D. E. (1989), Genetic Algorithms in Search, Optimization and Machine Learning, Addison-Wesley.

[3]Lewis J., Hart E., Graeme R. (1998), "A Comparison of Dominance Mechanisms and Simple Mutation on Non-Stationary Problems", Proceedings of Parallel Problem Solving from Nature, Springer.

[4]Ng K. P., Wong K. C. (1995), "A New Diploid Scheme and Dominance Change Mechanism for Non-Stationary Function Optimization", 6th International Conference on Genetic Algorithms.

[5]Ryan C. (1994), "The Degree of Oneness", in Proceedings of the 1994 ECAI Workshop on Genetic Algorithms, Springer Verlag.

[6]Smith R. E., Goldberg D. E. (1992), "Diploidy and Dominance in Artificial Genetic Search", Complex Systems, Vol. 6, pp.251-285.

[7]Uyar A. S., Harmanci A. E. (2002), "Preserving Diversity Through Diploidy and Meiosis for Improved Genetic Algorithm Performance in Dynamic Environments", LNCS Vol. 2457, pp. 314-323, Springer.

[8]Weaver R. F., Hedrick P. W. (1997), Genetics 3rd Ed., Wm. C. Brown Publishers.

New Heuristics for Flowshop Scheduling

Uday K. Chakraborty

Math & Computer Sc. Dept., Univ. of Missouri, St. Louis, MO 63121, USA
uday@cs.umsl.edu

1 The Problem

In the flowshop scheduling problem, n jobs are to be processed on m machines. The order of the machines is fixed. We assume that a machine processes one job at a time and a job is processed on one machine at a time without preemption. Let $t_p(i,j)$ denote the processing time of job j on machine i and $t_c(i,j)$ denote the completion time of job j on machine i. Let J_j denote the j-th job and M_i the i-th machine. The completion times of the jobs are obtained as follows. For $i = 1, 2, \ldots, m$ and $j = 1, 2, \ldots, n$, $t_c(M_1, J_1) = t_p(M_1, J_1); t_c(M_i, J_1) = t_c(M_{i-1}, J_1) + t_p(M_i, J_1); t_c(M_1, J_j) = t_c(M_1, J_{j-1}) + t_p(M_1, J_j); t_c(M_i, J_j) = max\{t_c(M_{i-1}, J_j), t_c(M_i, J_{j-1})\} + t_p(M_i, J_j)$. *Makespan* is defined as the completion time of the last job, $t_c(M_m, J_n)$. The goal is to obtain the n-job sequence that minimizes the makespan. The search space consists of $n!$ possible job sequences. The problem is NP-hard.

2 The Proposed Heuristics

2.1 Heuristic H

This heuristic improves upon NEH [4], the best-known deterministic heuristic for minimizing makespan. The algorithm is given below:

1. For each job i, find the total processing time T_i which is given by $T_i = \sum_{j=1}^{m} t_p(j,i)$ where $t_p(j,i)$ is the processing time of job i on machine j.
2. Sort the n jobs on descending order of their total processing times.
3. Take the first four jobs from the sorted list and form $4! = 24$ partial sequences (each of length 4). The best k (k is a parameter of the algorithm) out of these 24 partial sequences are selected for further processing. The relative positions of jobs in any partial sequence is not altered in any later (larger) sequence.

4. Set $z = 5$.
5. The z-th job on the sorted list is inserted at each of the z positions in each of the k $(z-1)$-job partial sequences, resulting in $z \times k$ z-job partial sequences.
6. The best k out of the $z \times k$ sequences are selected for further processing.
7. Increment z by 1.
8. If $z > n$, accept the best of the k n-job sequences as the final solution and stop. Otherwise go to step 5.

Step 1 of our algorithm computes a sum of m terms for each of the n jobs and is thus of complexity $\Theta(mn)$. Step 2 involves sorting n items and can be implemented using any good algorithm from the literature. Quicksort, with an average-case complexity of $\Theta(n \lg n)$, is a natural choice. Step 3 takes a constant amount of time. Steps 5 through 8 together take a total time given by $\sum_{z=5}^{n} k \times z \times \text{TMS}(z)$ where $\text{TMS}(z)$ denotes the time to compute the makespan for a z-job partial sequence. For a direct comparison with NEH [4], we note that the total number of enumerations (of partial and complete sequences) in the present method is $4! + \sum_{z=5}^{n} k \times z = 4! + k \times \sum_{z=5}^{n} z = \Theta(n^2)$. The number of enumerations in NEH was shown [4] to be $n(n+1)/2 - 1$ which, clearly, is $\Theta(n^2)$. Thus the asymptotic time complexity of our method is the same as that of NEH.

2.2 Heuristic H1

This algorithm modifies NEH by replacing the best partial sequence obtained at any stage of the algorithm by a randomly generated adjacent (i.e., neighboring) sequence. The neighbor is generated from the original by a single swap of a pair of (independently) randomly picked jobs.

1. For each job, find the total processing time of the job on all machines.
2. Sort the n jobs on descending order of their total processing times.
3. Take the first two jobs from the sorted list and compute the makespans of the two possible partial sequences of these two jobs. The better of the two partial sequences is selected for further processing. The relative positions of jobs in any partial sequence is not altered in any later (larger) sequence.
4. For $i = 3$ to n do
   - Without disturbing the relative positions of jobs in the best partial sequence obtained in the immediately preceding step, place the i-th job at all the i positions in the best partial sequence, and save the best of the i resulting (partial) sequences as the best i-job partial sequence for further processing.
   - Randomly interchange any two jobs in the best partial sequence obtained in the above step, find the makespan of the modified sequence, and accept that as the current best partial sequence if its makespan is better, otherwise continue working with the original partial sequence.
5. Output the current best sequence as the final solution.

2.3 Heuristic H2

Algorithm H2 is the same as H1 except for step 4 which is changed to:
For $i = 3$ to n do

- Without disturbing the relative positions of jobs in the best partial sequence obtained in the immediately preceding step, place the i-th job at all the i positions in the best partial sequence, and save the best of the i resulting (partial) sequences as the best i-job partial sequence for further processing.
- Randomly interchange any two jobs in the best partial sequence obtained in the above step, find the makespan of the modified sequence, and accept that as the current best partial sequence if its makespan is better. Otherwise accept it with probability $e^{-\Delta/T}$, where Δ = makespan (modified) - makespan (original), and T is a parameter of the algorithm that is held fixed at a predetermined value during the course of a run.

2.4 Heuristic H3

This heuristic applies simulated annealing to improve upon the best partial sequence obtained at any step of the algorithm. The algorithm is described below:

1. For each job, find the total processing time of the job on all machines.
2. Sort the n jobs on descending order of their total processing times.
3. Take the first two jobs from the sorted list and compute the makespans of the two possible partial sequences of these two jobs. The better of the two partial sequences is selected for further processing. The relative positions of jobs in any partial sequence is not altered in any later (larger) sequence.
4. For $i = 3$ to n do
   a) Without disturbing the relative positions of jobs in the best partial sequence obtained so far, place the i-th job at all the i positions in the best partial sequence, and save the best of the i resulting (partial) sequences as the best i-job partial sequence for further processing.
   b) Initialize Max-iterations, Temp-start. Set Count = 1, T = Temp-start. Let the best partial sequence obtained in the previous step be called the current sequence, x_c. Compute makespan(x_c).
   c) Randomly generate a neighboring sequence by using either the *interchange neighborhood* or the *shift neighborhood* (these neighborhoods are explained below). Let the neighboring sequence be called the adjacent sequence, x_a. Compute makespan(x_c).
   d) If makespan(x_a) < makespan(x_c) then set $x_c = x_a$;
   else
   Set Δ = makespan(x_a) - makespan(x_c); Set T = Temp-start / log(1+Count); With probability $e^{-\Delta/T}$ set $x_c = x_a$.
   e) Increment Count by 1; If Count < Max-iterations, go to step (c).

5. Output the current best sequence as the final solution.

The annealing schedule used in step 4(d) of the above algorithm is due to Hajek [1]. For performance comparisons with the Osman-Potts (OP) algorithm [6], we implemented two variants of neighborhood: "interchange" and "shift".

3 Experimental Results

Algorithms H, H1, H2 and H3 were run on 28 different problem sizes (n = 12, 18, 24, 30, 40, 50, 100 and m = 5, 10, 15, 20). For each problem size, 15 independent problem instances were created. Each problem instance corresponds to a new t_p matrix: each processing time ($t_p(.,.)$ value) was independently obtained from a uniform random u(1,99) discrete distribution. Table 1 shows makespan values (averaged over 15 independent instances) obtained by NEH and heuristic H for two values of k: 6 and 24. For space limitations, only a few entries are shown in each table. These values of k are only representative; we did not attempt any tuning for the parameter k. These results bring out the superiority of the proposed approach H. Table 2 shows results of statistical tests of significance for a representative case ($k = 24$). Each test suite gives us 15 pairs of makespan values and we thus have a paired comparison. We now test the hypothesis that the population corresponding to the differences has mean, μ, zero. If the hypothesis is true, $T = \sqrt{N}\frac{\bar{X}-\mu_0}{S}$ has a t-distribution with $N-1$ degrees of freedom. The critical value c is obtained from the relation $Prob(T > c) = \alpha = 0.05$.

Two additional metrics have been used to quantify the improvement. These are the average relative percentage deviation (ARPD) and the maximum percentage deviation (MPD). The best possible performance corresponds to both ARPD and MPD being zero.

To our knowledge, the HFC heuristic [3] is the only competitor of NEH to date, but as the author of HFC admits, the HFC is no better than NEH for permutation flowshop problems (HFC is better than NEH only for non-permutation problems). Thus heuristic H is better than HFC, too.

For H1, H2 and H3 the makespan of each instance is obtained by averaging five independent runs (these independent runs are taken by varying the pseudo-random number generator seed). The same averaging is done on the makespan produced by the Osman-Potts algorithm. In Tables 3 – 6 each test suite comprises 15 independent instances. For Tables 7 – 10 this number is 5. The makespan values appearing in Table 5 are averages of 15 makespans obtained from as many problem instances (this averaging is not to be confused with the averaging mentioned above *for each instance*). The makespans in Tables 3 and 4 are also averages over 15 instances, but unlike the values in Table 5, each of the 15 individual makespans is the best (*not* the average) of 5 independent runs. The results show that the both H1 and H2 produce solutions of a better quality than NEH for all problem sizes but one. Heuristic

H3 is superior to H1 or H2 and outperforms NEH in all cases. For each test suite, the mean and the standard deviation of the fifteen (or five) differences in makespan are used to compute the sample t. Almost all cases show that the proposed algorithms produce makespans that are statistically significantly better than those produced by the best-known existing heuristics. The critical t values, as obtained from tables of the t-distribution, corresponding to 14 and 4 degrees of freedom are 1.76 and 2.15, respectively. Comparisons between H3-interchange and Osman-Potts-interchange (Tables 9 and 10) and between H3-shift and Osman-Potts-shift (Tables 7 and 8) show the presence of a few cases where the sample t value is less than the critical t; however, even for these cases the ARPD and MPD metrics show the superiority of the proposed methods.

To our knowledge, the most recent simulated-annealing based heuristics for permutation flowshop problems are Ogbu and Smith [5] and Ishibuchi et al. [2]. The OP heuristic was shown [5, 2] to be slightly better than the Ogbu-Smith method. The results of Ishibuchi et al. are not statistically significantly better than those of OP; in fact, the main advantage of the algorithms of [2] is their robust performance with respect to the choice of a cooling schedule, and their algorithms have "comparable performance" [2] to OP.

Table 1. Performance comparison between NEH and heuristic H.

St#	n	m	Avg MS		
			NEH	H	
				$k=6$	$k=24$
1	12	5	838	834	831
2	24	20	2530	2492	2477
3	100	20	6660	6593	6580

References

1. Hajek B (1988) Cooling schedules for optimal annealing, *Math. Ops. Res.* 13:311-329.
2. Ishibuchi H, Misaki S, Tanaka H (1995) Modified simulated annealing algorithms for the flow shop sequencing problem, *European Journal of Operations Research* 81:388-398.
3. Koulamas C (1998) A new constructive heuristic for the flowshop scheduling problem, *European J. of Operations Research* 105:66-71.
4. Nawaz M, Enscore Jr EE, Ham I (1983) A heuristic algorithm for the m-machine n-job flowshop sequencing problem, *OMEGA* 11:91-95.
5. Ogbu FA, Smith DK (1990) The application of the simulated annealing algorithm to the solution of the n/m/Cmax flowshop problem, *Computers & Opr. Res.* 17:243-253.
6. Osman IH, Potts CN (1989) Simulated annealing for permutation flowshop scheduling, *OMEGA* 17:551-557.

Table 2. Results of statistical tests. Heuristic H is run with $k = 24$.

St#	Difference		t	ARPD		MPD	
	Mean	S.D.		NEH	H	NEH	H
1	6.8	7.103	3.71	0.853	0	2.570	0
2	52.93	21.96	9.335	1.837	0	3.955	0
3	79.53	58.51	5.26	1.034	0.028	2.415	0.341

Table 3. Performance comparison between NEH and heuristic H1.

St#	n	m	Avg MS		Difference		t
			NEH	H1	Mean	S.D.	
1	40	10	2689	2665	23.067	36.082	2.476
2	40	20	3459	3434	25.33	31.66	3.099
3	50	10	3157	3136	20.4	25.809	3.06
4	50	20	4040	4011	28.6	46.763	2.369
5	100	10	5820	5800	20.067	35.358	2.198
6	100	20	6660	6660	-0.267	64.713	-0.016

Table 4. Performance comparison between NEH and heuristic H2.

St#	n	m	Avg MS		Difference		t
			NEH	H2	Mean	S.D.	
1	40	10	2689	2666	22.0	28.498	2.99
2	40	20	3459	3442	16.6	30.849	2.084
3	50	10	3157	3130	26.3	31.176	3.27
4	50	20	4040	4007	32.8	44.934	2.827
5	100	10	5820	5798	21.67	41.587	2.018
6	100	20	6660	6653	6.73	56.739	0.459

Table 5. Performance comparison between NEH and heuristic H3-interchange.

Suite #	n	m	Average makespan	
			NEH	H3-in
1	12	5	838	831
2	24	20	2530	2511
3	100	20	6660	6624

Table 6. Results of statistical tests: NEH versus H3-interchange.

St#	Difference		t	ARPD		MPD	
	Mean	S.D.		NEH	H3	NEH	H3
1	6.533	7.269	3.48	0.821	0	3.202	0
2	19.733	24.953	3.063	0.886	0.099	2.411	0.756
3	35.53	65.928	2.087	0.716	0.187	2.399	1.022

Table 7. OP-shift vs. H3-shift.

Suite #	n	m	Average makespan	
			OP-shift	H3-sh
1	50	10	3104.4	3097.6
2	100	20	6589.4	6574.6

Table 8. Results of statistical tests: OP-shift versus H3-shift.

St#	Difference		t	ARPD		MPD	
	Mean	S.D.		OP-sh	H3-sh	OP-sh	H3-sh
1	6.84	8.344	1.83	0.042	0.022	0.611	0.11
2	14.9	20.092	1.658	0.227	0	0.749	0

Table 9. OP-interchange vs. H3-interchange.

Suite #	n	m	Average makespan	
			OP-in	H3-in
1	50	10	3142.6	3132.1
2	100	20	6636.8	6584.5

Table 10. Results of statistical tests: OP-interchange versus H3-interchange.

St#	Difference		t	ARPD		MPD	
	Mean	S.D.		OP-in	H3	OP-in	H3
1	12.28	9.624	2.853	0.0359	0	0.8317	0
2	52.28	21.052	5.553	0.729	0	1.174	0

Temporal Rule Discovery using Genetic Programming and Specialized Hardware

Magnus Lie Hetland[1] and Pål Sætrom[2]

[1] Norwegian University of Science and Technology, Dept. of Computer and Information Science,
Sem Sælands vei 9, NO–7491 Trondheim, Norway
magnus@hetland.org

[2] Interagon AS, Medisinsk-teknisk senter,NO–7489 Trondheim, Norway
paalsat@interagon.com

Abstract. Discovering association rules is a well-established problem in the field of data mining, with many existing solutions. In later years, several methods have been proposed for mining rules from sequential and temporal data. This paper presents a novel technique based on genetic programming and specialized pattern matching hardware. The advantages of this method are its flexibility and adaptability, and its ability to produce intelligible rules of considerable complexity.

1 Introduction

Discovering association rules is a well-established problem in the field of data mining, with many existing solutions. In later years, several methods have been proposed for mining rules from sequential and temporal data (see, for example, [1–3]). Quite a few of these methods are based on the common premise of counting the occurrences of viable rules or patterns. While this approach has the advantage of finding all highly frequent patterns, it constrains the set of possible solutions.

One promising alternative is to use evolutionary algorithms, as described in [4]. While this approach places fewer restrictions on the form of patterns and rules that can be discovered, the performance of the method relies heavily on a speedy evaluation of each candidate rule, and such an evaluation typically involves examining the entire training data set. When mining rules in relational databases, existing indexing methods make it possible to efficiently calculate the fitness of each rule. When mining sequences for complex patterns, this evaluation is not quite as straightforward. Efficient indexing methods for some forms of patterns exist, (for example, using Patricia trees, as in [5], or multigram indices, as in [6]), but in this paper we use a specialized co-processor designed to perform very advanced, high volume pattern matching.

The paper is structured as follows: Sect. 1.1 describes the problem we are trying to solve in more formal terms; Sect. 1.2 gives a brief overview of some related work; Sect. 2 describes the specifics of our method; Sect. 3 contains some empirical results; and, finally, Sect. 4 concludes the paper.

1.1 Problem Definition

Our problem is as follows: Given a sequence s, a predicate p over all the indices of s, and a delay δ, find a rule that estimates the value of $p(i+\delta)$ from the prefix $s_1, \ldots, s_i$ of s. The estimated predicate is written $\hat{p}$. In the terminology of [7] this is a problem of sequence recognition, although by letting the predicate p represent a type of event that may or may not occur at time $i + \delta$, the rules can also be used to make predictions. Note that we take all available history into consideration by using a full prefix of s, rather than a fixed-width sliding window.

1.2 Related Work

A survey of association rule mining algorithms can be found in [8]. The underlying principle in these algorithms is exploiting the lattice structure of the pattern space when searching for frequent patterns and rules. This principle has also been applied to sequence mining [1–3]. The assumptions about the pattern space makes this general approach unsuitable for mining more complex patterns and rules.

Even though the problem tackled in this paper is closer to that of sequence prediction than that of sequence mining (see [7] for several sequence prediction algorithms), the goal of our method is to find rules that are readable and understandable by a human expert. Since this is one of the fundamental goals of data mining and knowledge discovery, we have chosen to classify our method as a rule discovery method.

A problem similar to ours is tackled in [9], where Giles et al. use recurrent neural networks to predict fluctuations in foreign exchange rates. In addition to the prediction task, their method encompasses the extraction of deterministic finite state automata, which are equivalent to regular expressions. Like most current sequence learning methods, the algorithm works with a fixed-width sliding window. We have tested our method on the same data sets as [9] in Sect. 3.3.

2 Method

To evolve our predictor rules we use genetic programming, where each individual in the population represents a single rule. Since the consequent is a part of the problem definition, each rule is represented by its antecedent, an expression in a general query language that can be interpreted by our pattern matching hardware (see Sect. 2.1).

The training data consist of a discrete sequence s with elements from some finite alphabet, and a predicate p, represented by the set of indices for which p is true. The rules are evaluated by having the hardware interpret their antecedents as queries, and comparing the returned hit positions, that is, $\hat{p}$, with the correct positions given by p.

Compared to existing methods for discovering sequential rules, our method has two main advantages: (1) It can produce rules in a very rich rule language, which can be customized to each application as needed, and (2) each rule has the entire history of its sequence available when making a prediction. This means that the method is quite flexible, and that restrictions on the rule format are mainly dictated by the problem, rather than by the method itself.

The hardware and the process of fitness evaluation are discussed in the following subsections.

2.1 The Pattern Matching Chip

To make it possible to perform a full search in the training data for each fitness calculation, we used a specialized pattern matching chip (PMC). The PMC is a special purpose co-processor designed for locating complex patterns in unstructured data [10]. This section gives a brief description of how the PMC works.

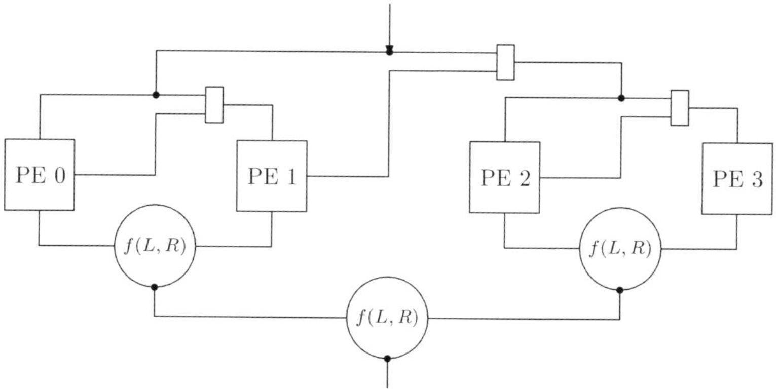

Fig. 1. A data distribution tree with four leaf nodes (PEs), and the corresponding result gathering tree, with $f(L, R)$ calculated from the above left (L) and right (R) results

The PMC consists of three functional units, as illustrated in Fig. 1: A data distribution tree (top), a set of processing elements, or PEs (middle), and a result gathering tree (bottom). The PEs monitor the data flowing through the chip. They receive the data from the data distribution tree, which can be configured so that single PEs and groups of PEs receive data either sequentially or in parallel. Each PE is configured with one byte (character) and a comparison operator, which it uses to look for bytes in the data stream. Matches are reported to the result gathering tree, which combines them and produces the final result, a Boolean value representing a hit or miss for the entire pattern. If the result is a hit, the hit location is reported back to the host computer.

The PMC is capable of performing searches at the rate of 100 MB/s and can handle several complex queries in parallel (from 1 to 64 depending on query complexity.)[3] Several PMCs may be combined for highly parallelized searches. The interface to the PMC is the special purpose query language IQL [11]. This language supports such functionality as regular expressions, latency (distance), alpha-numerical comparisons, and generalized Boolean operations. The regular expression syntax is similar to that of UNIX tools such as `grep`.

2.2 Rule Evaluation

In evolutionary algorithms (such as genetic programming) each individual of the population must be assigned a fitness score, which describes how well the individual solves the problem at hand. In rule mining there are several possible quality measures, including various measures of interestingness. In this paper we focus on predictive power, because the events to be predicted (given by p) are part of the problem definition, which makes most interestingness measures unsuitable.

Some measures of predictive power (precision, true positive rate, true negative rate, and accuracy rate, as well as some combinations) are described in [4, pp. 129–134]. In this paper we have used another measure, the correlation coefficient of the set of data points given by $(p(i), \hat{p}(i))$, for $1 \leq i \leq n$ (where n is the size of the training sequence). This can be interpreted as the cosine of the angle between two n-dimensional vectors p and $\hat{p}$, which means that we get a fitness of $+1.0$ for perfect prediction and -1.0 for completely erroneous prediction.

By classifying the hits reported by a given rule as true or false positives (correct or incorrect hits), the correlation can be expressed as

$$r(p, \hat{p}) = \frac{\text{TP} \cdot \text{TN} - \text{FP} \cdot \text{FN}}{\sqrt{(\text{TN} + \text{FN})(\text{TN} + \text{FP})(\text{TP} + \text{FN})(\text{TP} + \text{FP})}} \ . \tag{1}$$

Here TP and FP are the number of true and false positives, respectively. The number of true and false negatives (TN and FN) can easily be calculated from TP, FP, and the total number of positives and negatives (directly available from the training data).

[3] The prototype used in our experiments runs at 33 MB/s and handles up to 4 parallel queries.

3 Experiments

The method was tested on three data sets: synthetic data consisting of uniformly random symbols where certain positions were flagged according to predetermined rules; DNA sequence data, where the task was to predict the location of intron/exon splice sites; and foreign exchange rates, where the goal was to predict the trend for the next day (as in [9]). In the first two experiments we were mainly interested in the expressive power of our rule format, while in the last experiment we focused on predictive power.

For the first two data sets we used a "fuzzy" version of the problem definition from Sect. 1.1: For this fuzzy problem, the prediction predicate is the disjunction $p(i+\delta) \vee \cdots \vee p(i+\delta+\varepsilon)$ for some fixed window size ε. For these two data sets we used $\varepsilon = 10$. The original definition, which was used with the currency data, corresponds to $\varepsilon = 1$. For all the data sets we used $\delta = 1$.

For the last two data sets the technique of early stopping was used to prevent overfitting. This simply means that, in addition to a training set and a test set, we used a validation set, and the fitness calculated for this data set was used to select the best rules.

3.1 Synthetic Data

The synthetic data were constructed by repeatedly drawing symbols from the lowercase Latin alphabet (`a`–`z`) with a uniform probability distribution. The hit positions (representing the predicate p) were then found according to some predetermined antecedent patterns.

Two different antecedents types were used:

1. The regular expression `o[^n]*n`.
2. The letters `a`, `b`, `c`, `d` and `e` occurring in any order within a window of width 10.

The two sets consisted of 1 MB of sequence data with about 20000 and 1600 occurrences of antecedent types 1 and 2, respectively.

For the second antecedent type we wanted to simulate a search for fixed-width window rules, so we introduced a new operator into the language, composed from existing operators. The function of this operator was to match any number of arbitrary symbols (specified as parameters to the operator) that were all found within a window of width 10.

The system was able to generate expressions that recognized all occurences of both rules. Table 1 lists two of the expressions generated. As can be seen from this table, both problems were solved with a perfect correlation (100% prediction rate) for the test set. This correlation is certainly a result of the data being particularly well suited for our rule format.

Table 1. Results on synthetic data set

Type	Antecedent Expression	Corr.
1	([^n]+.+o[^n]+)n	1.0
2	{{a:d=9},{b:d=9},{c:d=9},{d:d=9},{e:d=9}:p=5}	1.0

3.2 DNA Sequence Data

The goal of this experiment was to find rules predicting intron/exon splice sites i DNA sequences. In addition to testing for predictive power, the rules were informally validated by comparing them to well-known splice site patterns.

The DNA sequences and exon locations were retrieved from NCBI [12]. The combined data set consisted of more than 6000 DNA sequences totalling 34 MB and about 20000 splice sites (of types $5'$ and $3'$, representing the beginning and end of an intron, respectively). The first 10 MB of this set were used in the training process (8 MB for training and 2 MB for early stopping.) The rest of the set was used for testing the generated rules.

Table 2 lists the rules produced for the $5'$ and $3'$ splice sites. The results are comparable to previously published splice site patterns (see, for example, [13]).

Table 2. Results on DNA data set

Type	Antecedent Expression	Corr.
$5'$	({t[ag]:d=52}{c:c=21})({(<=c):d=3}gg)t[ag]ag	0.266
$3'$	({[ct][ct][ct][cgt][ct][ct][ct][ct]:d=9}[ct]ag)&ag&(>=atctgt)	0.177

3.3 Foreign Exchange Rates

This experiment was performed on 5 sets of foreign exhange rates [14]. Because the number of data points was quite small (fewer than 4000 in total), tenfold cross-validation was used on each data set. The training data were discretized using the clustering method described in [2], and the resulting clusters were used to discretize the test and evaluation sets.

The average correlation and average prediction rate for each of the five currencies is listed in Table 3. Giles et al. [9] report a prediction rate of 52.9% on the same data set.

Due to the sequental nature of the data, the use of cross-validation is problematic. Therefore, we performed the same experiment on a larger data set (the exchange rate of BP to USD from 1971 to 2002 [15]) with about

Table 3. Results on currency data sets

Set	Avg. corr.	Avg. pred.	Max. pred.	Min. pred.
1	0.0465	54.5%	58.7%	50.5%
2	0.0446	54.3%	60.9%	45.1%
3	0.0845	54.8%	61.4%	48.4%
4	0.0527	56.1%	62.0%	52.2%
5	−0.0164	50.6%	57.6%	43.5%

8000 data points, using the first half as the training set. The experiment was repeated five times, giving prediction rates of 52.4%, 64.8%, 59.4%, 52.9%, and 65.1%.

4 Summary and Conclusions

The experiments performed so far seem promising: The method produces rules with good predictive power that are also readable by humans. For the synthetic data, we were able to reproduce the original rules, which contained regular expressions of varying complexity, without restriction on sequence history. For the DNA sequence data, our method produced rules with relatively good predictive power, which resemble well-known intron/exon splice site motifs. For the foreign exchange rates we were able to achieve a prediction rate comparable to that of Giles et al. in [9]. In each case, the rules were fairly easy to interpret, which means that it should be possible for a domain expert to find some structure in them, and perhaps modify them to a simpler or more general form, and to test them in real time, using the pattern matching hardware.

References

1. Agrawal, R., Srikant, R.: Mining sequential patterns. In Yu, P.S., Chen, A.S.P., eds.: Eleventh International Conference on Data Engineering, Taipei, Taiwan, IEEE Computer Society Press (1995) 3–14
2. Das, G., Lin, K.I., Mannila, H., Renganathan, G., Smyth, P.: Rule discovery from time series. In: Knowledge Discovery and Data Mining. (1998) 16–22
3. Mannila, H., Toivonen, H., Verkamo, A.I.: Discovery of frequent episodes in event sequences. Data Mining and Knowledge Discovery **1** (1997) 259–289
4. Freitas, A.A.: Data Mining and Knowledge Discovery with Evolutionary Algorithms. Springer-Verlag (2002)
5. Baeza-Yates, R.A., Gonnet, G.H.: Fast text searching for regular expressions or automaton searching on tries. Journal of the ACM **43** (1996) 915–936
6. Cho, J., Rajagopalan, S.: A fast regular expression indexing engine. In: ICDE. (2002)

7. Sun, R., Giles, C.L., eds.: Sequence Learning : Paradigms, Algorithms, and Applications. Number 1828 in Lecture Notes in Artificial Intelligence. Springer-Verlag (2000)
8. Hipp, J., Güntzer, U., Nakhaeizadeh, G.: Algorithms for association rule mining: A general survey and comparison. SIGKDD Explorations **2** (2000) 58–64
9. Giles, C.L., Lawrence, S., Tsoi, A.C.: Noisy time series prediction using a recurrent neural network and grammatical inference. Machine Learning **44** (2001) 161–183
10. Interagon AS: Digital processing device. PCT/NO99/00308 (2000)
11. Interagon AS: The Interagon query language: A reference guide. `http://www.interagon.com/pub/whitepapers/IQL.reference-latest.pdf` (2002)
12. National Center for Biotechnology Information: The NCBI Web site. `http://www.ncbi.nlm.nih.gov` (2002)
13. Burge, C.B., Tuschl, T., Sharp, P.A.: Splicing of precursors to mRNAs by the spliceosomes. In: The RNA World (Second edition). Cold Spring Harbor Laboratory Press, Cold Spring Harbor, New York (1999) 525–560
14. Weigend, A.S.: Daily exchange rates. `http://www-psych.stanford.edu/~andreas/Time-Series/Data/Exchange.Rates.Daily` (2002)
15. The Federal Reserve Board: Foreign exchange rates, historical data. `http://www.federalreserve.gov/Releases/H10/Hist` (2002)

Genetic Programming with Guaranteed Constraints

Colin G. Johnson

Computing Laboratory, University of Kent at Canterbury
Canterbury, Kent, CT2 7NF, England
email: C.G.Johnson@ukc.ac.uk

Abstract. Genetic programming is a powerful technique for automatically generating program code from a description of the desired functionality. However it is frequently distrusted by users because the programs are generated with reference to a training set, and there is no formal guarantee that the generated programs will operate as intended outside of this training set. This paper describes a way of including constraints into the fitness function of a genetic programming system, so that the evolution is guided towards a solution which satisfies those constraints and so that a check can be made when a solution satisfies those constraints. This is applied to a problem in mobile robotics.

Keywords. genetic programming, constraints, safety, mobile robotics, verification

1 Introduction and Background

1.1 Genetic Programming

Genetic programming [1, 7] is the application of genetic algorithms [3, 8] to the creation of program code. A set of programs are generated at random, tested on a set of training data, and their fitness for the desired task assessed by measuring their performance on those training data against some success metric. The better programs are selected from this set and a new set formed by combining the programs together in some randomized fashion which swaps routines between the programs, and small changes (mutations) are made to the programs. This process is then repeated starting from this new set and iterated until a good solution is found and/or the population stabilizes.

Clearly this cannot be carried out in an arbitrary fashion. The canonical way to ensure that the programs generated are executable is by swapping over components in a LISP parse tree [7]. Here we use the term to describe any application of genetic algorithms to the development of program code.

1.2 Trusting Genetic Programming

One problem with genetic programming is that the results cannot be formally trusted. At the end of a successful GP run the output is a program which performs on the training set, however there is no guarantee that it will perform adequately outside that set. In traditional programming this is not always a problem, because the program has been devised by a human programmer who has some knowledge of the problem which is to be solved, so testing is a confirmation that the problem has been adequately represented in the program logic. However in GP the final program will often be incomprehensible to human programmers so *post hoc* verification of its conformation to specification is difficult.

1.3 Towards Trustable GP—Guaranteed Constraints

One approach to solving this problem is to carry out some analysis of the program to determine whether the program satisfies certain properties. Several analysis techniques, such as static analysis [9] and model checking [2] provide ways to verify that properties of programs hold across all possible execution paths. The idea which is exploited in this paper is to use the output from such techniques as a component of program fitness; in the example below a component of fitness calculated using such an analysis is used alongside a fitness measure (for a different aspect of the problem) measured by execution of the programs on a training set. The benefit of this is that the end result, if the GP run is successful, is a program which works and has certain formally guaranteed properties. More details of this general approach are given in my recent papers [5, 6] and in a recent paper by Huelsbergen [4].

2 A Case Study from Robotics

2.1 Problem Details

The case study problem concerns the motion of a mobile robot reacting to two other moving objects in its environment. The environment is a square, 200×200 units in area. Within this a robot, represented by a point, moves under the control of the program evolved using the algorithm to be described. There are two other objects in the environment, which move around according to a random choice of direction each turn. The first is another point-robot, which shall be referred to as the *target*; the second is a barrier parallel to the y-axis. A snapshot from this is given in figure 1a.

The task for the main robot is to get to and remain as close as possible to the target robot. However the robot must always remain to the right of the barrier. This provides a problem which has both a data-driven component (remain as close as possible to the target) and a fixed constraint (remain to

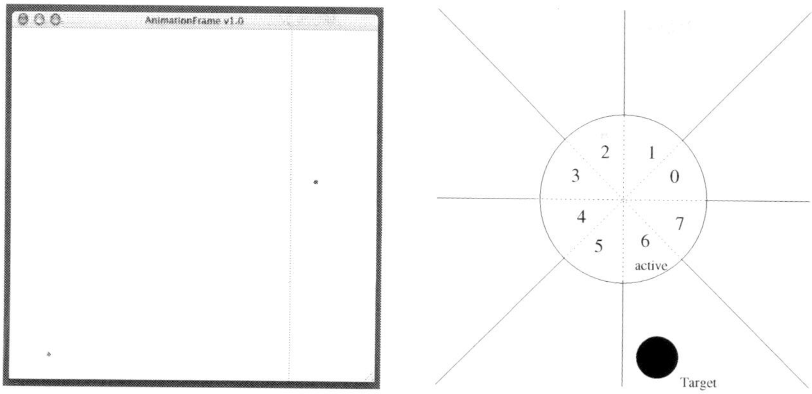

(a) (b)

Fig. 1. The experimental setup.

the right of the barrier). It is assumed that this constraint is a hard constraint, i.e. a solution which does not satisfy this is not acceptable.

The simulation works as follows. A test run of the program consists of 1000 discrete timesteps. Between timesteps the three objects in the environment make a move. At the beginning of the timestep the robot senses its environment. It has 8 sensors to sense the target (figure 1b), and a sensor which detects the shortest distance to the barrier.

The programs being evolved consist of a sequence of statements. The statements are each self-contained, e.g. there are no loops or `if` statements which in turn contain other loops or `if` statements. Each statement consists of either an unconditional piece of arithmetic (incrementing or decrementing either the x or y position of the robot), or the same type of statement conditional on an `if` statement which takes a condition on one of the sensor values or the distance to the barrier.

An individual in the population consists of a program containing 20 such statements. The population consists of 50 individuals. The population evolves by a generational genetic algorithm. Crossover is implemented using uniform crossover at the statement level, and mutation changes the calculation carried out in the statement or the condition in the `if` statements. Selection is roulette wheel selection with an elitist strategy ensuring that the top solution is migrated from one solution to the next. The robot begins each run at the position $(120, 100)$.

The fitness has two components. The first component is of traditional form, consisting of a measure based on the training data. At the beginning of each timestep the (euclidean) distance d from the robot to the target is measured, and the fitness calculated as follows:

$$\mathit{fitness} = \begin{cases} 0.0 & \texttt{if } d > 50.0 \\ 50.0 - d & \texttt{if } d \leq 50.0 \end{cases}$$

These fitnesses are summed over the 1000 timesteps of the run, and these summed over the 50 training cases.

The second component of the fitness ascertains whether the robot is always to the right of the barrier. This is done by carrying out an *interval analysis* of the change in the x position of the variable. An interval analysis consists of tracking a pair of numbers through the program, which represent the maximum and minimum values which the quantity can take at that current point in the program. This is set initially to a conservative approximation to the initial state of the quantity, and as each line is looked at in turn, the values are updated to represent the possible state *regardless* of what other data values were present.

In this example the analysis tracks the (dx_{min}, dx_{max}) values which represents a conservative approximation to the maximum and minimum values which the change in x can take. The initial value is $(0, 0)$—at the beginning of the program no change has been made. A simple arithmetic statement applies that calculation to both values; for example if the first statement in the program is `x++;`then the interval changes to $(1, 1)$. When a conditional statement is met then the interval is expanded to contain the outcome of executing either state. So for example if the next statement is

```
if (sensor[3] is active)
  {
    x++;
  }
```

then the interval becomes $(1, 2)$, because either the condition is true in which case the value can increase by one, or it is not in which case the value stays the same.

Two such interval analyses are done. One tracks the maximum and minimum x values in the case where the robot is "close" to the barrier, i.e. when it must move rightwards next turn or else there is some probability of it hitting the barrier; the second case is when the robot is elsewhere.

To generate the fitness the guaranteed minimum x distance which the robot moves when it is close to the barrier is examined for each program. If this value is greater than 5.0 then the constraint is considered to be satisfied, because the maximum the barrier can move in a turn is 5.0. This influences the GA in two ways. Firstly the fitness of any program which is guaranteed to satisfy the constraint is multiplied by three. Secondly the fact that the program satisfies the constraint is stored, and no program is reported to the user as successful unless it satisfies the constraint. To encourage the search towards solutions which satisfy the constraint, programs for which the guaranteed minimum is in $[0.0, 5.0]$ also get a fitness multiplier (an intermediate value of 1.8 was used).

This constraint analysis also affects the elitism strategy. Instead of simply choosing the program with the highest fitness, the fittest program which satisfies the constraint is chosen. Only if no programs in the population satisfy the constraint is the program with the absolute highest fitness chosen.

The training data are generated by making a random move of the barrier and the target each turn. The barrier moves with a speed between 0 and 5 units per timestep, whilst the target moves with a constant speed of one unit per timestep. The target starts at $(100, 100)$ and the barrier at $y = 100$.

2.2 Experimental Results

This section contains experimental results; these results are drawn from six experimental runs of the above system. Note that the figures given below are for the "raw" fitness value before the multipliers due to the constraint analysis are applied. Error bars give the standard error.

The first set of statistics (figure 2a) show the number of population members which satisfy the constraint at each generation. It can be clearly seen that this increases with successive generations. After a number of generations the algorithm is exploring almost solely within the space of solutions which satisfy the constraint.

The second set of statistics show the performance of the programs generated by the algorithm on the task of keeping close to the target robot. The first graph (figure 2b) gives the fitness across the set of those programs which satisfy the constraint; programs which do not satisfy the constraint are ignored. The second graph (figure 2c) gives the best fitness per run, and the final graph (figure 2d) gives the "best of the best", i.e. the best performance across all of the runs.

The maximum fitness attained is 1195559. By way of comparison the fitness of a basic hand written solution on the same test case was 910091; the best evolved solution is therefore somewhat better.

3 Conclusions and Future Work

This paper has presented a way of incorporating guaranteed constraints into the fitness function of a GP system. The effectiveness of this on a simple test problem from mobile robotics has been demonstrated. There are a number of more sophisticated tasks in mobile robotics which could be approached using this technique: one idea is to have multiple robots sharing a working area, yet guaranteeing that they are able to make use of an appropriate part of the area to carry out a particular task, or the idea of guaranteeing that a robot with react in a certain way to certain stimuli. Other ideas are suggested in [5].

References

1. Wolfgang Banzhaf, Peter Nordin, Robert E. Keller, and Frank D. Francone. *Genetic Programming: An Introduction.* Morgan Kaufmann, 1998.

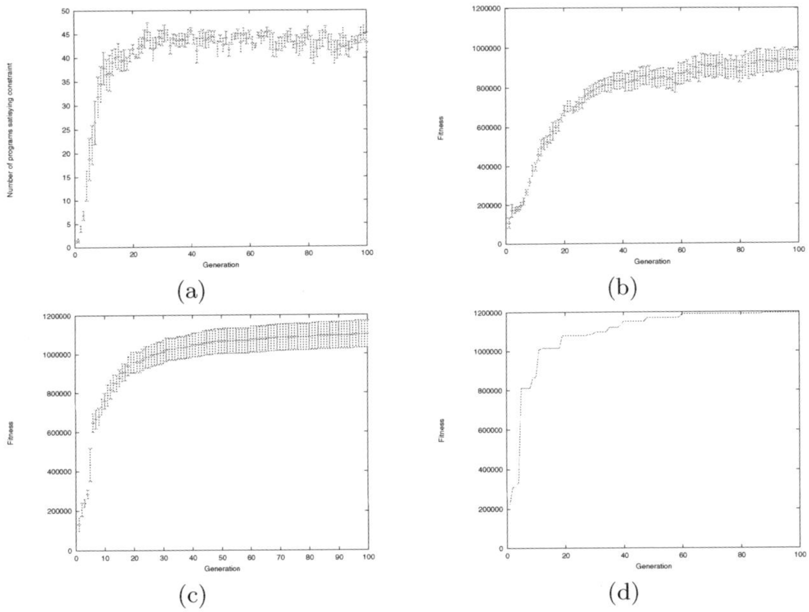

Fig. 2. Results: (a) The number of solutions which satisfy the constraint,per generation. (b) The average fitness of the constraint-satisfying solutions on the target-following task. (c) The fitness of the best of the constraint-satisfying solutions on the target-following task. (d) The best fitness over all runs of the constraint-satisfying solutions on the target-following task.

2. Edmund M. Clarke Jr., Orna Grumberg, and Doron A. Peled. *Model Checking.* MIT Press, 1999.
3. David E. Goldberg. *Genetic Algorithms in Search, Optimization and Machine Learning.* Addison-Wesley, 1989.
4. Lorenz Huelsbergen. Abstract program evaluation and its application to sorter evolution. In *Proceedings of the 2000 Congress on Evolutionary Computation,* pages 1407–1414. IEEE Press, 2000.
5. Colin G. Johnson. Deriving genetic programming fitness properties by static analysis. In James Foster, Evelyne Lutton, Conor Ryan, and Andrea Tettamanzi, editors, *Proceedings of the 2002 European Conference on Genetic Programming.* Springer, 2002.
6. Colin G. Johnson. What can automatic programming learn from theoretical computer science? In Xin Yao, Qiang Shen, and John Bullinaria, editors, *Proceedings of the 2002 UK Workshop on Computational Intelligence,* 2002.
7. John R. Koza. *Genetic Programming : On the Programming of Computers by means of Natural Selection.* Series in Complex Adaptive Systems. MIT Press, 1992.
8. Melanie Mitchell. *An Introduction to Genetic Algorithms.* Series in Complex Adaptive Systems. Bradford Books/MIT Press, 1996.
9. Flemming Nielson, Hanne Riis Nielson, and Chris Hankin. *Principles of Program Analysis.* Springer, 1999.

Part III

Fuzzy Systems

Precisiated Natural Language and Protoforms for Human Consistent Data Mining

Janusz Kacprzyk and Sławomir Zadrożny

Systems Research Institute, Polish Academy of Sciences
Newelska 6, 01-447, Warszawa
e-mail: kacprzyk@ibspan.waw.pl

Abstract. We show how Zadeh's concepts of a precisiated natural language (PNL) and protoforms can lead to a new direction in the use of natural language in data mining, linguistic data(base) summaries. We show that various types of linguistic data summaries may be viewed as items in a hierarchy of protoforms of summaries. We briefly present an implementation for a sales database of a computer retailer.

Keywords. precisiated natural language, protoform, linguistic database summary, fuzzy logic, fuzzy querying

1 Introduction

The basic philosophy and rationale of concepts discussed here is that since for a human being natural language is the only fully natural way of communication, then maybe it could be expedient to try to "directly" use (elements of) natural language in the formulation, processing and solution of problems considered to maintain a higher human consistence, hence a higher implementability in the domain of broadly meant information processing. A prerequisite for such an approach is to have some way to formally represent elements of natural language used. Zadeh proposed to use here the PNL (precisiated natural language). Basically, in PNL statements about values, relations, etc. between variables are represented by constraints. In the conventional case, a statement is, e.g., that the value of variable x belongs to a set X. In PNL, statements – written x isr R – may be different, and correspond to numeric values, intervals, possibility disctributions, verity distributions, probability distributions, usuality qualified statements, rough sets representations, fuzzy relations, etc. For our purposes, usuality qualified representation will be of special relevance. Basically, it says "x is usually R" that is meant as "in most cases, x is R". PNL may play various roles among which crucial are: description of perceptions, definition of sophisticated concepts, a language for perception based reasoning, etc.

Recently, Zadeh introduced the concept of a protoform. For our purposes, one should notice that most perceptions are summaries. For instance, a perception like "most Swedes are tall" is a summary. It can be represented in Zadeh's notation as "most As are Bs". This can be employed for reasoning under various assumptions. For instance, if we know that "x is A', we can deduce that, e.g. "it is likely that x is B", we can ask about an average height of a Swede, etc. One can go a step further, and define a protoform as an abstracted summary. In our case, this would be ""QAs are Bs". Notice that we now have a more general, deinstantiated form of our point of departure (most Swedes are tall), and also of "most As are Bs". Needless to say that most human reasoning is protoform based, and the availability of such a more general representation is vary valuable, and provides tools that can be used in many cases.

In this paper, we show that the concept of a protoform, viewed from the perspective of the computing with words, and precisiated natural language, can be of use in data mining. We show the idea of a linguistic data summarization. Data summarization is still in general unsolved a problem in spite of vast research efforts. Very many techniques are available but they are not "intelligent enough", and not human consistent, partly due to a little use of natural language. This concerns, e.g., summarizing statistics, exemplified by the average, median, minimum, maximum, 8-percentile, etc. which – in spite of recent efforts to soften them – are still far from being able to reflect a real human perception of their essence.

In this paper we will show the use of linguistic database summaries introduced by Yager [3, 4], and then considerably advanced by Kacprzyk [5], Kacprzyk and Yager [6], and Kacprzyk, Yager and Zadrożny [7, 8], and implemented in Kacprzyk and Zadrożny [9 - 15]. We will derive linguistic data summaries as linguistically quantified propositions as, e.g., "most of the employees are young and well paid", with a degree of validity (truth, ...), in case of a personnel database. Notice that this is related to statements used by Zadeh in the context of protoforms.

We follow Yager's [3, 4] idea, and present its implementation, mainly using Kacprzyk and Yager [6], and Kacprzyk, Yager and Zadrożny's [7, 8] extensions. We employ Kacprzyk and Zadrożny's [9, 13] idea of an interactive approach to linguistic summaries in which the determination of a class of summaries of interest is done via Kacprzyk and Zadrożny's [17, 18] FQUERY for Access, a fuzzy querying add-on to Access, extended to the querying over the Internet in Kacprzyk and Zadrożny [19 - 21].

We show that by relating various types of linguistic summaries to fuzzy queries, with various known and sought elements, we can arrive at a hierarchy of prototypical forms, or – in Zadeh's [1] terminology – protoforms, of linguistic data summaries.

Finally, we present a real application for a sales database of a computer retailer.

2 Linguistic data summaries

We follow Yager's [3, 4] concept of a linguistic summary of a set of data whose main elements are:

- V - a quality (attribute) of interest, e.g. salary in a database of workers,
- $Y = \{y_1, \ldots, y_n\}$ - a set of objects (records) that manifest V, e.g. the set of workers; $V(y_i)$ are values of quality V for object y_i.

A *summary* of data set consists of:

- a summarizer S (e.g. young),
- a quantity in agreement Q (e.g. most),
- truth (validity) T - e.g. 0.7,

as, e.g., "T(*most* of employees are *young*)=0.7".

Basically, the calculation of the truth (validity) of the basic type of a linguistic summary considered in this section is equivalent to the calculation of the truth value (from the unit interval) of a linguistically quantified statement (e.g., „*most* of the employees are *young*"). This may be done by using either Zadeh's [22] calculus of linguistically quantified statements (cf. Zadeh and Kacprzyk [23]) or Yager's [24] OWA operators [cf. Yager and Kacprzyk [25]). Evidently, we seek a linguistic summary that maximizes T.

Yager's [23] source approach to linguistic summaries is a conceptual point of departure, and recent developments in linguistic database summaries may be found in Kacprzyk and Yager [6] or Kacprzyk, Yager and Zadrożny [7].

3 A fuzzy querying add-on for formulating linguistic summaries

In Kacprzyk and Zadrożny's [9, 13] approach, *interactivity*, i.e. *user assistance*, is in the definition of summarizers (indication of attributes and their combinations). This proceeds via a user interface of a fuzzy querying add-on. In Kacprzyk and Zadrożny [17, 18], a conventional DBMS is used, and a fuzzy querying tool, FQUERY for Access, is developed to allow for queries with fuzzy (linguistic) elements.FQUERY for Access is embedded in the native Access's environment as an add-on.

It is obvious that fuzzy queries directly correspond to summarizers in linguistic summaries. Thus, the derivation of a linguistic summary may proceed in an interactive (user-assisted) way as follows:

- the user formulates a set of linguistic summaries of interest (relevance) using the fuzzy querying add-on described above,
- the system retrieves records from the database and calculates the validity of each summary adopted, and
- a most appropriate linguistic summary is chosen.

The use of fuzzy querying is very relevant because we can restate the summarization in the fuzzy querying context. The two propositions with fuzzy linguistic quantifiers (without and with importance) are now:

"Most records match query S" (1)
"Most records meeting conditions B match query S" (2)

In database terminology, *B* corresponds to a *filter* and (2) claims that *most* records passing through *B* match query *S*. Moreover, since the filter may be fuzzy, a record may pass through it to a degree from [0,1].

It may be noticed that the concept of a protoform in the sense of Zadeh [1] is highly relevant in this context. First of all, a protoform is defined as an abstract prototype, that is, in our context, for the query (summary) given by (1) and (2) as follows, respectively:

"*Most R's are S*" (3)
"*Most BR's* are *S*" (4)

where *R* means "records", *B* is a filter, and *S* is a query.

Evidently, as protoforms may form a hierarchy, we can define higher level (more abstract) protoforms, for instance replacing *most* by a general linguistic quantifier *Q*, we obtain, respectively:

"*QR's* are *S*" (5)
"*QBR's* are *S*" (6)

Basically, the more abstract forms correspond to cases in which we assume less about summaries sought. There are two limit cases, where we: (1) assume totally abstract protoform or (2) assume all elements of a protoform are given on the lowest level of abstraction as specific linguistic terms. In Table 1 5 basic types of linguistic summaries are shown, corresponding to protoforms of a more and more abstracted form.

Table 1: Classification of linguistic summaries

Type	Given	Sought	Remarks
1	S	Q	Simple summaries through ad-hoc queries
2	S B	Q	Conditional summaries through ad-hoc queries
3	Q $S^{structure}$	S^{value}	Simple value oriented summaries
4	Q $S^{structure}$ B	S^{value}	Conditional value oriented summaries
5	Nothing	S B Q	General fuzzy rules

where $S^{structure}$ denotes that attributes and their connection in a summary are known, while S^{value} denotes a summarizer sought.

Type 1 may be easily produced by a simple extension of FQUERY. Basically, the user has to construct a query – candidate summary, and it is to be determined what is the fraction of rows matching this query and what linguistic quantifier best denotes this fraction. A Type 2 summary is a straightforward extension of Type 1 by adding a fuzzy filter. Type 3 summaries require much more effort as their primary goal is to determine typical (exceptional) values of an attribute. So, query S consists of only one simple condition built of the attribute whose typical (exceptional) value is sought, the "=" relational operator and a placeholder for the value sought. For example, using: Q = *"most"* and S = "age=?" we look for a typical value of age. A Type 4 summary may produce typical (exceptional) values for some, possibly fuzzy, subset of rows. From the computational point of view. Type 5 summaries represent the most general form considered: fuzzy rules describing dependencies between specific values of particular attributes. Here the use of B is essential, while previously it was optional.

The summaries of Type 1 and 3 have been implemented as an extension to Kacprzyk and Zadrożny's [17 - 20] FQUERY for Access. Two approaches to Type 5 summaries producing has been proposed. Firstly, a subset of such summaries may be produced exploiting similarities with *association rules* concept and employing their efficient algorithms.

4 Implementation

As a simple illustration of Type 5 summaries, an implementation is shown for a sales database of a small computer retailer in Southern Poland. For illustration we will only show some examples of linguistic summaries for some interesting (for the user!) choices of relations between attributes.

Table 2. Linguistic summaries expressing relations between the group of products and commission

Summary	Degree of validity
About ½ of sales of network elements is with a high commission	0.3630
About ½ of sales of computers is with a medium commission	0.4753
Much sales of accessories is with a high commission	0.5713
Much sales of components is with a low commission	0.6707
About ½ of sales of software is with a low commission	0.4309
About ½ of sales of computers is with a low commission	0.4473
A few sales of components is without commission	0.0355
A few sales of computers is with a high commission	0.0314
Very few sales of printers is with a high commission	0.0509

Table 3. Linguistic summaries expressing relations between the groups of products and times of sale

Summary	Degree of validity
About 1/3 of sales of computers is by the end of year	0.2801
About ½ of sales in autumn is of accessories	0.4790
About 1/3 of sales of network elements is in the beginning of year	0.1957
Very few sales of network elements is by the end of year	0.0929
Very few sales of software is in the beginning of year	0.0958
About ½ of sales in the beginning of year is of accessories	0.4343
About 1/3 of sales in the summer is of accessories	0.3092
About 1/3 of sales of peripherals is in the spring period	0.2140
About 1/3 of sales of software is by the end of year	0.2258
About 1/3 of sales of network elements is in the spring period	0.2081
About 1/3 of sales in the summer period is of components	0.3081
Very few sales of network elements is in the autumn period	0.0955
A few sales of software is in the summer period	0.1765

Table 4. Linguistic summaries expressing relations between the attributes: size of customer, regularity of customer (purchasing frequency), date of sale, time of sale, commission, group of product and day of saleLinguistic summaries expressing relations between the groups of products and times of sale

Summary	Degree of validity
Much sales on Saturday is about noon with a low commission	0.3951
Much sales on Saturday is about noon for bigger customers	0.4430
Much sales on Saturday is about noon	0.4654
Much sales on Saturday is about noon for regular customers	0.4153
A few sales for regular customers is with a low commission	0.1578
A few sales for small customers is with a low commission	0.1915
A few sales for one-time customers is with a low commission	0.1726
Much sales for small customers is for non-regular customers	0.5105

5 Concluding remarks

We have presented an interactive, fuzzy logic based approach to the linguistic summarization of databases, and advocated it as a means to obtain human consistent summaries of (large) sets of data that can be easily comprehensible by human beings, hence that can be useful. We have indicated that Zadeh's [1] concept of protoforms, or a hierarchy of them, is a natural framework within which to consider various classes of linguistic summaries, and that this is particularly clear when we consider linguistic summaries from the perspective of their related fuzzy queries. The approach presented is implementable, and is a step towards an increase of the role of natural language in data mining. This can be viewed as another example that Zadeh's idea of *computing with words* as a new paradigm to represent and solve complicated problem may provide a viable alternative to the existing, traditional apprpoaches.

References

1. Zadeh L.A.: A prototype-centered approach to adding deduction capabilities to search engines – the concept of a protoform. BISC Seminar, 2002), University of California, Berkeley, 2002.
2. Zadeh L.A., J. Kacprzyk J., Eds.: Computing with Words in Information/Intelligent Systems. Vol 1: Foundations, Vol. 2: Applications, Physica-Verlag (Springer-Verlag), Heidelberg and New York, 1999.
3. Yager R.R.: A new approach to the summarization of data. Information Sciences, vol. 28, 1982, pp. 69-86.
4. Yager R.R.: On linguistic summaries of data. In W. Frawley and G. Piatetsky-Shapiro (Eds.): Knowledge Discovery in Databases. AAAI/MIT Press, pp. 347-363.
5. Kacprzyk J.: Intelligent data analysis via linguistic data summaries: a fuzzy logic approach. In R. Decker and W. Gaul (Eds.): Classification and Information Processing at the Turn of the Millennium. Springer-Verlag, Berlin, Heidelberg and New York, 2000. pp. 153-161.
6. Kacprzyk J. and Yager R.R.: Linguistic summaries of data using fuzzy logic, Int. J. of General Systems, vol. 30. 2001, pp. 133-154.
7. Kacprzyk J., Yager R.R.. and Zadrożny S.: A fuzzy logic based approach to linguistic summaries of databases", Int. J. of Applied Mathematics and Computer Science, vol. 10. 2000. pp. 813-834.
8. Kacprzyk J., Yager R.R. and Zadrożny S.: Fuzzy linguistic summaries of databases for an efficient business data analysis and decision support. In W. Abramowicz and J. Zurada (Eds.): Knowledge Discovery for Business Information Systems. Kluwer, Boston, 2001, pp. 129-152.
9. Kacprzyk J. and Zadrożny S.: Data Mining via Linguistic Summaries of Data: An Interactive Approach. In T. Yamakawa and G. Matsumoto (Eds.): Methodologies for the Conception, Design and Application of Soft Computing" (Proc. of IIZUKA'98), Iizuka, Japan, 1998, pp. 668-671.
10. Kacprzyk J. and Zadrożny S.: On combining intelligent querying and data mining using fuzzy logic concepts. In G. Bordogna and G. Pasi (Eds.): Recent Research Issues on the Management of Fuzziness in Databases. Physica - Verlag, Heidelberg and New York, 2000. pp. 67-81.
11. Kacprzyk J. and Zadrożny S.: On a fuzzy querying and data mining interface, Kybernetika, vol. 36, 2000. pp. 657-670.
12. Kacprzyk J. and Zadrożny S.: Computing with words in intelligent database querying: standalone and Internet-based applications. Infor. Sciences, vol. 134, 2001, pp. 71-09.

13. Kacprzyk J. and Zadrożny S.: Data mining via linguistic summaries of databases: an interactive approach. In L. Ding (Ed.): A New Paradigm of Knowledge Engineering by Soft Computing. World Scientific, Singapore, 2001, pp. 325-345.
14. Kacprzyk J. and Zadrożny S.: On interactive linguistic summarization of databases via a fuzzy-logic-based querying add-on to Microsoft Access, In B. Reusch (Ed.): Computational Intelligence – Theory and Applications. Springer, Berlin, 1999, pp. 462-472.
15. Kacprzyk J. and Zadrozny S.: Fuzzy linguistic summaries via association rules. In A. Kandel, M. Last and H. Bunke (Eds.): Data Mining and Computational Intelligence. Physica-Verlag (Springer-Verlag), Heidelberg and New York 2001, pp. 115-139.
16. George R. and Srikanth R.: Data summarization using genetic algorithms and fuzzy logic. In F. Herrera and J.L. Verdegay (Eds.): Genetic Algorithms and Soft Computing. Physica-Verlag, Heidelberg, pp. 599-611, 1996.
17. Kacprzyk J. and Zadrożny S.: Fuzzy querying for Microsoft Access, Proc. of FUZZ - IEEE'94 (Orlando, USA), 1994, vol. 1, pp. 167-171.
18. Kacprzyk J. and Zadrożny S.: FQUERY for Access: fuzzy querying for a Windows-based DBMS. In P. Bosc and J. Kacprzyk (Eds.): Fuzziness in Database Management Systems. Physica-Verlag (Springer-Verlag), Heidelberg, 1995, pp. 415-433.
19. Kacprzyk J. and Zadrożny S.: A fuzzy querying interface for a WWW-server-based relational DBMS. Proc. of 6th IPMU Conference (Granada), 1996, Vol. 1, pp. 19-24.
20. Kacprzyk J. and Zadrożny S.: Using fuzzy querying over the Internet to browse through information resources. In B. Reusch and K.-H. Temme (Eds.): Computational Intelligence in Theory and Practice. Physica-Verlag (Springer- Verlag), Heidelberg and New York, 2001, pp. 235-262.
21. Kacprzyk J. and Zadrożny S.: Data mining via fuzzy querying over the Internet. In O. Pons, M.A. Vila, J. Kacprzyk (Eds.): Knowledge Management in Fuzzy Databases. Physica - Verlag, Heidelberg, New York 2000. pp. 211-233.
22. Zadeh L.A.: A computational approach to fuzzy quantifiers in natural languages. Computers and Maths with Appls. Vol. 9, 1983, pp. 149-184.
23. Zadeh L.A. and Kacprzyk J., Eds.: Fuzzy Logic for the Management of Uncertainty, Wiley, New York, 1992.
24. Yager R.R.: On ordered weighted avaraging operators in multicriteria decision making. IEEE Trans. on Systems, Man and Cybern. SMC-18, 1988, 183-190.
25. Yager R.R. and Kacprzyk J. (Eds.): The Ordered Weighted Averaging Operators: Theory and Applications. Kluwer, Boston, 1997.

Design and Synthesis of Piecewise-Linear Fuzzy Systems

Inés del Campo, Javier Echanobe, and José M. Tarela

Department of Electricity and Electronics, University of the Basque Country

Abstract. The main advantages of piecewise linear approximations are the availability of mature and precise analysis methods, and their suitability for simple hardware implementations. The control surface of a large number of practical fuzzy controllers is well approximated by means of PWL functions. In this work a design methodology for PWL fuzzy controllers is presented. As an example, the development of a two-input single-output fuzzy controller is presented and its stability in the phase plane is investigated. The closed-loop system trajectories in the phase plane are simulated and the system stability is assured by means of a globally quadratic Lyapunov function. In addition, two efficient hardware implementations are presented.

1 Introduction

Three main aspects have to be taken into account in the development of fuzzy logic controllers (FLCs) for practical purposes: the approximation capability of the fuzzy model, the availability of structured analysis methods, and the suitability of algorithms for efficient hardware implementations. The first aspect is concerned with the capability of fuzzy models to approximate any input-output mapping to arbitrary accuracy. In this respect, the approximation capability of different type of FLCs has been addressed by many authors in the last decade [2, 3, 6, 7]. The well known ability of fuzzy logic to construct effective controllers is attributable to the verification of this property. Therefore, fuzzy models with universal approximation capability have to be selected. The second aspect requires formal design methods that assure the stability and the performance of the system. Although important advances in this sense have been reported in the past few years, most of them concerning the Takagi-Sugeno model [5, 9, 11], the problem of fuzzy system stability requires further investigation. Finally, fuzzy control require specific hardware suitable to achieve high-speed for real-time operation, (see [1] and references therein).

On the other hand, it is well known that piecewise-linear (PWL) methods have been widely used in modeling and synthesis of nonlinear electronic circuits. The main advantages of PWL approximations are the availability of well known analysis methods, as well as their suitability for simple hardware implementations. As in the case of some electronic devices (diodes, transistors, or operational amplifiers) which are PWL in essence, the control surface of a large number of practical fuzzy controllers is well approximated by means of PWL functions. Moreover, Rovatti [7] proposes an universal approximator fuzzy system, based on a non-classical "AND" operator, whose output is in fact PWL.

In this work a design methodology for efficient fuzzy controllers is presented. The proposed methodology is based on PWL fuzzy models with universal approximation capability. As an example, the development of a two-input single-output fuzzy controller is presented and its stability in the phase plane is investigated. In addition, two efficient hardware implementations are presented.

2 Fuzzy system representation

Consider a rule based fuzzy system in which the *j*-th rule can be expressed as:

$$R_j : \text{IF } x^1 \text{ is } M_j^1 \text{ and } x^2 \text{ is } M_j^2, \ldots, x^n \text{ is } M_j^n, \text{ then } y \text{ is } c_j,$$

where $\mathbf{x} = (x^1, x^2, ..., x^n) \in X = X_1 \times X_2 \times ... X_n$ is the input vector; $M_j^1, M_j^2, ..., M_j^n$ are fuzzy sets, $y \in Y$ is the output variable, and $c_j \in \mathrm{R}$ is a crisp consequent. Assume that M_j^i are triangular, normalized, and overlapped in each input dimension with an overlapping degree of two. Rovatti [7] demonstrated that under the above restrictions on the membership functions, and using the center-of-gravity defuzzification method, a PWL fuzzy system is obtained. Moreover, the resulting system is a first order universal approximator. The key of this important result is the PWL operator denoted Θ_n, which is the *n*-dimensional extension of the t-norm proposed by Sasaki and Ueno [10]. The above PWL fuzzy model induces a partition of the input space into polyhedral cells. In each cell the system dynamic is only concerned with the consequents of the local rules.

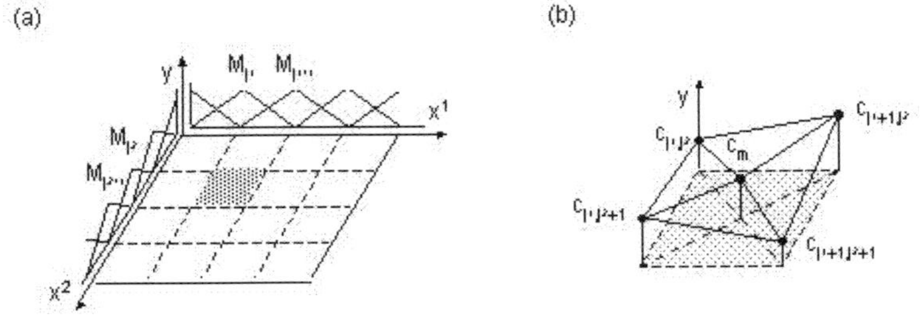

Fig. 1. (a) Fuzzy set definition and input partition of the proposed example. **(b)** Polyhedral (triangular) cells generated by the PWL fuzzy model.

To investigate the geometry of the input partition, consider a two-input single-output fuzzy system with 25 rules:

R_{j_1,j_2} : IF x^1 is $M^1_{j_1}$ and x^2 is $M^2_{j_2}$, then y is c_{j_1,j_2}, $1 \le j_1, j_2 \le 5$.

Fig. 1(a) shows the partition of the input universes into 5 triangular membership functions per input dimension $M^1_1(x^1),...,M^1_5(x^1)$ and $M^2_1(x^2),...,M^2_5(x^2)$. For simplicity, membership functions with the same width have been selected.

The PWL operator $\Theta_2 : [0,1]^2 \mapsto [0,1]$ is given by:

$$\Theta_2(t^1,t^2) = \frac{1}{4}\left\{\min(t^1,t^2) + \max(t^1 + t^2 - 1, 0)\right\} \tag{1}$$

where (t^1,t^2) are fuzzy true values. The system output is:

$$y = \sum_{j_1,j_2} c_{j_1,j_2} \Theta_2\left(M^1_{j_1}(x^1), M^2_{j_2}(x^2)\right) \tag{2}$$

the arguments of Θ_2 are the activation degrees of the rule antecedents. Note that the rule interaction determines 16 rectangular regions (see Fig. 1(a)). The system dynamic inside each region is governed by the 4 rules whose antecedents are non null in that region.

Consider now a generic rectangular region with unity width as in Fig. 1(b). The associated membership functions (half triangles) are: $M^1_{j_1}(x^1) = 1 - x^1$, $M^1_{j_1+1}(x^1) = x^1$, $M^2_{j_2}(x^2) = 1 - x^2$, and $M^2_{j_2+1}(x^2) = x^2$. Therefore, the system output (Eq. (2)) in this region is:

$$\begin{aligned} y = {} & c_{j_1,j_2}\Theta_2(1-x^1, 1-x^2) + c_{j_1+1,j_2}\Theta_2(x^1, 1-x^2) \\ & + c_{j_1,j_2+1}\Theta_2(1-x^1, x^2) + c_{j_1+1,j_2+1}\Theta_2(x^1, x^2). \end{aligned} \tag{3}$$

In Eq.(1), the conditions $t^1 = t^2$ and $t^1 + t^2 = 1$ delimit regions with different solution. Applying both conditions to the arguments of Θ_2 in Eq. (3), the geometry of the input partition can be deduced. The sub-regions with different solution are triangles delimited by the diagonals of the region, $x^1 = x^2$ and $x^1 = 1 - x^2$. It is easy to demonstrate that the system output in each triangular sub-region is given by the plane that contains the 2 associated consequents and the average of all the 4 consequents c_m, see Fig. 1(b). Finally, note that the system geometry depicted in Fig. (1) can also be used as a general implementation methodology to approximate any fuzzy model.

3 Closed-loop system

Consider now the system depicted in Fig. 2 where the fuzzy control law is given by a PWL fuzzy controller and the plant is modeled as a second order linear system.

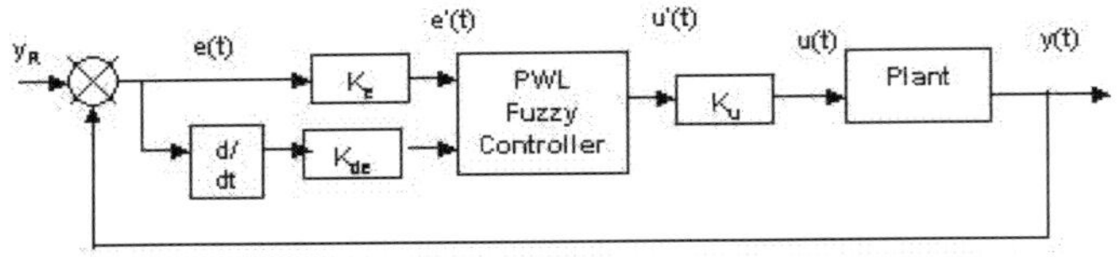

Fig. 2. Block diagram of a feedback system composed of a PWL fuzzy controller and a linear plant

To derive a state-space model for the closed-loop system, the following state variables are defined, $x_1 = y$ and $x_2 = \dot{y}$. The fuzzy controller inputs are $e' = K_e e$ and, $\dot{e}' = K_{de}\dot{e}$ where K_e and K_{de} are scaling factors, $e = y_R - y$ and $\dot{e} = -\dot{y}$. Let $\{P_i'\}_{i\in I'} \subseteq \mathbf{R}^2$ denote a partition of the input space $(e', \dot{e}')$ into regular triangles like in Fig. 1, with I' the set of cell indexes. This partition induces a partition of the state space (x_1, x_2) which can be obtained by applying the following transformation $x_1 = y_R - {e'}/{K_e}$ and $x_2 = -{\dot{e}'}/{K_{de}}$. Note that the transformed partition, $\{P_i\}_{i\in I} \subseteq \mathbf{R}^2$, preserves the triangular shape of the original cells. It can be demonstrated that in each cell the controller dynamics is of the form:

$$\mathbf{u}(t) = \mathbf{D}_i\mathbf{x}(t) + \mathbf{d}_i, \qquad \mathbf{x}(t) \in \{P_i\}_{i\in I}, \tag{4}$$

with $\mathbf{D}_i = [-\alpha_{1i}K_eK_u \quad -\alpha_{2i}K_{de}K_u]$ and $\mathbf{d}_i = K_u(\alpha_{1i}y_RK_e + \alpha_{3i})$, where α_{1i}, α_{2i}, and α_{3i} are the coefficients of the controller planes $u'(t) = \alpha_{1i}e'(t) + \alpha_{2i}\dot{e}'(t) + \alpha_{3i}$.

Consider next a second order linear plant $\dot{\mathbf{x}}(t) = \mathbf{B}\mathbf{x}(t) + \mathbf{C}\mathbf{u}(t)$, then, the dynamic of the closed-loop system can be modeled as follows:

$$\dot{\mathbf{x}}(t) = \mathbf{A}_i \mathbf{x}(t) + \mathbf{a}_i, \qquad \text{for } \mathbf{x}(t) \in P_i, \tag{5}$$

with $\mathbf{A}_i = \mathbf{B} + \mathbf{C}\mathbf{D}_i$, $\mathbf{a}_i = \mathbf{C}\mathbf{d}_i$. The above state-space equation defines a piecewise affine system. Therefore, the general results on the stability analysis of this kind of nonlinear systems can be applied [4]. As a numerical example consider the following plant: $\mathbf{B} = \begin{bmatrix} 0 & 1 \\ 0 & -10 \end{bmatrix}$, $\mathbf{C} = \begin{bmatrix} 0 \\ 300 \end{bmatrix}$, and a PWL fuzzy controller with the triangular membership functions shown in Fig. 1. The control surface has been represented in Fig. 3(a); the crisp consequents can be directly obtained from the surface value in the vertex of the cells. Then, Eq. (5) has analytical solution in each sub-region of the state space. The full system trajectory can be predicted by superposition of the trajectory pieces. Fig. 3(b) depicts a family of trajectories starting at different initial states (x_{10}, x_{20}); this family of curves has been generated with $K_e = K_{de} = 1$, and $K_u = 0.2$. The system reaches an equilibrium point (stable node) at $x_2(t) = \dot{y}(t) = 0$ and $x_1(t) = y(t) = 1 + \alpha_{3f} / \alpha_{1f} = 1$, where α_{1f} and α_{3f} are the coefficients of the last plane involved in the system trajectory; $\alpha_{3f} = 0$ for these planes.

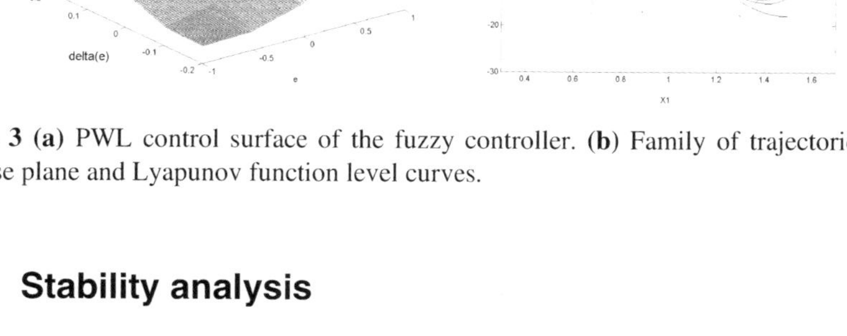

Fig. 3 **(a)** PWL control surface of the fuzzy controller. **(b)** Family of trajectories in the phase plane and Lyapunov function level curves.

4 Stability analysis

The problem of the construction of Lyapunov functions for stability analysis is of great interest in general nonlinear systems theory. Particularly, the search of quadratic Lyapunov functions for fuzzy systems is nowadays an active area of research. Recently, a novel approach based on piecewise quadratic Lyapunov functions, has been proposed [5]. This approach, which holds for both linear and affine systems, exploits the close connection between Takagi-Sugeno fuzzy systems and PWL systems. The same argument will be used in the following.

It has been seen that the dynamic of the system depicted in Fig. 1 is piecewise affine (see Eq. (5)). The stability of this class of systems can be proved using a globally quadratic Lyapunov function $V(\mathbf{x}) = \mathbf{x}^T \mathbf{P} \mathbf{x}$. The conditions for quadratic stability of piecewise affine systems are usually formulated as follows [4]:

Proposition 1: If there exists a matrix $\mathbf{P} = \mathbf{P}^T > 0$ such that $\mathbf{A}_i^T \mathbf{P} + \mathbf{P}\mathbf{A}_i < 0$, $i \in I$, then every trajectory $\mathbf{x}(t)$ of (1) with $\mathbf{a}_i = 0$ tends to zero exponentially.

However, the above condition is unnecessarily restrictive because the dynamic given by $\mathbf{A}_i$ is only valid in a sub-region of the state-space. Therefore, it is sufficient to require that:

$$\mathbf{x}^T (\mathbf{A}_i^T \mathbf{P} + \mathbf{P}\mathbf{A}_i)\mathbf{x} < 0 \text{, for } \mathbf{x} \in X_i . \tag{6}$$

In view of the simulations results, the fuzzy system shown in Fig. 3(b) seems to be stable. The search for a Lyapunov function has been successfully carried out. The candidate function $V(\mathbf{x}) = \mathbf{x}^T \mathbf{P} \mathbf{x}$ with $\mathbf{P} = \begin{bmatrix} 100 & 0.1 \\ 0.1 & 0.03 \end{bmatrix}$ verifies Eq. (6), assuring the stability of the system. A set of level curves can be seen in Fig. 3(b). In some cases, no globally quadratic Lyapunov function can be found. For these systems, piecewise quadratic Lyapunov functions can be considered [4].

5 Hardware implementation

It has been seen that using the fuzzy model (2) the input space of the FLC is split into a finite number of sub-regions or cells. Only one of these cells become active each time the input is updated. Therefore, efficient cellular architectures can be developed for high-speed hardware implementations [8]. In addition, the PWL nature of the inference scheme reduces the complexity of the implementations. The proposed architecture comprises three main blocks: the preprocessing module, the hyperplane generator, and the memory block. The system inputs are b-bit binary words generated by means of analog-to-digital (A/D) converters. The b_h most significative bits of each input determine the active cell, and the b_l least significative bits represent the input coordinates referred to that cell. First, the preprocessing module locate the triangular sub-region where the input falls; this task is done by a sorting procedure. This information is used to address the corresponding crisp consequents in the memory block. Then, the hyperplane generator reads the data in memory and implements the system output.

Two different implementation technologies have been selected to evaluate the performance of the proposed method: a digital signal processor (DSP) and a programmable logic device (PLD). In both cases the implementation consists in the two-input single-output FLC analysed in the previous sections. The internal architecture of DSPs is specially designed for multiply-accumulate operations, typical of signal processing algorithms. This DSP feature have been exploited to achieve a straightforward implementation of the proposed method. The DSP used in this application is the Motorola's DSP-56002 which is a general purpose DSP with a 24-bit data bus. The full system algorithm occupies 83 words in the 24-bit program memory and its execution requires only 160 clock cycles. The PWL system can also be implemented using a single PLD. In this approach the design has been programmed in a field programmable gate array (FPGA) from Altera (EPF10K20). This SRAM-based device provides up to 20.000 gates. It can also provide up to 12K bits of memory which can be configured with different wordlengths. As a case example, a 4-bit two's complement architecture has been implemented. The design software reports the utilization of the 25% of the device resources and a maximum clock frequency of 20MHz. The developed circuit provides a maximum performance nearly 600.000 samples per second.

6 Conclusions

In this work a design methodology for PWL fuzzy controllers has been presented. Although the proposed methodology is based on the PWL fuzzy model proposed by Rovatti, it can also be used as a general implementation methodology to approximate any fuzzy model. The main advantages of PWL approximations are the availability of analysis methods and their suitability for simple hardware implementations. As an example, the development of a two-input single-output FLC has been reported. The closed-loop system trajectories in the phase plane have been simulated and the system stability has been guaranteed by means of a globally quadratic Lyapunov function. Finally, two efficient hardware implementations have been presented. The first one based on a digital signal processor (DSP) and the other one based on a programmable logic device (PLD). Both developments show high operation speed and very low hardware requirements.

Acknowledgement

This work has been partially supported by the Spanish MCYT, European FEDER funds (TIC2002-02458), and the Basque Country University (UPV224.310-E-14029/2001).

References

1. Baturone I et. all (2000) Microelectronic Design of Fuzzy Logic Based Systems. International Series on Computational Intelligence, CRC Press LLC, Boca Raton, Florida
2. Buckley JJ (1993) Sugeno Type Controllers are Universal Controllers. Fuzzy Sets and Systems 53:299-304
3. Castro JL (1995) Fuzzy Logic Controllers are Universal Approximators. IEEE Trans. on Systems, Man, and Cybernetics 25:629-635
4. Johansson M, Rantzer A (1998) Computation of Piecewise Quadratic Lyapunov Functions for Hybrid Systems. IEEE Trans. on Automatic Control 43:555-559
5. Johansson M, Rantzer A, Arzén K (1999) Piecewise Quadratic Stability of Fuzzy Systems. IEEE Trans. on Fuzzy Systems 7:713-722
6. Kosko B (1994) Fuzzy Systems as Universal Approximators. IEEE Trans. on Computers 43:1329-1333
7. Rovatti R (1998) Fuzzy Piecewise Multilinear and Piecewise Linear Systems as Universal Approximators in Sobolev Norms. IEEE Trans. on Fuzzy Systems 6:235-249
8. Rovatti R, Fantuzzi C, Simani S (2000) High-speed DSP-based Implementation of Piecewise-affine and Piecewise-quadratic Fuzzy Systems. Signal Processing 80:951-963
9. Tanaka K, Ikeda T, Wang HO (1996) Robust Stabilization of a Class of Uncertain Nonlinear System via Fuzzy Control: Quadratic Stabilizability, H^{∞} Control Theory, and Linear Matrix Inequalities. IEEE Trans. on Fuzzy Systems 4:1-13
10. Sasaki M, Ueno F (1994) A Novel Implementation of Fuzzy Logic Controller Using New Meet Operation. In Proc. of the Third IEEE International Conference on Fuzzy Systems, IEEE Press, pp 1676-1681
11. Wang LX, Tanaka K, Griffin MF (1996) An Approach to Fuzzy Control of Nonlinear Systems: Stability and Design Issues. IEEE Trans. on Fuzzy Systems 4:14-23

A New Ranking Method For Fuzzy FMEA

Muhammad Suzuri Hitam*, Suresh Gobee* and Khairanum Subari**

* Department of Computer Science, University College of Science and Technology Malaysia, Mengabang Telipot, 21030 Kuala Ter-engganu, Malaysia.
e-mail: suzuri@kustem.edu.my
** School of Mechanical Engineering, Universiti Sains Malaysia, Engineering Campus, Nibong Tebal, Penang, Malaysia, e-mail: khairanu@usm.my

Abstract. In this paper, we extend the method introduced by Bowles and Peláez (1995) for the risk ranking in Failure Mode and Effect Analysis (FMEA). This method called as multi-ranking method allows combination of both the quantitative method (numeric ranking and fuzzy number) and the qualitative method (linguistic ranking) to be employed in the ranking of risk. It employs two types of aggregation method namely an input ranking aggregation and an output ranking aggregation method. Experimental investigations were carried out in actual industrial environment on a paste printing process of printed circuit board at Motorola, Penang, Malaysia. Comparisons were made between traditional FMEA Risk Priority Number (RPN) risk ranking and the newly proposed method. The results show that the use of both the qaulitative and quantitave information in the risk ranking could provides almost similar outcomes with that of the Fuzzy FMEA.

1 Introduction

Failure Mode and Effect Analysis (FMEA) is a popular tool used to identify failure modes of the process and it's effects to the system's functionality. The failure modes are prioritized according to the risk by experts and prevention action is taken to reduce the risk accordingly. In general, there are three main stages in the FMEA process. The first stage is to identify the potential failure modes in the process. The second stage is to rank the risk and the final stage is to take preventive action to reduce risk.

At present, the method of risk ranking is very subjective in nature due to the information for risk ranking is solely depends on the expert's experiences and knowledge of the process. In practice, the problem arises when the experts are having different opinions to rank the risk. This paper describes methods to further improve the aspect of ranking risk from failure modes based on the Risk Priority Number method (RPN) [7]. In short, the RPN method involves the multiplication of 3 parameters; occurrence (O) of failure, severity (S) of failure and detection (D) or detectability of failure, where each parameter having the scale in the range between 1 to 10. Failure modes having a high RPN value are assumed to be more important and given a higher priority than those having a lower value [7]. This method is commonly employs in industries such as in electronics and automobiles.

Bowles and Peláez [1] introduced a method whereby fuzzy logic technique is intergrated into the FMEA risk ranking. Their approach resolves some of the problems in traditional methods of evaluation and it has several advantages over strictly numerical method: 1) ambiguous, qualitative or imprecise information, as well as quantitative data can be used in assessment and they are hendled in a consistant manner; 2) the risk associated with item failure modes can be evaluated directly using linguistic terms that are employed in making the criticallity assessment; 3) it give more flexible structure for combining the severity, occurance, and detectability parameters.This method allows the analyst to evaluate the risk associated with item failure modes directly using linguistic terms that are employed in making the criticality assessment.

Despite such advantages, approach reported in [1] did not address several aspects: 1) the process of risk ranking is a group decision process where a method to aggregate the experts' ranking (numerical and linguistic) is not clearly described; 2) the different degree of contribution of an expert is not taking into account during assessment of risk; 3) the different degree of important in the S, O and D parameters is not described.

In this paper, a new method of ranking risk called multi–ranking is proposed to improve the aforementioned weaknesses. This method combines both the quantitative method (numeric ranking and fuzzy numbers) and the qualitative method (linguistic ranking) to be employed in the ranking of risk. In both aspects, the concept of fuzzy logic [9] is employed. In the qualitative method, ranking is done by using linguistic variable whereas in the quantitative method, ranking could be made either numerically (traditional RPN method) or by employing fuzzy numbers [3].

The main objective of this method is to provide the experts with more flexibility in expressing their opinions on ranking and also to give the expert choices to choose the ranking method to use according to their preference. In this way, it allows experts opinions to be combined in a flexible way without bias of other expert opinions; noting that FMEA risk ranking is a team work process whereby ranking is given by several experts having different experiences and backgrounds.

2 Quantitative Versus Qualitative Ranking

Combining quantitative and qualitative make the process of risk ranking to be more flexible and more representative of the real world decision making. Flexibility in ranking could be achieved in several ways. A way is to allow experts to use linguistic ranking when there is uncertainty in the information regarding the ranking value. Otherwise, numerical ranking could be employed. It should be noted that the risk ranking is very subjective in nature where experts' knowledge and experience were consulted when statistical or reliability data are unavailable [1] and

[4]. Hence, combination of both the numerical and linguistic ranking is expected to be more robust and could resemble the overall expert opinions.

Another way which could provide more realistic resemblence of the real world decision making process is to incorporate weights to the different experts performing the risk ranking. This process is known as peer ranking [3]. In this process, each expert ranks other expert's weightage. In other words, the contribution of an expert is determine by other experts in a group of experts. The higher the weight of a particular expert, it is assumed that the more important the expert's contribution to the process.

Finally, more flexibility in the FMEA risk ranking process could be achieved by allowing all of the expert to come up with their own rule base instead of using a common rule base. In other words, experts are allowed to design the rule base according to their policies and experiences in their related areas [2].

2.1 Fuzzy FMEA Methodology

To test the effectiveness of the proposed method, a case study was preformed at Motorola, Penang, Malaysia in the solder paste printing process. The FMEA team consists of 3 experts; a quality engineer, a process engineer and a design engineer. It should be noted that the Motorola, Penang had already carried out the traditional process FMEA. Therefore, all the numerical data are already available from the existing records. In this study, the linguistic ranking for severity, occurrence and detection were reconstructed by using similar data.

Rule base in the form of *IF ... Then ...* is extracted from the 3 experts to define the risk. Examples of fuzzy rule base are as follows:

If *occurrence* is moderate **and** *detection* is low **and** *severity* is moderate **then** *risk* is moderate

If *occurrence* is moderate **and** *detection* is remote **and** *severity* is moderate **then** risk is important

The rule base is defined in such a way that priority is given to severity followed by occurrence and lastly detection so that the different degree of contribution from these 3 parameters is automatically accommodated. The rule will produce a very high risk category if any of the ranking values of severity (S), occurrence (O) or detection (D) exceeds 7. It should be noted that the experts have to agree in consensus regarding the parameters and shape of the fuzzy sets.

3 Ranking Methods

In this paper, 2 methods were used to evaluate the risk ranking; the traditional numerical RPN method and the multi–ranking method. Finally a comparison is made between these two methods to evaluate the effectiveness of each of the method employed.

3.1 Numerical Ranking

Ranking for severity, detection and occurrence is carried out numerically using ranking values in the range between 1 to 10. It should be noted that this method of ranking is employed only in situations where the experts are more than 90% certain of the ranking value for any particular parameters. For example, *severity* **very high** is ranked as 10 or *detection* **moderate** is ranked as 6.

3.2 Fuzzy Ranking

Fuzzy ranking could be carried out in two ways; ranking using fuzzy numbers and ranking with linguistic variables. The former method is proposed to be employed when the expert is less than 90% certain of the numeric value in which a degree of certainty is associated with the ranking values from 1 to 10. In such cases, percentage is employed to express the degree of certainty about the ranking and is represented in the form of a triangular fuzzy number (T.F.N). The later ranking method is employed when the percentage of certainty is less than 50%. Fig. 3.1 illustrates the fuzzy sets labelled *low*, *moderate* and *high* associated with linguistic ranking which are used in this study.

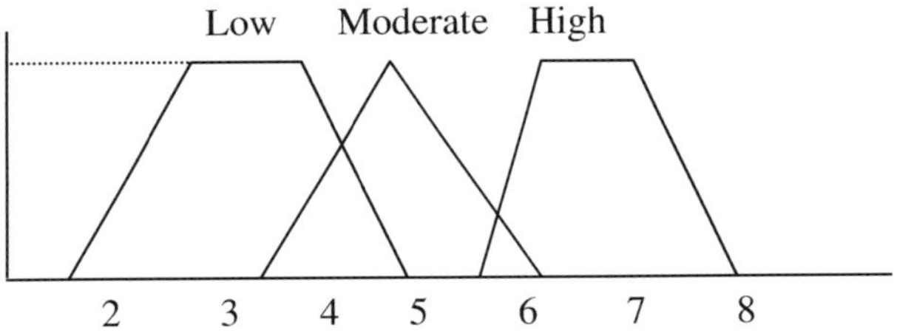

Fig. 3.1. Linguistic ranking

3.2 Multi-Ranking Method

The multi-ranking can be divided into two main methods; input ranking aggregation method and output risk ranking aggregation method. The later could further be devided into 3 sub-methods; with common rule-bases, with peer-ranking and common rule-bases and with peer-ranking and different rule bases.

Multi-ranking using input ranking aggregation method aggregates the ranking input of S, O and D from several experts using either linguistic or numerical values into a single fuzzy set according to their respective parameters before the input goes through fuzzy inference system (FIS). In this paper, the FIS used for all the multi-ranking methods is the Mamdani min-max method [6]. In multi-ranking with risk output aggregation the input from several experts are evaluated separately through the FIS without aggregation. In this method the difference using a common rule base for the all the experts and different rule base for each experts are studied. The effect of using peer ranking on the expert is studied for the output risk aggregation ranking.

3.3.1 Input Ranking Aggregation

The process of aggregation mentioned in the multi-ranking methodology consists of two stages. The first stage is the ranking stage and the second is the aggregation stage. In the ranking stage, the expert linguistically rank O, S and D as a fuzzy set which can be either linguistic variable, fuzzy numbers or numeric values. In the aggregation stage, the ranked fuzzy sets and numeric values are aggregated to find a single fuzzy set.

For example, ranking for Severity (S) by the expert t which ranks S as *High* can be represented with a trapezoidal fuzzy number as $M_S^t = (l_S^t, m_S^t, n_S^t, r_S^t)$. Let n be the number of experts involved and k is the fuzzy set under study, therefore the aggregated trapezoidal fuzzy number can be calculated by using the Eq. (1), (2), (3) and (4) [8].

$$m_k = \frac{1}{n}\sum_{t=1}^{n}\left(m_k^t\right) \tag{1}$$

$$n_k = \frac{1}{n}\sum_{t=1}^{n}\left(n_k^t\right) \tag{2}$$

$$l_k = \underset{t=1,\ldots n}{Min}\left\{l_k^t\right\} \tag{3}$$

$$r_k = \underset{t=1,\ldots n}{Max}\left\{r_k^t\right\} \tag{4}$$

3.3.2 Output Risk Ranking Aggregation

This method could be carried out in three different ways that is by using a single common rule-base, with common rule-base and peer ranking and finally using different rule–base and peer ranking. The third method allows each expert to come up with their own rule base individually or by groups and the output risk is combined and defuzzified to get the risk outcome. The third method also allows peer to rank his/her expert colleagues by giving weightage of importance to certain experts as compared to others. It should be noted that all the methods use the same fuzzy universe of discourses for ranking.

Another important factor that has to be highlighted here is that in these new multi-ranking methods the input from the experts are evaluated separately and not aggregated as in the earlier section. After the input from experts are evaluated separately, the output risk are combine in an additive manner. Therefore, the aggregation is only performed on the output risk values.

3.3.3 Peer Ranking

Input from the 3 experts either in the form of linguistic labels, numerical or T.F.N are evaluated according to expert's defined rule base. In the aggregation process all the output risk information from the experts are combined to come up with a single risk area. The output region of risk is first scale according to the experts peer ranking before aggregation. Scaling of the input is done with the Eq. 5 [2].

$$\mu_{si}[x] = \underset{i=1}{\forall} \mu_{si}(x) \times \mathbf{P}_i \tag{5}$$

Where μ_s is the risk membership function, P_i is the peer ranking which could takes the value in the range from 0 to 1 and and N is the maximum number of fuzzy set defined.

3.3.4 Output Aggregation

An important factor that must be observed in the aggregation process is to reduce or minimize the loss of information from each expert in the aggregation process. So the aggregation method must be able to maintain the information. Consequently the additive aggregation method is used to combine the output risk. This output is later defuzzified with MOM method. Eq. 6 shows the additive aggregation operator [2].

$$\mu_{outi}[x] = \overset{N}{\underset{i=1}{\forall}} (\min(\mu_{outi}[x] + \mu_{si}[x]), 1) \tag{6}$$

Where N, μ_{out} and μ_{si} is the number of expert involved, the output risk after aggregation and the output risk from the *i*-th expert, respectively.

4 Results and Discussions

Comparisons are made between traditional RPN risk ranking, multi-ranking with input aggregation, multi-ranking methods with common rule base, different rule base and with peer ranking. Comparison is also made between these methods to analyse which among them is the most suitable and flexible method. To compare, a cut-off value where the risk is important for the ranking method has to be defined. This process has to be carried out so that preventive action can be taken to remove or minimize the risk. In this project, the cut off value is decided using consensus among the experts. It is decided that for traditional RPN ranking and the multi–ranking methods, the cut off value is ≥ 90 and ≥ 7, respectively.

A comparison of the number of failure mode that has risk important (i.e. ≥ 90) for traditional RPN and ≥ 7 for both the multi-ranking and fuzzy numerical ranking methods are shown in the Table 1. From this table, it can be seen that the number of failure mode for multi-ranking using input aggregation method is 13 which far exceeds the number of failure mode from traditional RPN ranking. This is due too the definition of risk in the fuzzy rule base. It should be noted that, in this case study, the value for severity, occurrence and detection if exceeds 7, then risk is considered important. However, for traditional RPN method, only the RPN multiplication value is considered. Subsequently, even though the RPN value is low for traditional ranking, the risk may turn out to be important or high for both the multi–ranking and numerical fuzzy ranking, if either severity, occurrence or detection is more than 7.

It can also be seen form Table 1, the multi-ranking using common rule base and input aggregation has the most number of failure modes when risk is very high. This is because the input range is largest compared to other methods because input from several experts are aggregated together before evaluation. Furthermore in input aggregation the input is not modified by peer ranking and this will also contribute to the large input range.

In the output aggregation method, inputs from several experts' are evaluated through FIS without aggregation. This will cause the input range used for evaluation become smaller and therefore will only allow fewer number of rules to be fired. Hence, the risk outcome will be reduced as compared to multi-ranking using input aggregation method. For instance let's look at the comparison made between input aggregation and output aggregation multi-ranking using common rule base without peer ranking. It is found that the input aggregation gives higher number of failure mode (13) as compared to the output aggregation method, which gives only 1 high-risk failure mode.

Peer ranking provides higher risk values compared to without peer ranking because ranking from expert has different weights. Let's say, if an expert with a high weightage ranking gives a high risk ranking, a high output risk is likely be produced and this is not affected much by experts with low weightage ranking giving a risk ranking input of low or moderate. Therefore, if FMEA team consist of experts of different levels of experience, peer ranking can be used to reduce the bias from a less experience expert. However, if the FMEA team has experts of the same level of experience, the input ranking could be used instead. In other words, this study shows that multi-ranking method using input aggregation could be utilised when the experts are at the same level of experience and multi-ranking using output risk aggregation with peer ranking method could be applied when the experts experience are at different level.

Table 4.1. Comparing number of failure modes with high risk

Output Aggregation	Number of failure mode with high risk	Common failure mode between the method marked (*)		
Multi-ranking (common rule base and no peer ranking)	1			
Multi ranking (common rule base and with peer ranking)	5	*	*	*

Table 4.1. (cont.)

Multi-ranking (different rule base and with peer ranking)	8	*		*	*
Input Aggregation Multi-ranking (common rule base)	13	*	*	*	
Traditional Ranking	5	*			
Common failure modes		4	4	8	5

4 Conclusions

In this paper, a new ranking methods based on fuzzy logic technique for FMEA ranking called multi-ranking method had been introduced. This method allows more flexibility for the experts to define risk in the form of a fuzzy rule base. This method also allows a greater degree of freedom for the expert to express their uncertainty in the ranking of risk either linguistically or numerically depending on their preference. If the group of experts are from different level of experience, peer ranking is recommended to provide better outcomes as it considers the expert's expertise.

References

1. Bowles JB, Peláez CE (1995) Fuzzy logic prioritization of failure in a system failure mode, effects and criticality analysis. Reliability Engineering and System Safety, 50: 203-213
2. Cox E (1995) Fuzzy logic for business and industry. Charles River Media
3. Cox E (1998) The fuzzy system handbook. AP Professional
4. Gilchrist W (1993) Modeling failure modes and effect analysis. International Journal of Quality and Reliability Management.10:16-23
5. Kaufman A, Gupta M (1988) Fuzzy mathematical models in engineering and management science. Elsivier Science
6. Lee CC (1990) Fuzzy logic in control systems. Fuzzy Logic Controller: Part II, IEEE Trans. on Syst. Man and Cybernetics, 20: 419-435
7. Stamatis DH (1998) FMEAfrom Theory to Execution, Milwaukee. ASQ Quality Press
8. Wang J (1999) Fuzzy outranking approach to prioritise design requirement in quality function deployment. International Journal of Production Research, 37: 899-916
9. Zadeh LA (1973) Outline of a new approach to the analysis of complex system and decision process. IEEE Trans Systems, Man and Cybernetics. 3: 28-44

Representing Reliability and Hesitation in Possibility Theory: A General Framework

Chris Cornelis, Martine De Cock, and Etienne Kerre

Fuzziness and Uncertainty Modelling Research Unit, Applied Mathematics and Computer Science Dept., Krijgslaan 281 (S9), B–9000 Gent, Belgium
{chris.cornelis,martine.decock,etienne.kerre}@UGent.be
http://fuzzy.UGent.be

1 Introduction

Possibility theory, introduced by Zadeh in [5], is a formalism that tries to capture in mathematical terms imprecise (typically, linguistic) information about the more or less plausible values that a variable may assume. For instance, a statement like "decent houses in Gent do not come cheap" flexibly restricts plausible buying prices by pointing out that a low budget will not get very you far in this town. Compare this to "I found a nice place in Gent for about 100.000 EUR", which gives you a guarantee (by explicit example) that properties in a given price range can be found. The examples we quoted are instances of what is called negative or constraint–based and positive or observation–based information respectively in the literature [3, 4]: the first kind rules out certain values for the buying price X, while the second kind only designates certain observed values as "definitely possible" while saying nothing about the rest.

To mathematically represent both types of information, possibility distributions and guaranteed possibility distributions are introduced. Formally, a possibility distribution π_X on a variable X (e.g. buying price) in a universe U (e.g. a domain $[0, 200.000]$ of prices in euro) is a $U \to [0,1]$ mapping such that $\pi_X(u) = p$ means that it is (at most) possible to degree p that X can take the value u. Possibility degrees typically emerge by evaluating a fuzzy set like "expensive" and subsequently imposing

$$\pi_X(u) \leq expensive(u) \tag{1}$$

The inequality allows for the existence of other information items restricting X (specifically, new constraints can make the possibility degrees decrease). On the other hand, a guaranteed possibility distribution δ_X on U is another $U \to [0,1]$ mapping such that $\delta_X(u) = d$ means that it is (at least, or guaranteed) possible to degree d that X can take the value u. In our example, δ_X would be obtained by setting

$$\delta_X(u) \geq \text{about-100.000-EUR}(u) \tag{2}$$

Analogously to (1), the inequality is meant to imply that later on new observations can make the guaranteed possibility degrees increase. Constraint–based and observation–based information induce different types of inequalities; it is generally assumed that $\delta_X(u) \leq \pi_X(u)$ for $u \in U$, an integrity constraint expressing that impossible values cannot be guaranteed.

(Guaranteed) possibility distributions have been applied fruitfully in a wide range of knowledge–based systems. To increase their expressivity, it is worthwhile to take into account the following ideas:

- It is not always realistic to pin down exact (guaranteed) possibility degrees, since the person assessing them is hesitant about them to a given extent.
- We sometimes may have doubts about the truthfulness of information, i.e. claims can have varying degree of credibility.

In this paper, we study extensions, in terms of interval–valued fuzzy sets and intuitionistic fuzzy sets, of classical (guaranteed) possibility distributions that are able to deal with the facets described above. We will illustrate their usefulness by some concrete examples.

2 Preliminaries

2.1 Interval–Valued and Intuitionistic Fuzzy Sets

Interval–valued fuzzy sets (IVFSs, for short) form an extension of fuzzy sets in the sense that crisp, $[0,1]$–valued membership degrees (say, $\mu_A(u)$) are replaced by intervals $[\mu^l_A(u), \mu^u_A(u)]$ in $[0,1]$. As a general principle, IVFSs emerge whenever fixing exact membership degrees is conceived as unrealistic, unjustified or too restrictive.

Intuitionistic fuzzy sets (IFSs, for short) defy the claim that from the fact that an element u "belongs" to a given degree $\mu_A(u)$ to a fuzzy set A, naturally follows that u should "not belong" to A to the extent $1 - \mu_A(u)$, an assertion implicit in the concept of a fuzzy set. On the contrary, IFSs assign to each element of the universe both a degree of membership $\mu_A(u)$ and one of non–membership $\nu_A(u)$ such that $\mu_A(u) + \nu_A(u) \leq 1$, thus relaxing the enforced duality $\nu_A(u) = 1 - \mu_A(u)$ from fuzzy set theory.

It is instructive from a syntactical perspective to study IVFSs and IFSs as particular cases of L–fuzzy sets. To recall, let $(L, \leq_L)$ be a complete lattice. An L–fuzzy set in X is an $X \to L$ mapping. Putting e.g. $L^* = \{(x_1, x_2) \in [0,1]^2 \mid x_1 + x_2 \leq 1\}$, $(x_1, x_2) \leq_{L^*} (y_1, y_2) \Leftrightarrow x_1 \leq y_1$ and $x_2 \geq y_2$, one can easily verify that $(L^*, \leq_{L^*})$ is a complete lattice, and also that the class of IFSs is isomorphic to that of L^*–fuzzy sets. Hence, IFSs emerge, syntactically, as a specific subclass of L–fuzzy sets. On the other hand, defining $(L_I, \leq_{L_I})$ as $L_I = \{(x_1, x_2) \in [0,1]^2 \mid x_1 \leq x_2\}$, $(x_1, x_2) \leq_{L_I} (y_1, y_2) \Leftrightarrow x_1 \leq y_1$ and

$x_2 \leq y_2$, it is clear that IVFSs are also specific kinds of L–fuzzy sets. Moreover, IFSs are formally equivalent to IVFSs: indeed, a couple $(x_1, x_2) \in L^*$ may be mapped bijectively to an interval $[x_1, 1 - x_2]$.

For a more detailed account of the semantical aspects of IVFSs and IFSs, we refer to [2]. What should be clear is that IVFSs and IFSs correspond to distinct paradigms: the former acknowledge the lack of justification for precise membership degrees, while the latter address positive (membership) and negative (non–membership) assessments separately.

2.2 Selected Notions from Possibility Theory

As explained in the introduction, the essence of (guaranteed) possibility distributions is that they provide us with information about the values a variable X may assume. In the constraint–based framework, this information comes in a negative guise, expressing which values u are more or less impossible for X; this is done by associating with X a $U \to [0,1]$ mapping π_X called a possibility distribution on X. π_X can be used to evaluate to what extent it is possible that X takes a value in a crisp subset A of U. To this aim, we calculate $\Pi_X(A) = \sup\limits_{u \in A} \pi_X(u)$, the possibility of A. On the other hand, we may also be interested to know to what extent we can be sure that X takes its values in A; $N_X(A) = \inf\limits_{u \notin A} 1 - \pi_X(u)$ is called the certainty or necessity of A. It is clear that $N_X(A) = 1 - \Pi_X(coA)$; this equality is known as possibility–necessity duality.

In the observation–based framework information emerges as soon as some value u receives a strictly positive degree $\delta_X(u)$ of guaranteed possibility, where the guaranteed possibility distribution δ_X on X is again a $U \to [0,1]$ mapping δ_X. It can be used to define two dual measures Δ_X and ∇_X of guaranteed possibility and potential necessity. For a crisp subset A of U, $\Delta_X(A) = \inf\limits_{u \in A} \delta_X(u)$ calculates the extent to which it is guaranteed possible that X takes a value in A, while $\nabla_X(A) = \sup\limits_{u \notin A} 1 - \delta_X(u)$ expresses the potential ("best–case") necessity or certainty that X indeed takes its values in A. It can be verified that $\nabla_X(A) = 1 - \Delta_X(coA)$.

3 Extensions

In this section, we explore two kinds of situations which the traditional patterns of possibility theory have difficulties coping with. To each problem, an extension of the original definitions is proposed to mend their shortcomings.

3.1 Interval–Valued Extension

Recall the example statement from the introduction that houses in Gent are expensive. Compare it to the following conversation: "This house costs 250.000

EUR."— "Wow, that's expensive!" Whereas in the latter situation there is no uncertainty (the use of a linguistic term is just a way of abstracting from numerical data, in other words an instance of vagueness), in the former there is usually plenty; it seems therefore unnatural to suppress this uncertainty by enforcing the rigid delimitation that a fuzzy set representation of "expensive" implies. In other words, while people may be able to draft a membership function for "expensive" (evaluating prices on a given scale), when they assign the label to a given situation, usually this association is pervaded with some uncertainty. As a result, people may be reluctant to pin down exact (guaranteed) possibility degrees decisively, because they are to some extent hesitant about such an assignment (which would involve a strong commitment); what they often are prepared to do, is to fix thresholds α and β which they consider a sure lower and upper bound for it, respectively. In figure 1, this is illustrated for our example. The left hand side picture presents a π_X which is defined by the fuzzy set *expensive*, while in the right hand side picture, a margin of hesitation is introduced by introducing lower and upper possibility degrees $\pi_X^1(u)$ and $\pi_X^2(u)$.

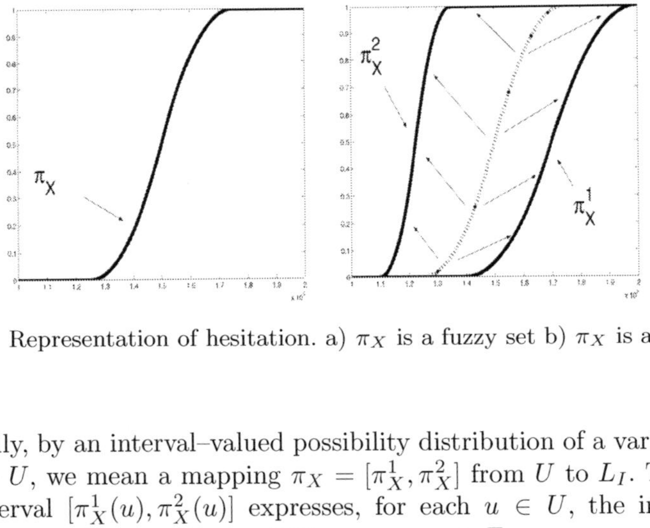

Fig. 1. Representation of hesitation. a) π_X is a fuzzy set b) π_X is an IVFS

Formally, by an interval–valued possibility distribution of a variable X on a universe U, we mean a mapping $\pi_X = [\pi_X^1, \pi_X^2]$ from U to L_I. The length of the interval $[\pi_X^1(u), \pi_X^2(u)]$ expresses, for each $u \in U$, the information source's hesitation w.r.t. the possibility assessment. From π_X, we may derive generalized versions of the possibility and necessity of a crisp subset A of U:

$$\Pi_X(A) = \sup_{u \in A} \pi_X(u) = \left[\sup_{u \in A} \pi_X^1(u), \sup_{u \in A} \pi_X^2(u)\right]$$

$$N_X(A) = \inf_{u \notin A} \mathcal{N}(\pi_X(u)) = \left[\inf_{u \notin A} 1 - \pi_X^2(u), \inf_{u \notin A} 1 - \pi_X^1(u)\right]$$

where the complement operation $\mathcal{N}$ on L_I is defined by $\mathcal{N}([u_1, u_2]) = [1 - u_2, 1 - u_1]$. It can be verified that $N_X(A) = \mathcal{N}(\Pi_X(coA))$. Analogously, we can derive an interval–valued extension of guaranteed possibility distributions [3].

3.2 Intuitionistic Fuzzy Extension

With the traditional machinery of (guaranteed) possibility distributions, there is no way to cast doubt about a given statement. This becomes obvious when one takes a closer look at the relationship between (guaranteed) possibility and (potential) necessity. When $\pi_X(u) = 0$, then from the definition of necessity it is clear that $N_X(co\{u\}) = 1$, a statement that means that we are certain that X cannot take the value u. Unfortunately, this forces us into a position of unconditional faith in the truthfulness of the information provider. A similar remark holds when $\delta_X(u) = 1$ for some u; in that case, $\nabla_X(co\{u\}) = 0$, implying that the potential (and hence the actual) certainty that X cannot take the value u does not surpass 0. But what if we found out that whoever provided the information were lying? It should be a sign for us to cancel all commitments (i.e. assessments of non–trivial certainties) based on the information.

For this reason, we propose to opt out of the possibility–necessity duality (resp. the guaranteed possibility–potential necessity duality); to see how this can be done, recall the example "I found a place for about 100.000 EUR". The left–hand picture in figure 2 shows the corresponding guaranteed possibility distribution δ_X. Its complement (let us call it ω_X) is also shown; recall that $\omega_X(u) = \nabla_X(co\{u\})$ expresses, for $u \in U$, the potential certainty that X is different from u. Now let us assume that we got this information from someone who is always bragging about what good deals he has been able to strike, but in general pays much more than he is willing to admit. To account for this scepticism, we should indicate that the guaranteed possible degrees for certain low amounts of money are not actually credible in our eyes. We do this raising (increasing) ω_X, such that e.g. $\omega_X(100000) = 1$, i.e. the potential certainty that X differs from u is equal to one. This is shown in the right hand side picture of figure 2.

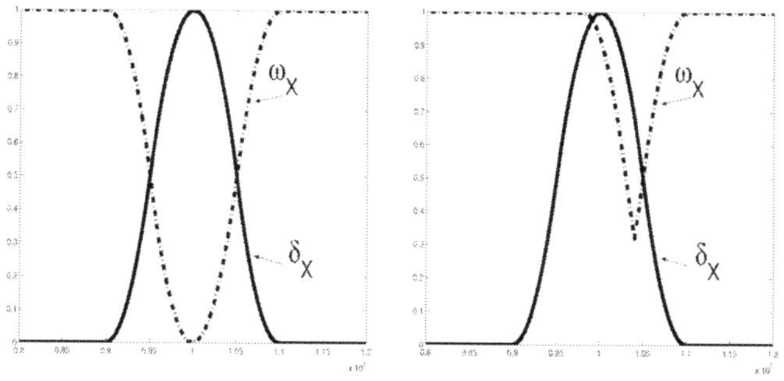

Fig. 2. Representation of unreliable information. a) δ_X and ω_X are each other's complement. b) δ_X and ω_X are partially independent distributions

Formally, by an intuitionistic fuzzy guaranteed possibility distribution of a variable X on a universe U, we mean a couple (δ_X, ω_X) where δ_X and ω_X are two $U \to [0,1]$ mappings such that $\delta_X(u) + \omega_X(u) \geq 1$ holds for all $u \in U$. Completely analogously, an intuitionistic fuzzy possibility distribution of a variable X on a universe U is a couple (π_X, ν_X) where π_X and ν_X are two $U \to [0,1]$ mappings such that $\pi_X(u) + \nu_X(u) \leq 1$ for all $u \in U$, which explains the terminology "intuitionistic fuzzy". The newly introduced concepts also give rise to altered definitions of (guaranteed) possibility and (potential) necessity measures. Given (π_X, ν_X) and (δ_X, ω_X), we may define, for a crisp subset A of U:

$$\Pi_X(A) = \inf_{u \in A} \pi_X(u) \qquad \Delta_X(A) = \inf_{u \in A} \delta_X(u)$$
$$N_X(A) = \sup_{u \notin A} \nu_X(u) \qquad \nabla_X(A) = \sup_{u \notin A} \omega_X(u)$$

It is clear that $N_X(A) \leq 1 - \Pi_X(coA)$ and $\nabla_X(A) \geq 1 - \Delta_X(coA)$.

4 Conclusion

This paper described two extensions to the classical setting of possibility theory: an interval–valued extension that draws π_X and δ_X from L_I instead of from $[0,1]$, and an intuitionistic fuzzy extension that gives the concepts of (guaranteed) possibility and (potential) necessity (up to a certain point) independent status. The resulting formalism can be used to express, in a quantitative manner, the reservation of the parties involved: hesitation on the part of the information source, and skepticism on the part of its recipient. The material in this paper is exposed in a more detailed fashion in [3].

References

1. Cornelis C, De Cock M, Kerre E (2002) Assessing degrees of possibility and certainty within an unreliable environment. In: Lotfi A, Garibaldi J, John R (eds) Proceedings of Fourth International Conference on Recent Advances in Soft Computing.
2. Cornelis C, Atanassov K, Kerre E (2003) Intuitionistic Fuzzy Sets and Interval–Valued Fuzzy Sets: a Critical Comparison, To appear in: Proceedings of EUSFLAT 2003.
3. Cornelis C, Kerre E (2003) Generalized (Guaranteed) Possibility Distributions for Hesitation and Reliability Representation. Submitted to Fuzzy Sets and Systems.
4. Dubois D, Prade H, Ughetto L(2002) A New Perspective on Reasoning with Fuzzy Rules. In: Pal N, Sugeno M (eds) Proceedings of 2002 International Conference on Fuzzy Systems, LNAI 2275, Springer.
5. Zadeh L (1978) Fuzzy Sets as a Basis for a Theory of Possibility. Fuzzy Sets and Systems 1:3–28

Noise Clustering with a Fixed Fraction of Noise

Frank Klawonn

Department of Computer Science, University of Applied Sciences, Salzdahlumer Str. 46/48, D-38302 Wolfenbuettel, Germany
f.klawonn@fh-wolfenbuettel.de

1 Introduction

Cluster analysis is an exploratory data analysis technique that is designed to group data and to detect structures within data. Exploratory techniques are applied as first steps in data analysis, immediately after elementary cleaning and visualisation of the data has been carried out. This means that a specific model for the data is not available at this state. However, exploratory methods usually incorporate control parameters that influence the result of this early data analysis step. Therefore, it is desirable to design methods that are robust w.r.t. the variation of such control parameters. In this paper, we modify the so-called noise clustering technique making it more robust against a wrong choice of its main control parameter, the noise distance. Section 2 briefly reviews the necessary background in fuzzy clustering. Section 3 introduces our modified noise clustering approach including a computationally efficient algorithm. We finish the paper by an example and some concluding remarks.

2 Objective Function-Based Fuzzy Clustering

Fuzzy clustering is suited for finding structures in data. A data set is divided into a set of clusters and – in contrast to hard clustering – a datum is not assigned to a unique cluster. In order to handle noisy and ambiguous data, membership degrees of the data to the clusters are computed. Most fuzzy clustering techniques are designed to optimise an object function with constraints. The most common approach is the so called probabilistic clustering with the objective function

$$f = \sum_{i=1}^{c}\sum_{j=1}^{n} u_{ij}^{m} d_{ij} \quad \text{constrained by} \quad \sum_{i=1}^{c} u_{ij} = 1 \quad \text{for all } j = 1,\ldots,n \qquad (1)$$

that should be minimized. It is assumed that the number of clusters c is fixed. We will not discuss the issue of determining the number of clusters here and refer for an overview to [2, 4]. The set of data to be clustered is $\{x_1, \ldots, x_n\} \subset R^p$. u_{ij} is the membership degree of datum x_j to the ith cluster. d_{ij} is some distance measure specifying the distance between datum x_j and cluster i, for instance the (quadratic) Euclidean distance of x_j to the ith cluster centre. The parameter $m > 1$, called fuzzifier, controls how much clusters may overlap. The constraints lead to the name probabilistic clustering, since in this case the membership degree u_{ij} can also be interpreted as the probability that x_j belongs to cluster i. The parameters to be optimised are the membership degrees u_{ij} and the cluster parameters that are not given explicitly here. They are hidden in the distances d_{ij}. Since this is a non-linear optimisation problem, the most common approach to minimize the objective function (1) is to alternatingly optimise either the membership degrees or the cluster parameters while considering the other parameter set as fixed.

In this paper we are not interested in the great variety of cluster shapes (spheres, ellipsoids, lines, quadrics,...) that can be found by choosing suitable cluster parameters and an adequate distance function. (For an overview we refer again to [2, 4].) We only concentrate on the aspect of the membership degrees. Interpreting the membership degrees in terms of probabilities, u_{ij} specifies the probability that datum x_j belongs to cluster i, under the assumption that it must be assigned to a cluster. As a consequence, we obtain the following effect, which can lead to undesirable results. If we have, for instance, only two clusters and a datum has approximately the membership degree 0.5 to both clusters, it means either that the datum fits to both clusters equally well (the datum is near the border between the two clusters) or equally bad (the datum is noise and far away from both clusters).

In order to avoid this effect, possibilistic clustering was introduced [5], dropping the probabilistic constraint completely and introducing an additional term in the objective function to avoid the trivial solution $u_{ij} = 0$. However, the aim of possibilistic clustering is actually not to find the global optimum of the corresponding objective function, since this is obtained, when all clusters are identical. [6] describes an improved approach for the price of solving an additional non-linear optimisation problem in each iteration step.

Noise clustering [3] is another approach extending probabilistic clustering. The principle of probabilistic clustering is maintained, but an additional noise cluster is introduced. All data have a fixed (large) distance to the noise cluster. In this way, data that are near the border between two clusters still have a high membership degree to both clusters as in probabilistic clustering. But data that are far away from all clusters will be assigned to the noise cluster and have no longer a high membership degree to other clusters. The crucial point in noise clustering is the choice of the noise distance. If the noise distance is chosen too small, most of the data will simply be assigned to the noise cluster, if the noise distance is too high, the results are more or less identical to standard probabilistic clustering.

3 Noise Clustering with an Expected Fraction of Noise

The considerations in the previous section inspired the idea to introduce a new parameter into noise clustering that specifies the number or the fraction of noisy data expected or admitted in the data. At first sight, it seems that this approach makes the situation even more complicated, since in addition to the noise distance the additional parameter must also be specified. However, it turns out that a very rough (usually much too large) estimation of the noise distance in combination with a rough estimation of the number of noisy data leads to very good clustering results.

The basis of noise clustering with a fixed number $0 \leq N < n$ of noisy data is the following objective function

$$f = \sum_{i=1}^{c+1} \sum_{j=1}^{n} u_{ij}^2 d_{ij} \quad \text{constrained by} \quad \sum_{i=1}^{c+1} u_{ij} = 1 \quad \text{for all } j = 1, \ldots, n. \qquad (2)$$

$$\text{with the additional constraint} \qquad \sum_{j=1}^{n} u_{c+1,j} = N. \qquad (3)$$

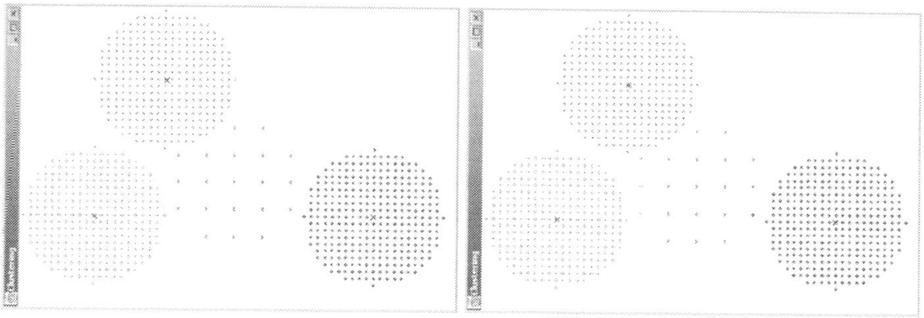

Fig. 1. Detection of noise with modified (left) and standard (right) noise clustering.

We assume that cluster number $(c+1)$ is the noise cluster and therefore the distance $d_{c+1,j} = d_{\text{noise}}$ is the fixed noise distance d_{noise}. In order to obtain an analytical solution for the alternating optimisation scheme, we have chosen a fuzzifier of $m = 2$. The constraint (3) reflects the requirement that N data are accepted or considered as noise. In order to obtain the update equations for the u_{ij}, we must find the global minimum of this objective function satisfying the constraints specified in (2) and (3), while we consider the distance values d_{ij} as fixed. Therefore, we compute the corresponding partial derivatives of the Lagrange function

$$f_L = \sum_{i=1}^{c+1} \sum_{j=1}^{n} u_{ij}^2 d_{ij} + \sum_{j=1}^{n} \lambda_j \left(1 - \sum_{i=1}^{c+1} u_{ij}\right) + \lambda \left(N - \sum_{j=1}^{n} u_{c+1,j}\right). \qquad (4)$$

We obtain
$$\frac{\partial f_L}{\partial u_{ij}} = \begin{Bmatrix} 2u_{ij}d_{ij} - \lambda_j & \text{if } 1 \le i \le c \\ 2u_{ij}d_{\text{noise}} - \lambda_j - \lambda & \text{if } i = c+1 \end{Bmatrix} = 0. \tag{5}$$

Writing down these equations and the constraints specified in (2) and (3), we obtain the following system of $(n(c+2)+1)$ linear equations.

$u_{1,1}$	$\ldots$	$u_{1,n}$	$\ldots$	$u_{c,1}$	$\ldots$	$u_{c,n}$	$u_{c+1,1}$	$\ldots$	$u_{c+1,n}$	λ_1	$\ldots$	λ_n	λ	RHS
$2d_{1,1}$										-1				
	$\ddots$										$\ddots$			
		$2d_{1,n}$										-1		
			$\ddots$								$\vdots$			
				$2d_{c,1}$						-1				
					$\ddots$						$\ddots$			
						$2d_{c,n}$						-1		
							$2d_{\text{noise}}$			-1			-1	
								$\ddots$			$\ddots$		$\vdots$	
									$2d_{\text{noise}}$			-1	-1	
1				1			1							1
	$\ddots$				$\ddots$			$\ddots$						$\vdots$
		1				1			1					1
							1	$\ldots$	1					N

Empty entries correspond to zeros. RHS stands for the right hand side of the equation. An ad hoc solution of this equation would not be feasible for large data sets. But we can see that the corresponding matrix is almost an upper triangular matrix. Only the last $(n+1)$ rows disturb the triangular structure. We can use the equations with $2d_{\text{noise}}$ to eliminate the coefficients 1 in the last row, thus replacing the last row by

$$\frac{1}{2d_{\text{noise}}}\lambda_1 + \ldots + \frac{1}{2d_{\text{noise}}}\lambda_n + \frac{n}{2d_{\text{noise}}}\lambda = N$$

or, equivalently,

$$\lambda_1 + \ldots + \lambda_n + n\lambda = 2Nd_{\text{noise}}. \tag{6}$$

From (5), we know

$$u_{ij} = \frac{\lambda_j}{2d_{ij}} \quad (\text{for } i \le c) \quad \text{and} \quad u_{c+1,j} = \frac{\lambda_j + \lambda}{2d_{\text{noise}}}. \tag{7}$$

The constraint in (2) together with (7) yields

$$\frac{\lambda_j + \lambda}{2d_{\text{noise}}} + \sum_{i=1}^{c} \frac{\lambda_j}{2d_{ij}} = 1, \tag{8}$$

so that we have

$$\lambda_j = \frac{2 - \frac{\lambda}{d_{\text{noise}}}}{\frac{1}{d_{\text{noise}}} + \sum_{i=1}^{c} \frac{1}{d_{ij}}}. \tag{9}$$

Inserting (9) into (6) and solving for λ, we obtain

$$\lambda = \frac{2Nd_{\text{noise}} - 2\sum_{j=1}^{n} \frac{1}{\frac{1}{d_{\text{noise}}} + \sum_{i=1}^{c} \frac{1}{d_{ij}}}}{n - \frac{1}{d_{\text{noise}}} \sum_{j=1}^{n} \frac{1}{\frac{1}{d_{\text{noise}}} + \sum_{i=1}^{c} \frac{1}{d_{ij}}}}.$$

After having computed λ, we can use (9) to determine $\lambda_1, \ldots, \lambda_n$. Now we can directly compute the membership degrees u_{ij} from (7).

For this new type of noise clustering, the update equations for the membership degrees are no longer as simple as they are in probabilistic, possibilistic or standard noise clustering. However, the scheme we have derived is computationally efficient and is not significantly slower than the other clustering algorithms. If we had simply solved the system of linear equations in a naive way, the computation would not be feasible for larger data sets.

4 An Example

In this section we briefly illustrate with a simple data set, how our new algorithm works. Figure 1 shows the result of applying the well known fuzzy c-means clustering algorithm [1] with standard noise clustering on the right hand side and with our new approach on the left hand side. In both cases the cluster centres are positioned correctly and the data obviously belonging to a cluster are assigned with the highest membership degree to the corresponding cluster. The good result for standard noise clustering could only be obtained, by tuning the noise distance manually, finally to a value of 1.7. The data assigned to the noise cluster with the highest membership degree are marked as small vertically oriented rectangles. We can see that still some of the noisy data are not assigned to the noise cluster with the highest membership degree in the case of standard noise clustering. With our modification all noisy data are assigned to the cluster. For practical purposes, we do not specify the expected number of noisy data N, but the expected percentage P_{noise} or fraction of noisy data, so that the parameter N is determined by $N = \frac{P_{\text{noise}}}{100} \cdot n$. It should be noted that we can and should overestimate the percentage of noisy data. The nature of fuzzy clustering is that zero membership degrees nearly never occur. Therefore, even for a very large noise distance all data will have at least a small membership degree to the noise cluster. And all these small membership degrees contribute to the value of N of noisy data. Figure 2 (left) shows the clustering result with heavily overestimated noise where we have assumed 60% noisy data: $P_{\text{noise}} = 60$. Even for this case, the clustering result is still acceptable. In addition to the data that should be considered as noise, only data at the very boundary of the clusters are assigned to the noise cluster. If we apply standard noise clustering and decrease the more or less optimum noise distance of 1.7 in figure 1 to 1.6, the effect for the clustering is disastrous as the right hand side of figure 2 shows.

5 Conclusions

We have introduced an extension of noise clustering that allows the specification of the fraction of expected noisy data. A very rough value for this additional parameter frees the user from an accurate estimation of the noise distance. Therefore, our approach can be seen as a further step for making exploratory data analysis techniques more robust against tedious parameter selections.

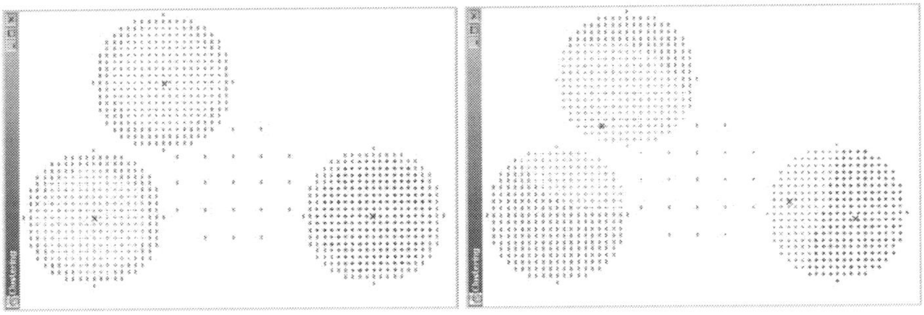

Fig. 2. Clustering result with heavily overestimated noise (left) and standard noise clustering with a slightly decreased noise distance.

References

1. Bezdek JC (1981) Pattern recognition with fuzzy objective function algorithms. Plenum Press, New York
2. Bezdek JC, Keller J, Krishnapuram R, Pal NR (1999) Fuzzy models and algorithms for pattern recognition and image processing. Kluwer, Boston
3. Davé RN (1991) Characterization and detection of noise in clustering. Pattern Recognition Letters 12: 657–664
4. Höppner F, Klawonn F, Kruse R, Runkler T (1999) Fuzzy cluster analysis. Wiley, Chichester
5. Krishnapuram R, Keller J (1993) A possibilistic approach to clustering. IEEE Trans. on Fuzzy Systems 1: 98–110.
6. Timm H, Borgelt C, Kruse R (2002) A modification to improve possibilistic cluster analysis. IEEE Intern. Conf. on Fuzzy Systems, Honululu

Selective Sampling for Combined Learning from Labelled and Unlabelled Data

Lina Petrakieva[1] and Bogdan Gabrys[2]

[1] Applied Computational Intelligence Research Unit, School of Information and Communication Technologies, University of Paisley, UK
petr-ci0@wpmail.paisley.ac.uk

[2] Computational Intelligence Research Group, School of Design, Engineering & Computing, Bournemouth University, bgabrys@bournemouth.ac.uk

Abstract. This paper examines the problem of selecting a suitable subset of data to be labelled when building pattern classifiers from labelled and unlabelled data. The selection of representative set is guided by a clustering information and various options of allocating a number of samples within clusters and their distributions are investigated. The experimental results show that hybrid methods like Semi-supervised clustering with selective sampling can result in building a classifier which requires much less labelled data in order to achieve a comparable classification performance to classifiers built only on the basis of labelled data.

Keywords. Combined Learning, Labelled and Unlabelled Data, Clustering, Selective Sampling.

1 Introduction

In many domains labelled data is difficult or expensive to obtain as it may require manual work or be computationally expensive. Therefore it is not surprising that there has been much interest in applying techniques that incorporate knowledge from unlabelled data into a supervised learning system [2, 3, 9, 1, 5, 10, 6, 12]. The task is to minimize the overall cost of the classification process, which depends both on the classifier accuracy and the cost of obtaining labelled data as discussed in [7].

In a number of publications discussing hybrid methods for coping with labelled and unlabelled data, the use of additional unlabelled data has been shown to offer improvements in comparison to classifiers generated only on the basis of labelled data [2, 3, 9, 1, 5, 6]. However it was not clear whether the improved performance of the classifiers supplemented by unlabelled data was mainly due to representativeness of the original labelled set or to the specific combined methods.

In order to answer this question we had conducted extensive experimental analysis of different combined approaches (shortly summarised in Section 4) for different ratios of labelled to unlabelled samples [3]. One of the main findings of that study was that the final classification performance depends more on the specific labelled subset used rather than on the combined classification method. Additionally, as a result of random selection of samples to be labelled from the initial pool of unlabelled data, a high variability of the classifier performance was commonly observed.

In this paper a continuation of the previous study is presented with a focus on methods for selecting samples to be labelled rather than choosing them randomly. It is hoped that a suitable selection of samples to be labelled could reduce the variance of the final solutions and improve mean classification performance obtained from random selection.

The rest of the paper is organised as follows: In Section 2 a general problem statement is given. This is followed by description of two selective sampling methods in Section 3 and a summary of the investigated combined methods in Section 4. Section 5 presents the experimental results. And finally conclusions are given in the last section.

2 General Problem Statement

Let $D = \{L, U\}$ be the training data set with $L = \{(\mathbf{x}_i, t_i) \mid i = 1 \ldots M\}$, representing a set of M labelled samples and $U = \{(\mathbf{x}_j, 0) \mid j = 1 \ldots N\}$, representing a set of N unlabelled samples where $\mathbf{x} = (x_1, x_2, \ldots, x_n) \in R^n$ is an n-dimensional feature vector and $t \in \{1, \ldots, p\}$ is a class label representing one of p classes with 0 used to denote an unlabelled sample. As in the conventional cases of designing a classifier on the basis of a training data set the main goal is to find a function transforming a feature vector $\mathbf{x}$ into one of the p classes, which can be formally written as:

$$C_D : \mathbf{x} \rightarrow t \qquad \text{or} \qquad t = C_D(\mathbf{x}) \tag{1}$$

where C_D is a classifier C designed on the basis of the data set D. However, depending on the ratio

$$r = \frac{M}{(M + N)} \tag{2}$$

of the labelled samples to the total number of samples in D the problem ranges from the pure supervised learning for $r = 1$ to the pure unsupervised learning for $r = 0$. In [3] various hybrid methods for coping with cases for $r \in (0, 1)$

with random selection of the labelled samples, in order to obtain L have been examined.

But since the use of specific labelled subset has been found to be of crucial importance our main effort in this study will concentrate on selecting representative samples to be labelled rather than creating sophisticated methods for coping with both types of data.

3 Selective Sampling Methods

In the context of pattern classification systems selective sampling techniques have been most frequently used in active learning approaches [11], where samples for labelling are selected in a dynamic manner (one at a time) - active selection. In the research presented here the static selection techniques will be examined. In contrast to the active selection, the static selection operates on the basis of selecting whole batches of data to be labelled (i.e. all M samples forming the labelled subset L).

Trying to find representative samples when working with unlabelled data means that one has to make decisions based only on clustering information. If the clusters are already available one needs just to select the samples from the clusters. However, the immediate question is: How many samples and from which clusters? The following two distinctive approaches of allocating the number of samples per cluster have been investigated: a) *proportional* allocation - samples for labelling are allocated proportionally to the cardinality of the cluster which means more samples for bigger clusters and some of the small clusters may have no samples selected; b) *consecutive* allocation - samples for labelling are allocated uniformly disregarding clusters' sizes. Furthermore, the actual selection of the samples to be labelled per cluster have been done in two ways - a) by selecting cluster prototypes and b) by trying to describe a cluster by selecting samples close to its boundary. A short description of both methods is presented bellow.

Cluster Mean Selection - The subset of samples to be labelled is created from the prototypes of the clusters of the dataset. The prototypes are selected as the closest samples to the means of the clusters. If there are more than one sample per cluster to be selected the clusters are divided into subclusters and their prototypes are selected. In case when the clusters are predefined the number of data points for labelling per cluster has to be calculated as discussed above. Then the process of selecting the samples can be applied to each cluster separately. If there are no defined clusters the whole dataset can be considered as one cluster or the dataset can be divided into b clusters where

b is the number of samples to be selected.

Cluster Boundary Selection - The process of selection begins with a set of randomly picked b samples. Then the algorithm is optimising this initial set by removing from it the samples that are too close to each other and by selecting outermost samples. Thus by maximizing the minimum distance between the selected data points the algorithm is selecting them around the boundary of the cluster. If there are many samples to be selected the method is placing some of them at the boundary and when they become too close to each other it is selecting the rest of the samples spread within the cluster.

4 Combined Learning Methods

Once the labelled subset has been generated, either using random selection or the approaches discussed in the previous section, one of a number of combined learning algorithms can be used to design a classifier. A short description of the combined approaches used in the experiments and formally defined in [3] follows.

The first most obvious way of dealing with situations when labelled and unlabelled data is available is to ignore the unlabelled data (referred to as Labelled Only method) and build the classifier C_L using just the labelled subset L from D completely ignoring U. The classification process from the Eq.(1) in this case becomes:

$$t = C_L(\mathbf{x}) \tag{3}$$

Labelled Only method is implemented for comparison purposes. However, since using the unlabelled data can be advantageous we used various combined learning methods falling into one of the following three major groups (see [3] for more details):

Pre-labelling approaches - a) Static methods - The first of the considered approaches to utilising the unlabelled data, referred to as Static Labelling approach in the later sections, is based on generating an initial classifier on the basis of the labelled data only C_L and labelling the remaining unlabelled data U by applying the initial classifier so that W is the newly labelled set U. Finally the classifier is redesigned using both the original L and the newly labelled W data sets; b) Dynamic methods - This approach is a modification of the above whereas an initial classifier is generated on the basis of the labelled data only C_L but the unlabelled data U are iteratively labelled one sample at a time. The newly labelled sample is added to the pool of labelled data and the classifier is redesigned at each step. The samples which can be most confidently classified are chosen first. The process is continued until all unlabelled samples have been labelled and the final classifier obtained. This will be referred to as a Dynamic Labelling approach.

Post-labelling approaches - Majority Clustering - The considered method is based on clustering all the data and using the labelled data for labelling the whole clusters by applying the majority principle i.e. the label of the cluster is assigned on the basis of the largest number of samples from a given class represented in the cluster. We will refer to this method as the Majority Clustering method.

Semi-supervised approaches - Semi-supervised Clustering - The detailed algorithm is presented in [3]. In contrast to the standard clustering the labels are actively used for guiding the clustering process. In result the algorithm is more robust in the sense of the number of created clusters and their sizes which to a large extent is dependant on the relative placement of the labelled samples in the input space. The idea is to split the initial clusters until there is an overwhelming presence of one type of labelled samples in each of newly created sub-clusters. After splitting the clusters the labelling process is carried out using Majority Clustering method.

5 Experimental Results and Analysis

Methods described in Section 4 are used in selective sampling experiments. As described in Section 3, both proportional and consecutive distribution as well as Mean and Boundary Selection methods are used. The nearest neighbour (NN) and pseudo-fisher support vector (PFSV) classifiers implemented in [8] have been used as the base classifiers for labelling and testing purposes. A complete-linkage hierarchical clustering has been used for Majority Clustering ([3]) and Semi-supervised Clustering ([3]) with the shortest Euclidean distance adopted for the cluster similarity measure. The user defined parameter Θ used in the Semi-supervised Clustering has been set to 0.3.

Due to the space limitations only selected results are presented for one artificial (Normal mixtures) and one real (Glass) datasets. Normal mixtures dataset (available at www.stats.ox.ac.uk/~ripley/PRNN) represents a two dimensional problem with two highly overlapping classes. The training set consists of 250 samples and a separate testing set has 1000 samples. This dataset has been constructed in such a way as to allow the best possible performance of around 8%. The Glass dataset [4] consists of 214, 10 dimensional samples representing 6 classes of different glass types found at the crime scenes and used during forensic investigations. Testing for Glass dataset is performed using 5-fold cross-validation.

The experiments have been performed for different ratios r of labelled data to the total number of data samples ranging from virtually unlabelled sets ($\sim 0\%$ of labelled data is one sample per class for "random per class" selections and 3 samples for selective sampling methods) to the fully labelled data sets (100% of labelled data). For the random selection the experiments have been repeated many times at each level for collecting reliable statistical information. The same sets of labelled samples have been used in all the experiments with different classification methods. Two types of random selection have been performed - a) random selection but ensuring that at least one sample per class is selected (referred to as "random per class") and b) completely random selection. The results for random selection are compared with the results for selective sampling methods.

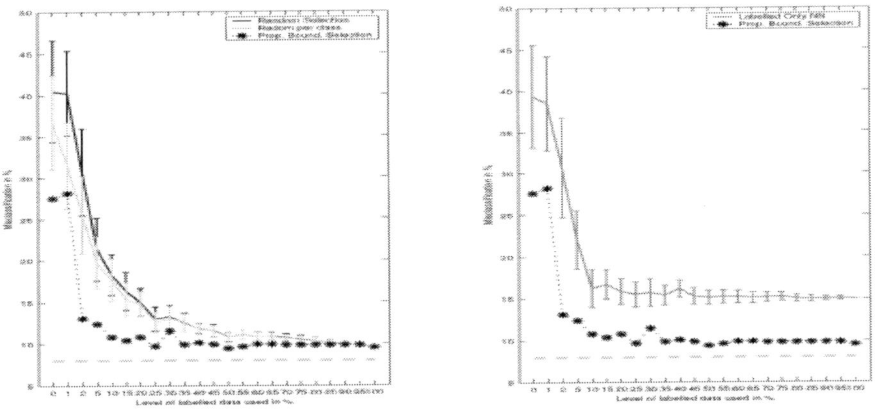

Fig. 1. Normal mixtures dataset - Left: Semi-supervised clustering - mean classification error and standard deviation of Random selection method compared to Random selection per class and to Proportional Boundary selection; Right: Proportional Boundary selection v Labelled Only NN.

As illustrated in the left parts of Fig.1 and Fig.2 the combined methods using selective sampling have shown an improved performance in comparison to completely random selection methods. This is especially evident for small values of r. However, it can also be noted (Fig.2 left) that the prior information about the number of classes used in the "random per class" selection method for the Glass dataset resulted in much better performance for small r than when using selective sampling where no information about the number of classes is used. This is common feature in multiclass problems with uneven distribution (prior class probabilities) of samples from different classes. The right parts of Fig. 1 and Fig. 2 illustrate the better performance when using selective sampling together with Semi-supervised clustering in comparison to classifiers generated on the basis of labelled data only selected randomly. In

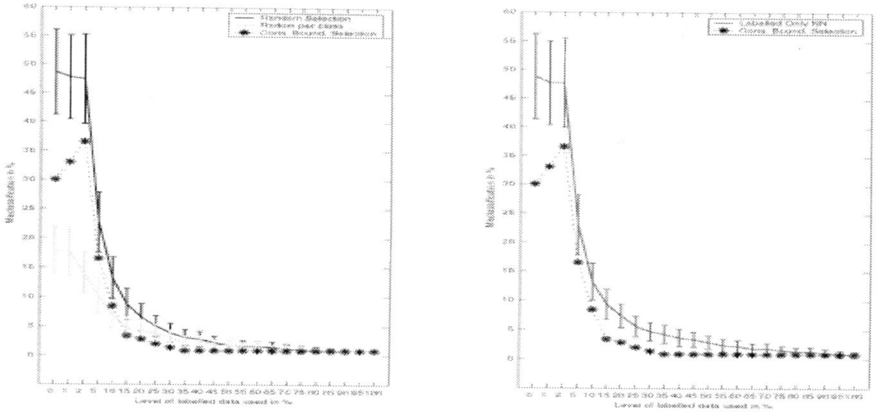

Fig. 2. Glass dataset - Left: Semi-supervised clustering - mean classification error and standard deviation of Random selection method compared to Random selection per class and to Proportional Boundary selection; Right: Proportional Boundary selection v Labelled Only NN.

all the cases a high classification errors are observed when only a very limited number of labelled data is used (small r). No consistent significant difference have been noted when comparing the Boundary Selection with Mean Selection and/or consecutive and proportional allocation methods. The results depend on suitable choice of the number of clusters for different levels of labelled data. In general, the better results have been obtained when using smaller number of clusters for small r and increased number of clusters with an increase of available labelled samples.

In the patterns of change illustrated in Fig.1 and Fig. 2 the level after which there is no significant improvement of the classifier performance is referred to as a *sufficient* level (SLLS - Sufficient Level of Labelled Samples). This level is different for different datasets. In general, the more complex the dataset distribution, the more labelled samples the algorithm needs to describe it so the SLLS will be at a higher ratio r, i.e. when more labelled samples are used. This level indicates that generally much less labelled data is needed for constructing a reliable classifier when unlabelled data is used in addition. When using selective sampling methods the stable performance related to the the SLLS is often achieved at lower values of r compared to the random sampling methods.

6 Conclusions

The random sampling methods analysed in [3] show that selection of a representative labelled subset is more important than combining learning from labelled and unlabelled data. Therefore in this paper we have concentrated our investigations on selective sampling methods. The preliminary results presented here indicate an improvement of both the mean classifier performance and reduction of the classification variance when using selective sampling methods in comparison to random selection of samples to be labelled.

A distinct disadvantage of the discussed methods is that they assume static selection. The algorithms used that way cannot take advantage of any available class information in contrast to the active selection approaches. Therefore our future research will extend to active selection as an alternative to overcoming the disadvantages of the static selection methods presented in this paper.

References

1. Klose A and Kruse R. Enabling neuro-fuzzy classification to learn from partially labeled data. In *IEEE World Congress on Computational Intelligence, IEEE International Conference on Fuzzy Systems*, page 4232, 2002.
2. Gabrys B and Bargiela A. General fuzzy min-max neural network for clustering and classification. *IEEE Transactions on Neural Networks*, 11(3):769, 2000.
3. Gabrys B and Petrakieva L. Combining labelled and unlabelled data in the design of pattern classification systems. In *Proceedings of EUNITE2002 conference, Hybrid Methods for Adaptive Systems (HMAS2002) workshop*, 2002.
4. Blake CL and Merz CJ. Uci repository of machine learning databases. In *[http://www.ics.uci.edu/ mlearn/MLRepository.html]*, 1998.
5. Larsen J, Szymkowiak A, and Hansen LK. Probabilistic hierarchical clustering with labeled and unlabeled data. *International Journal of Knowledge-Based Intelligent Engineering Systems*, 6(1):56–62, January 2002.
6. Nigam K and Ghani R. Analyzing the effectiveness and applicability of co-training. In *Ninth International CIKM*, pages 86–93, 2000.
7. Duda RO, Hart PE, and Stork DG. *Pattern Classification.* A Wiley-Interscience Publication, 2nd edition, 2001.
8. Duin RPW. Pattern recognition tools for matlab. In *[ftp://ftp.ph.tn.tudelft.nl/pub/bob/prtools/]*, 2000.
9. Goldman S and Zhou Y. Enhancing supervised learning with unlabeled data. In *Proceedings of the Seventeenth ICML*, pages 327–334, 2000.
10. Mitchell TM. The role of unlabeled data in supervised learning. In *Proceedings of the Sixth International Colloquium on Cognitive Science*, 1999.
11. Iyengar VS, Apte C, and Zhang T. Active learning using adaptive resampling. In *ACM SIGKDD*, 2000.
12. Pedrycz W and Waletzky J. Fuzzy clustering with partial supervision. *IEEE Transactions on Systems, Man and Cybernetics - Part B:Cybernetics*, 27(5):787, 1997.

Part IV

Hybrid and Intelligent Systems

New Results for a Hybrid Decision Tree/Genetic Algorithm for Data Mining

Deborah R. Carvalho* and Alex A. Freitas**

* Computer Science Department, Universidade Tuiuti do Paraná, Av. Comendador Franco, 1860, Curitiba PR. 80215-090. Brazil, email: deborah@utp.com.br
** Computing Laboratory, University of Kent at Canterbury, Canterbury, Kent, CT2 7NF, UK email: A.A.Freitas@kent.ac.uk

Abstract. This paper proposes a hybrid decision tree/genetic algorithm for solving the problem of small disjuncts in the classification task of data mining. It reports computational results comparing the proposed algorithm with two versions of C4.5 (one of them also specifically designed for solving the problem of small disjuncts) in 22 data sets.

Keywords. data mining, classification, genetic algorithm, decision trees.

1 Introduction

This paper addresses the data mining task of classification [4]. Classification rules are often expressed in disjunctive normal form, where each rule represents a disjunct and each rule condition represents a conjunct – i.e., the conditions in a rule antecedent are connected by a logical conjunction (AND) operator. In this context, a small disjunct can be defined as a rule which covers a small number of training examples (records, or cases) [5].

We have recently proposed a hybrid decision tree/genetic algorithm (GA) method to cope with the problem of small disjuncts [1]. This paper extends that previous work in two ways: (a) in [1] we reported results for only 8 data sets, whilst in this paper we report computational results for 22 data sets; (b) in [1] we reported results concerning only the predictive accuracy of the hybrid method, whereas in this paper we report results concerning both the predictive accuracy and the simplicity of the discovered rules.

2 The Hybrid Decision-Tree/Genetic Algorithm (GA)

Our hybrid decision tree/GA method discovers classification rules in two training phases. First, it runs C4.5, a decision-tree induction algorithm [7]. The induced, pruned tree is transformed into a set of rules (disjuncts). Each of these rules is considered either as a "small" or a "large" disjunct, depending on whether or not the number of examples covered by the rule is smaller than or equal to a given threshold. Second, it uses a GA to discover rules covering examples belonging to small disjuncts. Examples belonging to large disjuncts are classified by the decision tree produced by C4.5. The rationale of the method is as follows. Decision-tree algorithms have a bias towards generality that is well suited for large disjuncts, but not for small disjuncts. On the other hand, GAs tend to cope with attribute interaction better than most greedy rule induction algorithms [3]. This makes them a promising solution for the problem of small disjuncts, since attribute interactions are believed to be one of the causes of small disjuncts [2].

The first step consists in creating a suitable training set for the GA component of the hybrid method, as follows. All the examples belonging to all the leaf nodes considered small disjuncts are grouped into a single training set, called the "second training set" (to distinguish it from the original training set used to build the decision tree). This second training set is provided as input data for the GA.

Each individual represents the antecedent (IF part) of a small-disjunct rule, i.e., a conjunction of conditions (attribute-value pairs) [1]. The consequent (THEN part) of each rule is not represented in the genome of individuals. Rather, the consequent of each rule is dynamically chosen as the most frequent class in the set of examples covered by that rule's antecedent.

In addition to standard crossover and mutation, the GA uses a new rule-pruning operator. Each condition in the genome is associated with a flag, the active bit, which takes the value 1 or 0 to indicate whether or not, respectively, the associated condition occurs in the decoded rule antecedent. This allows the GA to use a fixed-length genome to represent a variable-length antecedent. The heuristic for rule pruning is based on the idea of using the decision tree built by C4.5 to compute a classification accuracy rate for each attribute, according to how accurate were the classifications performed by the decision tree paths in which that attribute occurs. That is, the more accurate were the classifications performed by the decision tree paths in which a given attribute occurs, the higher the accuracy rate associated with that attribute, and the smaller the probability of removing a condition with that attribute form a rule. (See [1] for more details.)

In order to discover a diverse set of rules the GA uses an iterative niching method which is similar to the sequential covering approach used by some rule induction algorithms. The basic idea is that the GA is iteratively applied to the second training set in such a way that each iteration (i.e., each GA run) discovers a single rule, and each iteration discovers a rule covering examples which are different from examples covered by rules discovered in previous iterations. In the first iteration the GA has access to all the examples in the second training set, and it discovers a single rule. Then the examples correctly covered by this rule are removed from the second training set. An example is "correctly covered" by a rule if the example's attribute values satisfy all the conditions in the rule antecedent and the example belongs to the same class as predicted by the rule. This process is iteratively performed while the cardinality of (number of examples in) the second training set is greater than 5.

The fitness function is given by the formula: Fitness = (TP / (TP + FN)) * (TN / (FP + TN)), where TP, FN, TN and FP stand for the number of true positives, false negatives, true negatives and false positives [4].

3 Computational Results

We have evaluated the performance of our hybrid decision tree/GA method across 22 data sets, out of which 12 are public domain data sets (available from the University of California at Irvine (UCI)'s data repository at: http://www.ics.uci.edu/~mlearn/MLRepository.html). The other 10 are proprietary data sets contain data about the scientific production of researchers [8]. These 10 data sets are denoted by DS-1 through DS-10.

The experiments used C4.5 [7] as the decision-tree component of our hybrid method. The performance of the hybrid decision tree/GA method was compared with the performance of two versions of C4.5 alone, as follows. The first version consists of running C4.5 with its standard set of parameters, using the tree constructed by C4.5 to classify all test examples – both large-disjunct and small-disjunct examples. The second version is a variant of C4.5, which we call double C4.5, specifically designed for coping with small disjuncts. This variant consists of two runs of C4.5. The first run considers all examples in the original training set, producing a first decision-tree. Once all the examples belonging to small disjuncts have been identified by this decision tree, C4.5 is run again on the "second training set" (the same training set used as input to the GA), producing a second decision tree. A test example is classified as follows. If it belongs to a large disjunct of the first decision tree (i.e., the tree built using the entire training set), it is classified by the corresponding leaf node of that tree. Otherwise – i.e., it belongs to a small disjunct of the first tree – it is classified by the second decision tree (i.e., the tree built using the second training set). In our experiments a decision-tree leaf if considered a small disjunct if and only if the number of examples belonging to that leaf is smaller than or equal to a fixed size S. Due to limitation of space, we report only the results for S = 10 here.

In three data sets – Adult, Connect and Letter - we used a single partitioning of the data into a training and test sets, since these data sets are relatively large. In the other data sets we used a well-known 10-fold stratified cross-validation procedure [4]. In the case of the hybrid C4.5/GA, the GA was executed ten times, with a different random seed each time. The GA parameters were: population of 200 individuals, 50 generations, one-point crossover probability: 80%, mutation probability: 1%, tournament selection with tournament size of 2. We made no attempt to optimize these parameters, in order to make a fair comparison with C4.5 and double C4.5, since we are using standard, non-optimized parameters for C4.5 as well.

Table 1. Accuracy rate (%) on the test set

Data Set	C4.5	Double C4.5	C4.5/AG
Connect	72,60 ± 0,5	76,19 ± 0,6 +	**76,95 ± 0,1** +
Adult	78,62 ± 0,5	76,06 ± 0,5 –	**80,04 ± 0,1** +
Crx	**91,79 ± 2,1**	90,78 ± 1,2	91,66 ± 1,8
Hepatitis	80,78± 13,3	82,36 ± 18,7	**95,05 ± 7,2**
House	93,62 ± 3,2	89,16 ± 8,0	**97,65 ± 2,0**
Segmentat.	**96,86 ± 1,1**	72,93 ± 5,5 –	78,68 ± 1,1 –
Wave	75,78 ± 1,9	64,93 ± 3,9 –	**83,95 ± 3,0** +
Splice	65,68 ± 1,3	61,51 ± 6,6	**70,70 ± 6,3**
Covertype	**71,61 ± 1,9**	68,64 ± 14,8	68,71 ± 1,3
Letter	**86,40 ± 1,1**	82,77 ± 1,0 –	79,24 ± 0,2 –
Nursery	95,40 ± 1,2	**97,23 ± 1,0**	96,77 ± 0,7
Pendigits	96,39 ± 0,2	**96,86 ± 0,4**	95,72 ± 0,9
DS-1	60,71 ± 3,0	**63,82 ± 5,2**	63,43 ± 1,4
DS-2	65,55 ± 1,5	72,52 ± 5,9	**73,77 ± 2,5** +
DS-3	75,65 ± 2,4	82,27 ± 1,3 +	**84,15 ± 0,9** +
DS-4	**92,97 ± 0,9**	92,58 ± 1,0	92,72 ± 1,0
DS-5	82,7 ± 2,8	83,01 ± 1,9	**83,36 ± 2,1**
DS-6	57,78 ± 2,1	60,68 ± 3,2	**61,69 ± 1,6** +
DS-7	65,18 ± 1,0	70,29 ± 2,4 +	**71,27 ± 1,6** +
DS-8	75,57 ± 1,4	81,03 ± 1,9 +	**82,63 ± 1,9** +
DS-9	93,00 ± 0,5	93,72 ± 1,2	**93,80 ± 1,4**
DS-10	82,80 ± 1,7	85,60 ± 1,4	**86,88 ± 1,6** +
# significantly better results		4	9
# significantly worse results		4	2

Table 1 reports the accuracy rate of standard C4.5, double C4.5 and our hybrid C4.5/GA in the 22 data sets. For each data set, the highest accuracy rate among the three methods is shown in bold. The numbers after the "±" symbol denote standard deviations. In the columns referring to double C4.5 and C4.5/GA, the cells where the corresponding method achieved a significantly larger (smaller) accuracy rate than C.45 (considered a baseline method) are indicated by the symbol "+" ("–") after the standard deviations. The accuracy rate of double C4.5 or C4.5/GA is considered significantly larger or smaller than the accuracy rate of standard C4.5 if the corresponding accuracy rate intervals, taking into account the standard deviations, do not overlap. The results are summarized in the last two rows of the table. More precisely, the last but one row shows the number of data sets were double C4.5 and C4.5/GA obtained an accuracy rate significantly better than the baseline standard C4.5, whereas the last row shows the number of data sets were double C4.5 and C4.5/GA obtained an accuracy rate significantly worse than standard C4.5.

Note that the hybrid C4.5/GA outperforms the two other algorithms in 14 of the 22 data sets. Double C4.5 outperforms the other two algorithms in only 3 data sets. Not all results are statistically significant. The hybrid C4.5/GA is significantly better than C4.5 in 9 data sets and the reverse is true in only 2 data sets, which indicates that the hybrid method is a good solution for the problem of small disjuncts. Double C4.5 is significantly better than C4.5 in 4 data sets and reverse is true in 4 data sets as well, which indicates that double C4.5 does not seem to be an improvement over standard C4.5, with respect to predictive accuracy. Overall, the results indicate that the hybrid C4.5/GA achieved considerably better results than both standard C4.5 and double C4.5, with respect to predictive accuracy.

With respect to simplicity (number of discovered rules and number of conditions per rule), in almost all the 22 data sets the hybrid C4.5/GA discovered a rule set considerably simpler (smaller) than the rule set discovered by standard C4.5. (Each path from the root to a leaf node of the decision tree is considered a rule.) More precisely, the number of rules discovered by C4.5/GA was significantly smaller (considering the standard deviations) than the number of rules discovered by standard C4.5 in 21 of the 22 data sets. The number of rules discovered by double C4.5 was significantly smaller than the number of rules discovered by standard C4.5 in 18 data sets, but the former was significantly larger than the latter in 2 data sets.

With respect to the average number of conditions per discovered rule, once again the hybrid C4.5/GA obtained the best results. The number of conditions per rule discovered by C4.5/GA was significantly smaller (taking into account the standard deviations) than the number of conditions per rule discovered by standard C4.5 in 16 of the 22 data sets, i.e., in 72% of the cases. The former was significantly larger than the latter in only one data set (Hepatitis). The number of conditions per rule discovered by double C4.5 was significantly smaller than the number of conditions per rule discovered by standard C4.5 in 8 of the 22 data sets. The former was significantly larger than the latter in only one data set (Hepatitis).

4 Conclusions and Future Work

Concerning predictive accuracy, the hybrid C4.5/GA can be considered a good solution for the problem of small disjuncts. It obtained results considerably better than both standard C4.5 and double C4.5. Another advantage of the hybrid C4.5/GA is that in general it discovers a rule set considerably simpler (smaller) than the rule set discovered by standard C4.5.

One possible research direction consists of performing a "meta-learning" [6] on the computational results obtained in our experiments. That is, one could use a classification algorithm to discover rules predicting, for each data set, which method designed for coping with small disjuncts will obtain the best result. In this case different methods for coping with small disjuncts (such as C4.5/GA and double C4.5) would correspond to the classes to be predicted, and the predictor attributes would be characteristics of the data sets. We are currently working towards this direction.

References

1. Carvalho, D.R; Freitas, A.A. A genetic algorithm with sequential niching for discovering small-disjunct rules. *Proc. Genetic and Evolutionary Computation Conf. (Gecco-2002),* 1035-1042. Morgan Kaufmann, 2002.
2. Freitas, A.A. Understanding the crucial role of attribute interaction in data mining. *Artificial Intelligence Review 16(3),* Nov. 2001, 177-199.
3. Freitas, A.A. *Data Mining and Knowledge Discovery with Evolutionary Algorithms.* Springer-Verlag, 2002.
4. Hand, D.J. *Construction and Assessment of Classif. Rules*, John Wiley 1997.
5. Holte, R.C.; Acker, L.E.; Porter, B.W. Concept Learning and the Problem of Small Disjuncts, *Proc. IJCAI – 89*, pp.813-818. 1989.
6. Michie, D; Spiegelhalter, D.J. and Taylor, C.C. (Eds). *Machine Learning, Neural and Statistical Classification*, Ellis-Horwood, 1994.
7. Quinlan, J.R. *C4.5: Programs for Machine Learning*, Morgan Kaufmann 1993.
8. Romao, W.; Freitas, A.A. and Pacheco, R.C.S. A Genetic Algorithm for Discovering Interesting Fuzzy Prediction Rules: applications to science and technology data. *Proc. Genetic and Evolutionary Computation Conf. (GECCO-2002)*, pp. 1188-1195, 2002.

Embedding Support Vector Machines into Localised Mixtures of Experts[1]

Clodoaldo A. M. Lima, André L. V. Coelho, and Fernando J. Von Zuben

DCA-FEEC-Unicamp, Campinas, Brazil
e-mail: {moraes,coelho,vonzuben}@dca.fee.unicamp.br

Abstract. Aiming at distilling complex problems into simple subtasks, a mixture of expert (ME) may be implemented so that a gating network is in charge of learning how to softly divide the input space into overlapping regions to be each assigned to one or more expert networks. In this work, we focus mostly on the development and initial assessment of a new approach, named Localised Mixture of Support Vector Machine Experts (L-MSVME), that defines an SVM-based extension of localised ME models. Mixture models will then be used both to express prior belief in the existence of more than one process in the data generation model, and to provide an efficient method for increasing modelling power. An algorithm based on the maximum likelihood criteria (EM) was considered for training, and we demonstrate that it is possible to train each ME expert through an SVM perspective. This process also allows the decoupled training of gating and expert networks. Simulation results were then considered for performance assessment.

1 Introduction

Mixture of Experts (ME) models [4] consist of a family of modular neural network (NN) approaches that follow the "divide-and-conquer" strategy of distilling complex problems into simple subtasks. In a statistical sense, an ME should be regarded as a mixture model for estimating conditional probability distributions. In terms of a mixture model [6], expert networks (ENs) correspond to conditional component densities whereas the gating network (GN) generates input-biased mixture coefficients. Such interpretation is interesting for enabling an ME to be trained with the Expectation-Maximisation (EM) algorithm [2].

The standard ME model, as introduced by Jordan and Jacobs [4], has a single-layer perceptron with soft-max activation function as GN and ENs with linear activation functions. Amongst the ME variants proposed in the literature, the *localised ME*, formulated by Xu *et al.* [10], uses normalised Gaussian kernels as activa-

---

[1] **CNPq** has sponsored the authors via scholarships n° 141394/00-5 and n° 140719/99-7, and via grant n° 300910/96-7.

tion function for the GN in such a way as to divide the input space via soft hyper-ellipsoids (localised regions) to be assigned to the non-linear/linear experts.

By other means, there has been much interest nowadays in the study of support vector machines (SVMs) for regression and classification problems. SVMs are based on the minimisation of the structural risk [9], which asserts that the generalisation error is delimited by the sum of the training error and a parcel that depends on the Vapnik-Chervonenkis classifier dimension. By minimising this sum, high generalisation performance may be obtained. Besides, the number of free parameters in an SVM does not explicitly depend upon the input dimensionality of the problem at hand, which suggests that it may be applicable to untangle complex input data relationships.

In this paper, we propose an SVM-based extension of localised ME models in order to combine the benefits exhibited by both models, denoted *Localised Mixture of Support Vector Machine Experts* (L-MSVME, for short). In this context, mixture models are used both to express prior belief in the existence of more than one process in the data generation model, and to provide an efficient method for increasing modelling power.

The idea of merging (some authors also use the term "mixture" in this sense) many SVM instances into the same setting is not novel. For instance, Kwok [5] demonstrated earlier how to employ single SVM instances in the place of previously combined experts (an approach he called "Support Vector Mixtures") and also presented a learning algorithm based on SVM methodology for training the whole framework. In another work [7], Rida *et al.* proposed (i) to firstly divide the training set using an unsupervised algorithm for data clustering; then (ii) to train an expert (SVM) on each data subset corresponding to a cluster; and finally (iii) to recombine the outputs of all experts. A different scheme was proposed in [1] in which each SVM is trained upon a random subset of the training set. According to the performance shown by the expert, some points of its assigned subset may be reallocated to other peers, and the training process is restarted. In contrast to our work, the aforementioned approaches are not fully compliant with the idea behind mixture models, since they do not provide any probabilistic interpretation of the expert outcomes. Besides, the decoupled training of the gate and the experts is accomplished in a very arbitrary way, which may sometimes lead to non-optimal solutions.

Similarly to standard mixture models, there are many algorithms for ME training. In this work, an algorithm based on the maximum likelihood criteria (EM) will be considered. In the development of L-MSVME, we start from the standard ME [4] and demonstrate that, by applying a regularisation term (to control the model complexity and to improve the generalisation performance by minimising the overfitting problem) in the cost function of each expert, we can find, after some algebraic manipulations, a set of equations that is similar to those found in the development of support vector machines [9]. In this way, we can train each ME expert through an SVM perspective (that is, via a specific quadratic programming problem), as well as also allowing the decoupling in the training of gate and experts. To demonstrate the potentials of our proposal, some experiment results on a complicated approximation problem [3] are presented.

2 Overview of Support Vector Machines

Given a training data set of N points $\{x_k, y_k\}_{k=1}^{N}$ with data $x_k \in \Re^n$ and output data $y_k \in \Re$, we assume that the function f, which is desirable to approximate the input/output mapping, can be written as

$$f(x) = w^T \varphi(x) + b \tag{1}$$

where $\varphi(.) : \Re^n \rightarrow \Re^{n_h}$ is a function which maps the input space into a so-called higher dimensional feature space, and weight vector $w \in \Re^{n_h}$ is in primal weight space. Parameters w and b can be obtained by solving the following optimisation problem in the primal weight space

$$\Phi(w, \xi, \xi^*) = \frac{1}{2}(w^T w) + \frac{C}{k}\left(\sum_{i=1}^{N}(\xi_i)^k + \sum_{i=1}^{N}(\xi_i^*)^k\right) \tag{2}$$

(with a given C value) subject to constraints in (3), where $k = 1$ for the linear ε-insensitive loss function and $k = 2$ for the quadratic ε-insensitive loss function (here, we consider the latter).

$$\begin{array}{ll} y_i - w^T\varphi(x_i) - b \le \varepsilon_i + \xi_i \ \text{ and } \ \xi_i \ge 0 & i = 1, \cdots, N \\ w^T\varphi(x_i) + b - y_i \le \varepsilon_i + \xi_i^* \ \text{ and } \ \xi_i^* \ge 0 & i = 1, \cdots, N \end{array} \tag{3}$$

For simplicity, we construct (2) in the dual space. Then, the parameters of the function f can be retrieved by solving the following optimisation problem

$$L(\lambda, \lambda^*) = -\frac{1}{2}\left(\sum_{i,j=1}^{N}(\lambda_i^* - \lambda_i)(\lambda_j^* - \lambda_j)K(x_i, x_j) + \frac{1}{C}\sum_{i=1}^{N}(\lambda_i^*)^2 + \frac{1}{C}\sum_{i=1}^{N}(\lambda_i)^2\right) - \sum_{i=1}^{N}\varepsilon_i(\lambda_i + \lambda_i^*) - \sum_{i=1}^{N}y_i(\lambda_i^* + \lambda_i)$$

subject to the constraints: $\sum_{i=1}^{N}\lambda_i^* = \sum_{i=1}^{N}\lambda_i$, $\lambda_i^* \ge 0$, $\lambda_i \ge 0$, $i = 1, \cdots, N$, where the function $\varphi(x_i)$ above is then related to $K(x,x_i)$ by imposing $K(x, x_i) = \varphi(x)^T \varphi(x_i)$ according to Mercer's Theorem.

3 Mixture of Experts

In the standard ME framework, we have a set of expert networks $j=1,...,m$, all of which looking at the input vector x to form mapping outputs y_j. There is only one gating network, also looking at the input, but that, instead, produces outputs $g_j(x) \ge 0$, $\sum_{j=1}^{m} g_j(x) = 1$. The output vector of an ME is given by

$$y(x) = \sum_{j=1}^{m} g_j(x) y_j(x) \tag{4}$$

Each output g_j, thus, should be viewed as the probability of assigning input x to expert j. To be compliant with such interpretation, the activation functions for the GN outputs (known as soft-max functions) are given by

$$g_j(x) = \frac{\exp(z_j)}{\sum_{i=1}^{m} \exp(z_i)} \tag{5}$$

where z_i's (normalised exponentials) are gating outputs before threshold. This choice makes the experts more competitive and brings about the constraints that the outputs should be positive and sum to unity. Here, we consider a modified GN with normalised kernels (known as localised ME [10]), given by

$$g_j(x) = P(j \mid x) = \frac{\gamma_j P_j(x)}{\sum_{i=1}^{m} \gamma_i P_i(x)} \tag{6}$$

where $\sum_{i=1}^{m} \gamma_i = 1$, $\gamma \geq 0$ and P_i's are probability density functions.

The typical Expectation-Maximisation formulation [4][2] for the parameter estimation of both GN and EN comprehends two well-defined steps defined over a log-based error function and working upon the whole training set (indexed by t). The steps are:

- **E-Step**: $h_j^{(k)}(y^t \mid x^t) = \dfrac{g_j^{(k)}(x,v)\phi_j(y^t \mid x^t)}{\sum_{i=1} g_i^{(k)}(x,v)\phi_i(y^t, x^t)}$;
- **M-Step**: Minimise $E_{gate} = -\sum_{t=1}^{N}\sum_{j=1}^{m} h_j^{(k)}(y^t, x^t)\ln(g_j(x))$ for the GN.

The parameters of the gating network μ, Σ_j^{-1}, and γ_j can be obtained by taking the partial derivatives of the gating error function, that is,

$$\gamma_j^{k+1} = \frac{1}{N}\sum_{t=1}^{N} h_j^{(k)}(y^t \mid x^t) \qquad \Sigma_j^{(k+1)} = \frac{\sum_{t=1}^{N} h_j^{(k)}(y^t \mid x^t)[x^t - m_j^{(k)}][x^t - m_j^{(k)}]^T}{\sum_{t=1}^{N} h_j^{(k)}(y^t \mid x^t)} \tag{7}$$

$$m_j^{(k+1)} = \frac{1}{\sum_{t=1}^{N} h_j^{(k)}(y,x)} \sum_{t=1}^{N} h_j^{(k)}(y^t, x^t)x^t$$

Conversely, for expert j the term to be minimised is:

$$E_{\exp ert_j} = -\sum_{t=1}^{N} h_j^{(k)}(y^t, x^t)\ln(\phi_j(y^t, x^t)) \tag{8}$$

where ϕ is the conditional probability density, given by

$$\phi_j(y^t \mid x^t) = \frac{1}{(2\pi)^{d/2}(\sigma_j^2)^{1/2}} \exp\left\{-\frac{0.5(y - y_j)^T(y - y_j)}{\sigma_j^2}\right\} \tag{9}$$

where d is the dimension of y, and y_j is a function of some parameters of expert j (usually, its weights).

Substituting (9) in (8), we obtain the following expression

$$E_{\exp ert_j} = \sum_{t=1}^{N}\left(\frac{h_k(x^t, y^t)}{2\sigma_j^2}\left(y^i - y_j^t\right)^T\left(y^i - y_k^t\right)\right) - \sum_{t=1}^{N} h_j(x^t, y^t)(\ln\sigma_j^2 + \ln 2\pi) \tag{10}$$

The M-step minimises the cost function in (10) for each EN by adjusting the parameters. The parameters under consideration are the variances of the expert, as well as the weights of all experts and the gating network. The variance updates can be computed directly by setting $\partial E_j / \partial \sigma_j^2$ to zero.

4 Localised Mixture of Support Vector Machine Experts

We assume that the function y_j of each expert can be represented as:

$$y_j = f_j(x) = \sum_{i=1}^{\infty} c_n \varphi(x) + b \tag{11}$$

where $\{\varphi_i(x)\}_{i=1}^{\infty}$ is a given set of linearly independent basis functions, and c_n and b are parameters to be estimated from the data. Since it is possible to obtain an expression for calculating variances σ_j^2 in only one step, the cost function in (10) may be rewritten to emphasise its dependency on the weights:

$$E_{\exp ert_j} = \sum_{t=1}^{N} \left(\frac{h_k(x^t, y^t)}{2\sigma_j^2} \left(y^i - y_j^t\right)^T \left(y^i - y_k^t\right) \right) \tag{12}$$

The problem of recovering the coefficients c_n and b from the data set is clearly ill-posed, since it has an infinite number of solutions. In order to turn it well-posed, we follow the approach of regularisation theory [8] and impose an additional smoothness constraint on the solution of the approximation problem. Therefore, we choose as a solution the function that solves the following problem

$$\min_{f_j \in \mathrm{H}} E[f_j]_{\exp ert} = \frac{C}{2} \sum_{t=1}^{N} \left(\frac{h_k(x^t, y^t)}{2\sigma_j^2} \left(y^i - f_j(x^t)\right)^T \left(y^i - f_j(x^t)\right) \right) + \frac{1}{2} \Psi[f] \tag{13}$$

where C is a positive number, $\psi[f]$ is a smoothness functional and **H** is a set of functions over which the smoothness functional $\psi[f]$ is well defined. The first term is enforcing closeness to the data, the second term gauges the smoothness, while C controls the trade-off between these two terms. Expression (13) can be replaced by the following equivalent problem, in which an additional set of variables is introduced:

$$\min_{f_j \in \mathrm{H}} E[f_j]_{\exp ert} = \frac{C}{2} \sum_{t=1}^{N} \pi_j^t \left((\xi_i)^2 + (\xi_i^*)^2\right) + \frac{1}{2} \psi[f], \text{ where } \pi_j^t = \frac{h_k(x^t, y^t)}{2\sigma_j^2} \tag{14}$$

subject to

$f_j(x^t) - y^t \le \varepsilon + \xi_i \qquad i = 1, \cdots, N \qquad \xi_i \ge 0 \qquad i = 1, \cdots, N$

$f_j(x^t) - y^t \le \varepsilon + \xi_i \qquad i = 1, \cdots, N \qquad \xi_i^* \ge 0 \qquad i = 1, \cdots, N$

The equivalence is established just noticing that in the problem above a penalty is paid only when the absolute value of the interpolation error exceeds ε. To solve the above constrained minimisation problem, we have worked on dual space. The Lagrangian corresponding to the problem above is

$$L(f,\xi,\xi^*,\alpha,\alpha^*,\beta,\beta^*)=\frac{C}{2}\sum_{t=1}^{N}\pi_j^t(\xi_t)^2+\frac{C}{2}\sum_{t=1}^{N}\pi_j^t(\xi_t^*)^2+\frac{1}{2}\psi[f]-\sum_{t=1}^{N}\alpha_t\left(y^t-f(x^t)+\varepsilon+\xi_t^*\right)$$
$$-\sum_{t=1}^{N}\alpha_t^*\left(-y^t+f(x^t)+\varepsilon+\xi_t^*\right)-\sum_{t=1}^{N}\left(\beta_t\xi_t+\beta_t^*\xi_t^*\right) \tag{15}$$

where α, α^*, β, β^* are positive Lagrange multipliers. The solution of the constrained problem in (14) is now obtained by minimising the Lagrangian (15) with respect to f (that is, with respect to c_n and b), ξ and ξ^* and maximising with respect α, α^*, β, β^*. Since the minimisation step is now unconstrained, we set the derivatives with respect to c_n, b, ξ and ξ^* to zero, obtaining

$$\frac{\partial L}{\partial c_n}=0\rightarrow c_n=\lambda_n\sum_{t=1}^{N}(\alpha_i^*-\alpha_i)\varphi(x^t) \tag{16}$$
$$\frac{\partial L}{\partial \xi_i}=0\rightarrow \beta_i+\alpha_i\leq\pi_i C\xi_i \quad i=1,\cdots,N \tag{17}$$
$$\frac{\partial L}{\partial b}=0\rightarrow \sum_{t=1}^{N}\alpha_t^*=\sum_{t=1}^{N}\alpha_t \tag{18}$$
$$\frac{\partial L}{\partial \xi_i^*}=0\rightarrow \beta_i^*+\alpha_i^*\leq\pi_i C\xi_i^* \quad i=1,\cdots,N \tag{19}$$

Substituting the expansion (16) for the coefficients in the model in (11) we then conclude that the solution of the problem is a function of the form

$$f(x)=\sum_{t=1}^{N}(\alpha_i^*-\alpha_t)K(x,x^t)+b \tag{20}$$

Introducing Eq. (20) into the Lagrangian, we obtain a novel expression that should now be maximised with respect to α, α^*, β, β^*, and with the additional constraints listed above (17-19). We now obtain the following QP problem

$$L(\alpha,\alpha^*)=-\frac{1}{2}\left(\sum_{i,j=1}^{N}(\alpha_i^*-\alpha_i)(\alpha_j^*-\alpha_j)K(x_i,x_j)+\frac{1}{C}\sum_{i=1}^{N}\frac{(\alpha_i^*)^2}{\pi_i}+\frac{1}{C}\sum_{i=1}^{N}\frac{(\alpha_i)^2}{\pi_i}\right)-\sum_{i=1}^{N}\varepsilon_i(\alpha_i+\alpha_i^*)-\sum_{i=1}^{N}y_i(\alpha_i^*+\alpha_i) \tag{21}$$

subject to constraints: $\sum_{i=1}^{N}\alpha_i^*=\sum_{i=1}^{N}\alpha_i$, $\alpha_i^*\geq 0$, $\alpha_i\geq 0$, $i=1,\cdots,N$.

The solution for the QP problem in (21) is equivalent to the solution achieved by a single SVM (see Section 2). This justifies our approach of modelling the training of each ME expert as a QP problem.

5 Experiment

In the following, we assess the L-MSVME approach with respect to a function approximation problem taken from Hwang *et al.* [3]. Three datasets were generated: one for training; another for the validation of the model; and another to test the L-MSVME performance. A data pre-processing was realised in our simulations by normalising the input values to the range [-1,1]. The motivation for restricting the analysis to data sets of low dimensionality is the possibility of a 3D visualisation of the non-linearities involved. Below, we present the expression for the applied function $s:[0,1]^2\rightarrow\Re$ and its graphical sketch is shown in Fig 1, with and without noise. The noise corresponds to a Gaussian distribution with mean zero and standard deviation 0.25. The function was assumed unknown in such a way that it was only possible to obtain samples of its values for input vectors $\boldsymbol{x}=[x_1\ x_2]^T$, defined to adequately cover the approximation space. We produced training data sets (N_{tr}) with 225 uniformly distributed points, and validation and test data sets of dimensions $N_v = 225$ and $N_{te} = 625$ by generating input vectors $\boldsymbol{x}_l=[x_{1l}, x_{2l}]^T$ via a random distribution. The analytical form of s is

$$s(x_1,x_2) = 1{,}3356[1{,}5(1-x_1)+e^{2x_1-1}\,\mathrm{sen}(3\pi(x_1-0{,}6)^2)+e^{3x_2-1{,}5}\,\mathrm{sen}(4\pi(x_2-0{,}9)^2)] \tag{22}$$

For assessing the E-SVM proposal, we adopted the following algorithm:

1. Initialise the gating and expert networks
- **1.1**. Use K-means or a random process for the initialisation of the gating parameters.
- **1.2**. Divide the training dataset in m partitions and assign each one to an expert.
- **1.3**. For each EN j, solve the maximisation problem in (21).

2. Calculate the ME likelihood function [4].

3. While the variation in the likelihood function is less than a given threshold (here, 0.001)
- **3.1**. Step E: For each pair (x^l, y^l), compute the posterior probabilities h_i^l
- **3.2**. Step M
  - **3.2.1**. For each expert network j, solve the maximisation problem in (21).
  - **3.2.2**. For the gating network, solve the optimisation problem in (7).
- **3.3**. Update the parameter values.

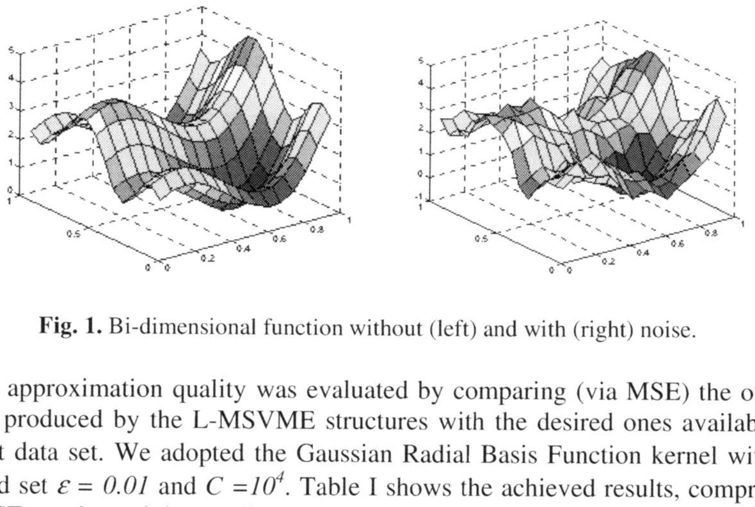

Fig. 1. Bi-dimensional function without (left) and with (right) noise.

The approximation quality was evaluated by comparing (via MSE) the output values produced by the L-MSVME structures with the desired ones available in the test data set. We adopted the Gaussian Radial Basis Function kernel with $\sigma^2 = 1$, and set $\varepsilon = 0.01$ and $C = 10^4$. Table I shows the achieved results, comprising the MSE on the training, validation and test sets. For both the single SVM (S-SVM) and L-MSVME approaches we employ the ε-quadratic loss function. L-MSVME configurations with 2, 3 or 5 expert networks were tested. From the results, it is possible to ascertain that L-MSVME invariably produced a significant improvement over the S-SVM approach, even when employing a small number of ENs. Taking into account the experiments without noise, the gain when employing L-MSVME over S-SVMs was near of ten times. Particularly for the noise-related results, our approach achieved only 14% of improvement over the S-SVM (with two experts) since the noise level introduced was too high for keeping up with.

6 Final Remarks

In this paper, we have presented a novel approach for combining SVMs and ME models into the same conceptual framework. In the development of L-MSVME, we start from the standard ME model and demonstrate that, by applying a regularisation term in the cost function of each expert, we can find a set of equations that is similar to those found in the development of support vector machines. After the theoretical formalisation, we have shown here, through a complicated approximation problem, that employing L-MSVME comes to be a good alternative when one wishes to achieve high approximation accuracy. The performance achieved (albeit the extra computational requirements underlying L-MSVME simulations) was remarkable, either with or without data noise. As future work, we will investigate the possibility of also setting up the training of the localised gate as a QP problem (that is, employing SVMs for coping also with the gating mapping).

Approach	m	EXPERIMENT					
		Without noise			With noise		
		Training	Validation	Test	Training	Validation	Test
S-SVM	--	0.01201	0.01323	0.01349	0.06252	0.02209	0.02175
L-MSVME	2	0.00099	0.00162	0.00160	0.05411	0.02013	0.01912
L-MSVME	3	0.00086	0.002	0.00211	0.053	0.01895	0.01842
L-MSVME	5	0.00110	0.00177	0.00137	0.05042	0.01806	0.01782

References

[1] R. Collobert, S. Bengio and Y. Bengio (2002). A Parallel Mixture of SVMs for Very Large Scale Problems, *Neural Computation*, vol. 14, pages 1105-1114.

[2] A. P. Dempster, N. M. Laird and D. B. Rubin, (1977). Maximum Likelihood from Incomplete Data via the EM Algorithm. *J .of the Royal Statistical Society B*, 39, 1-38.

[3] J-N. Hwang *et al.* (1994). Regression Modeling in Back-propagation and Project Pursuit Learning, *IEEE Trans on NNs*, vol. 5, pages 342-353.

[4] M. I. Jordan and A. R. Jacobs (1994). Hierarchical Mixtures of Experts and EM Algorithm, *Neural Computation*, vol. 6, pages 181-214.

[5] J. T. Kwok (1998). Support Vector Mixture for Classification and Regression Problems. In *Proceedings of ICPR*, pages 255-258, Brisbane, Queensland, Australia.

[6] G. J. McLachlan and K. E. Basford (1998). Mixture Models: Inference and Applications to Clustering. Marcel Deckker, Inc., New York.

[7] A. Rida, A. Labbi and C. Pellegrini (1999). Local Experts Combination through Density Decomposition. In *Proc. of Uncertainty' 99*, Morgan Kaufmann.

[8] A. Tikhonov and V. Arsenim (1977). *Solutions of Ill-posed Problems*. W. Winston.

[9] V. N. Vapnik (1995). *The Nature of Statistical Learning Theory*. Springer Verlag.

[10] L. Xu, M. I. Jordan, and G. E Hinton (1995). An Alternative model for Mixtures of Experts. *Advances in Neural Information Processing Systems (vol. 7)*, G. Tesauro, D. S. Touretzky, and T. K. Leen, (eds), MIT Press, Cambridge MA, pages 633-640.

Gaussian Function Approximation in Neuro-Fuzzy Systems

Koldo Basterretxea[*], José Manuel Tarela[**], Inés del Campo[**]

[*] Elektronika eta Telekomunikazio Saila, IITUE Bilbao, Euskal Herriko Unibertsitatea (UPV/EHU), e-mail: jtpbaoyk@lg.ehu.es
[**] Elektrika eta Elektronika Saila, Zientzi Fakultatea, Euskal Herriko Unibertsitatea (UPV/EHU), (The University of the Basque Country)

Abstract. Using smooth membership functions and activation functions presumably enhances the performance of neural, fuzzy and neuro-fuzzy systems. In this work we present some results based on the efficient generation of gaussian piecewise-linear approximations and its application to neural/fuzzy parallel computing systems. The application of approximations to the gaussian nodes of radial basis function networks (RBFN), and the observation of the approximation capabilities of the networks after applying various learning algorithms, is revealing. We use the equivalence theorem between RBFN's and certain fuzzy inference systems for extracting conclusions applicable to de fuzzy world.

Keywords. Gaussian Function, Membership Function, Activation Function, Centred Recursive Interpolation, Neuro-Fuzzy System.

1 Introduction

The cost of generating or storing smooth membership functions (MF's) like gaussian or bell-shaped functions, make fuzzy inference system (FIS) designers to choose simple MF's like triangles or trapezoids. This is specially true when specific hardware is used for fast operation. The choice of such functions is usually justified by referring to works on approximation theory that demonstrate the universal approximation property of mentioned FIS's [2]. Similar statements are made when handling activation functions (AF's) in artificial neural networks, but these are qualitative properties, and nothing is mentioned about the size of systems based on so simple functions.

We have designed a method for PWL function description based on *max* an *min* operators with lattice properties. The result is a recursive computing scheme capable of generating successively smoother functions from very simple initial piecewise linear (PWL) structures. We called this scheme, due to its properties, *centred recursive interpolation* or CRI, and it has been applied to the efficient generation of approximated non-linear AF's and MF's [1, 8]. Specifically, for neuro-fuzzy applications with learning capabilities, the widely used gaussian function (1) approximator must be programmable in two parameters: centre (c_x) and width (σ).

$$f(x)=1\Big/\exp\left(\left(x-c_x\right)^2/\sigma^2\right) \qquad (1)$$

The CRI algorithm allows for the programmability of both parameters with no rise of the memory demand. Fig. 1 shows optimised approximations through CRI for the first four recursions or *interpolation levels* (q).

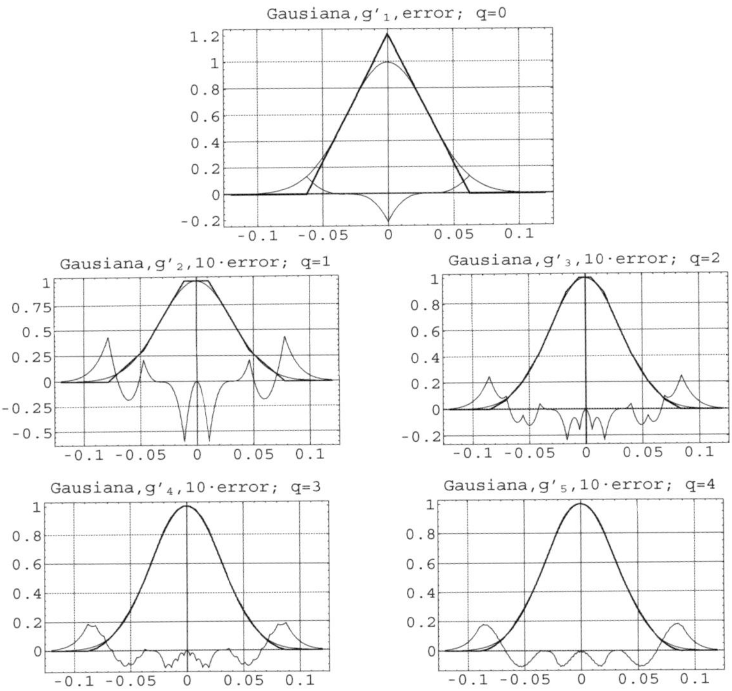

Fig. 1. Gaussian function, optimised approximation through CRI and error curves magnified by ten for interpolation levels from q=0 to q=4.

The aim of this work has been to analyse the influence of approximated gaussian functions, eventually generated by specific hardware based on the CRI algorithm, in the performance of neuro-fuzzy systems. To evaluate the performance of the systems, we have examined the approximation error and the learning evolution when applied to various benchmark function approximation problems. Some results are presented in this paper.

2 The RBFN as function approximator. Training for exact interpolation.

A simple one-dimensional function approximation problem has been analysed:

$$y(x) = 0.5\exp(-x)\sin(6x) \qquad x \in [-1,1] \tag{2}$$

This function has been used as a benchmark approximation problem in a previous work on RBFN's [3].

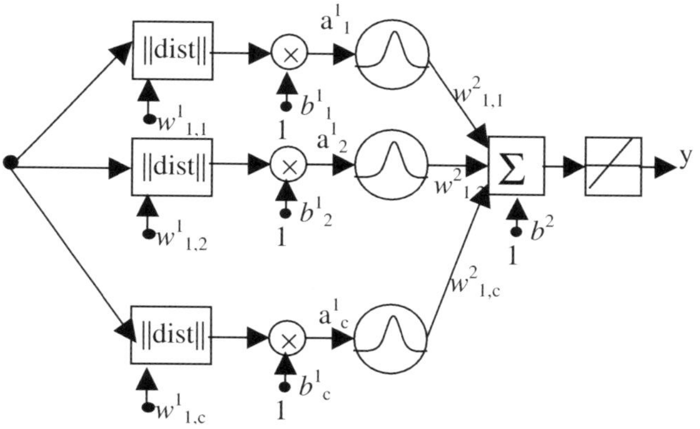

Fig. 2. RBFN structure with gaussian nodes – *Matlab*-style parameters –.

A RBFN with gaussian nodes (see Fig. 2) can be adjusted for exactly interpolate a data set obtained from (2). Such a net uses as many gaussian nodes or neurons as there are input vectors. The *Neural Network Toolbox* for *Matlab®* has been used for designing, training and simulating the behaviour of RBFN's. The applied training algorithms have been modified to allow the use of CRI approximations of the gaussian function. The exact fitting algorithm takes the input vectors and calculates the second layer matrix of weights $\mathbf{W}^2$ and the bias b^2. The elements of the weight matrix of the first layer $\mathbf{W}^1$ are set to the value of the training vectors, and biases of the first layer ($\mathbf{b}^1$) are set to 0.3862/*spread*, where $b_i^1 = 1/\sigma_i$.

As shown in Fig. 3, 21 equally distributed points have been taken to build the training data set. The training algorithm fixes the value of the spread parameter before it solves the fitting problem. This is irrelevant whenr solving the interpolation problem, but it is important for the resulting function approximation error – generalisation capability–. Consequently, we have programmed a sweeping of the spread parameter to obtain its optimum value. This procedure has been repeated for RBFN's with approximated gaussian nodes through CRI in the first four interpolation levels.

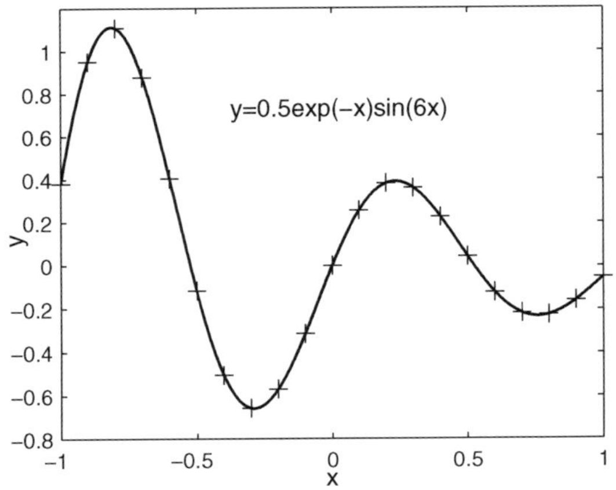

Fig. 3. Reference function to be approximated by the RBFN and training data-set.

The obtained results are summed up in Table 1, where the mean squared approximation error (MSE) has been evaluated at 1001 equally distributed points.

Table 1. Approximation error according to the neuron activation function

Neural function	Opt. spread	MSE_{1001}
Exact Gaussians	0.53	1.49×10^{-14}
CRI4 approx.	0.41	8.79×10^{-6}
CRI3 approx.	0.33	3.44×10^{-5}
CRI2 approx.	0.19	9.71×10^{-5}
CRI1 approx.	0.17	1.03×10^{-4}

As expected, even when the network interpolates training data exactly and the spread parameter is optimised, the successive gaussian function smoothing, provided by higher q values in the CRI algorithm, produces a better approximation of the objective function.

3 A growing training algorithm.

The training algorithm used for exact interpolation produces extremely big nets when many input vectors are needed to properly define the network. To overcome this problem, some more sophisticated algorithms have been designed. Usually, these algorithms train the net from a very simple initial structure –usually one neuron– and add more neurons when necessary for achieving an objective error [3, 6]. The *Neural Network Toolbox* incorporates such a training strategy for the definition of more efficient RBFN's. The training process adds new neurons until the sum of the squared errors (SSE) at the training points is reduced to the desired value.

Table 2. Net size according to the neuron activation function

Neural function	SSE = 0.01		SSE = 0.0041	
	Neurons	Opt. spread	Neurons	Opt. spread
Exact Gaussians	6	0.21	7	0.24
CRI4 approx.	6	0.30	11	0.26
CRI3 approx.	7	0.27	11	0.26
CRI2 approx.	7	0.27	12	0.24
CRI1 approx.	11	0.22	15	0.21

This time 41 points have been selected to build the training data set. The obtained results for two different objective errors are shown in Table 2, confirming our hypothesis: the smoother and more accurate the gaussian approximation is, the smaller the network generated is for a given approximation error.

The equivalence theorem between RBFN's and certain FIS's [5] has been applied to extract conclusions relating to the fuzzy world. Precisely, the equivalent gaussian MF based Sugeno-type FIS's, applied to the same approximation problem, have as many rules as neurons in the hidden layer of the RBFN. The neural structures obtained above can be then expressed linguistically. The network obtained for CRI4 approximated gaussians (SSE = 0.01), for instance, consists of 6 neurons; as many as for exact gaussians. The width parameter σ of each gaussian MF is defined by the vector of input biases of the equivalent RBFN:

$$\sigma_i = 0.30/0.8326 = 0.3603 \qquad i = 1,\ldots,6 \tag{3}$$

The value of input weights after training were

$$\mathbf{W}^{1^T} = \left[-0.85 \ -0.30 \ 0.20 \ -0.95 \ 0.75 \ 0.10\right] \tag{4}$$

that correspond to the positioning at the universe of discourse of the gaussian MF's (c_{xi}). The weights of the second layer, after training, were

$$\mathbf{W}^2 = \left[1.7823 \quad -0.6992 \quad 0.5069 \quad -0.8668 \quad -0.2613 \quad -0.1543\right] \tag{5}$$

that correspond to the values of the consequent of each Sugeno-type fuzzy rule. Consequently, we can now define the equivalent FIS, the input universe partition of witch is shown in Fig. 4. The knowledge base is composed of six linguistic rules:

If x is A1 *then* y(x) is 1,7820
If x is A2 *then* y(x) is -0,6992
If x is A3 *then* y(x) is 0.5069
If x is A4 *then* y(x) is -0,8668
If x is A5 *then* y(x) is -0,2613
If x is A6 *then* y(x) is -0,1543

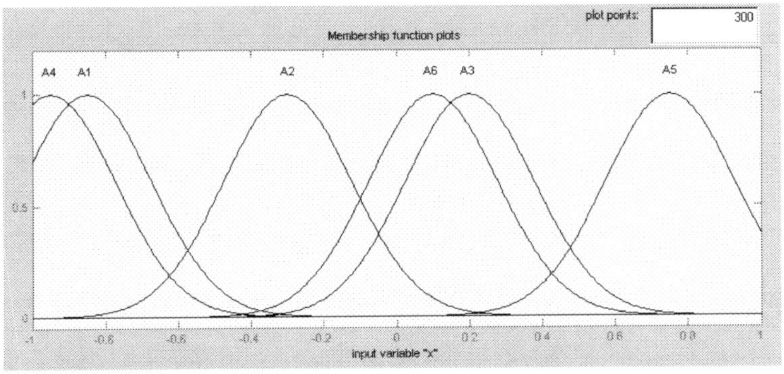

Fig. 4. Partition of the input universe for the FIS equivalent to the RBFN with CRI4 approximated gaussians

On the other edge, for CRI1 approximations, the equivalent FIS must contain 11 rules to achieve the same performance, almost double of the required rules when CRI4 approximations are used. This fact clearly shows the advantage of using smooth and accurately approximated gaussians when defining the input space partition of a FIS. The partition of the input domain for CRI1 is shown in Fig. 5.

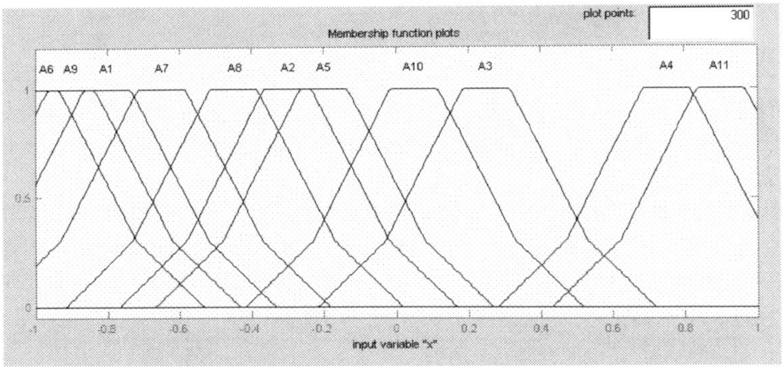

Fig. 5. Partition of the input universe for the FIS equivalent to the RBFN with CRI1 approximated gaussians.

4 A real world example: The Box&Jennkins gas furnace problem

The above described analysis has been applied to an example based on real word data, specifically to the well known Box&Jennkins gas furnace problem. This problem can be treated as a time series prediction problem by taking as inputs at a time t the inputs $u(t\text{-}i)$ and outputs $y(t\text{-}i)$ of previous sampling times. We have chosen as significant inputs the set $y(t\text{-}1)$ $y(t\text{-}2)$ $y(t\text{-}3)$ $u(t\text{-}1)$ $u(t\text{-}2)$ $u(t\text{-}3)$, first used in a work of Takagi and Sugeno [7]. 296 experimental data have been considered: 247 data for training and remaining 40 data to be predicted by the network. The training objective MSE has been 10^{-2}. Resulting network sizes after training are summed up in Table 3. Approximation and prediction of data for an exact gaussian network is shown in Fig. 6.a, and a learning evolution comparative for various CRI approximations is shown in Fig. 6.b. Here again, the reduction of the network size with the interpolation level of the CRI algorithm is clear and confirms above exposed results.

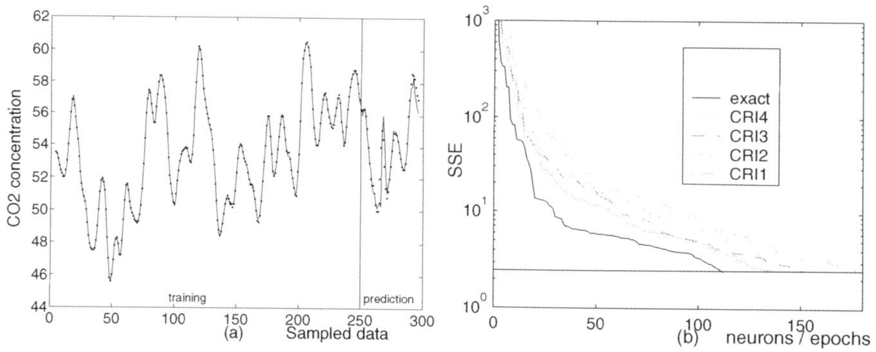

Fig. 6. Sampled data and systems output (a), and evolution of the training process for each CRI approximation (b).

Table 3. Net size according to the neuron activation function

Neural function	SSE = 2.47	
	Neurons	Opt. spread
Exact Gaussian	111	5.2
CRI4	140	4.1
CRI3	152	4
CRI2	152	3.9
CRI1	178	3.1

5 Conclusions

The hypothesis of how better approximations to non-linear smooth activation and membership functions –gaussian functions in this particular case– used in NFS's enhance the performance of such computational structures has been studied. An analytical inspection is, now a days, unapproachable, so previously used benchmark examples have been used in an experimental work. Obtained results suggest that the computational effort dedicated to the generation of AF's and MF's in NFS's, having in mind physical electronic circuitry for neuro-fuzzy applications, must be carefully considered. In this sense, the CRI recursive approximation algorithm, through the interpolation level parameter q, provides a powerful tool for flexible tuning of such effort. The conclusion is clear: smoother and more accurate approximations to AF's and MF's allow for defining smaller NFS's to obtain a given performance. This is very important as a consequence of the enormous complexity of NFS architectures when size grows –think of connections between neurons in NNA's for example–. Besides, it must be said that, not only same accuracy is achieved at training points with less nodes when using CRI approximations of higher interpolation levels, but greatest smoothness of the system output is obtained for the whole input domain. This is a specially valuable characteristic when FIS's are applied to the automatic control field, where high accuracy is required not only at a set of discrete points, but at a continuum of points over a large subset of the state space.

References

1. Basterretxea K., Tarela J.M, and del Campo I. (1999) PWL Approximation of Non-Linear Functions for the Implementation of Neuro-Fuzzy Systems, In: Proceedings of the CSCC'99 3rd World Multiconference on Circuits, Systems, Communications and Computers, Athens, Greece, pp.3001-3007.
2. Castro J. L. (1996) Fuzzy Logic Controllers Are Universal Approximators", IEEE Transactions on Systems, Man and Cybernetics, vol. 25, no. 4, pp. 629-635.
3. Chen C.-L., Chen W.-C. and Chang F.-Y. (1993) Hybrid Learning Algorithm for Gaussian Potential Function Networks", IEEE Proceedings-D, vol. 140, no. 6, pp. 442-448.
4. Esposito A., Marinaro M., Oricchio D. and Scarpetta S. (2000) Approximation of Continuous Mappings by a Growing Neural RBF-based Algorithm", Neural Networks, vol. 13, pp. 651-665.
5. Jang, J. S. R. and Sun, C. T. (1993) Functional Equivalence Between Radial Basis Function Networks and Fuzzy Inference Systems". IEEE Trans. on Neural Networks, vol. , no.1, pp. .
6. Lee S. and Rhee M. K. (1991) A Gaussian Potential Function Network with Hierarchically Self-Organizing Learning, Neural Networks, vol. 4, pp.207-224.
7. Takagi T., and Sugeno M. (1985), Fuzzy Identification of Systems and its Application to Modelling and Contro, IEEE Transactions on Systems Man and Cybernetics, vol. 15, pp. 116-132.
8. Tarela J.M., Basterretxea K., del Campo I., Martínez M.V., and Alonso E. (2002) Optimised PWL Recursive Approximation and its Application to Neuro-Fuzzy Systems, Mathematical and Computer Modelling, vol. 35, no. 7-8, pp. 867-883.

Automated Configuration of Multi-Spindle Drilling Gear Machines by Sub-Domain Knowledge Processing and Adaptive Algorithms

G. Klene*, A. Grauel*, H. J. Convey** and A. J. Hartley**

* University of Applied Sciences - South Westphalia, Soest Campus, Germany
{klene, grauel}@fh-swf.de
** Bolton Institute of Higher Education, Technology Development Unit, UK
{h.convey, a.j.hartley}@bolton.ac.uk

Abstract. A software concept and its realisation based on heuristic knowledge and pattern identification techniques for automated design of a multi-spindle drilling gear machine used in furniture production process is presented. The aim is to find an optimised design of the target-machine, this means to find a machine design with minimised number of drills and with the antagonistic goal to provide a fast production of the boards by minimising production-cycles per board. The design experience of a human expert was transferred to a design tool using his heuristic knowledge in combination with special developed pattern detection and recognition algorithms. Known and interpretable patterns are identified and used as information for a pre-design of the machine. The feasibility to manufacture each board is reached by analysing each single board to recognise known patterns for which drills are already equipped on the gears and the detection of new, uninterpretable patterns for which free spindle places can be equipped with suitable drills.

1 Introduction

The target machine for the research work is a large flexible machine consisting of up to eight drill supports and each drill support has one or two drilling gears each having up to 40 individual drill locations called spindles. Each machine has to be specifically designed with regard to the minimum number of drill supports, gears and spindles. The antagonistic goal is to minimise the production time by reducing the number of drilling cycles of a board during production. During one cycle the supports and gears are positioned and a selection or all of the holes of the board are drilled by moving up the spindles for selected drills. In other words, there are two antagonistic minimisation goals: the optimisation of the machine design and the optimisation of the board production. Each machine is designed to cover the customer's board specifications that are given by structural component engineering drawings.

The aim of the project is to automate the design of such multi-drilling gear machines. Dependent on the customer's requirements a few hundred different boards have to be processed on a machine. The time to design such a machine depends on the amount of boards and their complexity as well as the maximal number of drilling-cycles requested by the customer. A typical amount of time for the manual design is about three month for a machine to be designed for approximately 300 boards and maximal three cycles per board. An automated multi-spindle gear design based on intelligent database techniques using heuristics of the expert's design procedure [1-4] with pattern detection and pattern identification [5] is introduced to solve this multi-criteria optimisation problem [6-7] to reach a strong saving of time during the design process.

2 Target Machine and Board Data

The machine can be described by the set of supports, a set of gears on each support, and set of spindles on each gear (Fig. 1). Each spindle can be equipped with one drill whereby the type of drill-tool is a sink-drill, a through-drill or a drill that can be used for both kinds of holes. The used position field notation [8] denotes a filled spindle by a drill specific tool number and a free spindle by zero. The tool number gives information about the drills diameter and the drilling mode related to the kind of holes to be drilled like sink-holes or through-holes.

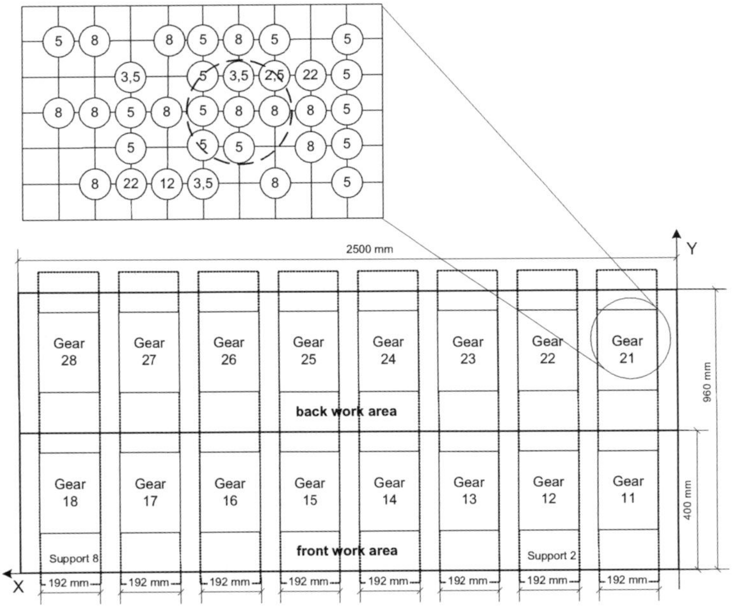

Fig. 1. Schematic of a multi-drilling gear machine and one specific gear.

During the design of the machine different constraint have to be considered rising from the construction of the machine, the control electronic or from production restrictions e.g. to under-run the maximal allowed time for a cycle. Important constraints [9] are for example the maximal number of spindles on one gear, the maximal distances of the feed-in piston, and stopper, the minimal distance between the gears and y = 0, two gears in y-direction, and the width of a support etc.

The board data are made up from the set of work pieces to be manufactured on the machine. Each work piece (Fig. 2) has a number of holes. The x- and y-position, the diameter, the depth of the hole and the drilling-mode determine each hole. The drilling-mode identifies each hole as a sink-hole or a through-hole.

3 Automated Design

Following the human expert the problem of automated configuration of the drilling gears can be divided in two major tasks [10-11]. The first step is to find a generalised pre-placement of drills. The second step is an iterative process, which processes each board by defining the placement of the board in the area of work of the machine, by finding optimised positions for each support and gear and by achieving the possibility to produce each board by defining cycles and suitable drills. A sequence for the consideration of boards is determined using the board complexity [8] before the iterative configuration process starts. During this configuration process restrictions related to the parameters of the machine have to be observed and each structural component has to be checked to ensure that the production is feasible. Boards containing holes for fittings are detected and planned in a previous step considering the requirements that the accessory holes are usually not in a grid distance and the main holes are too big to place them on spindles in grid distance.

3.1. Pattern Detection and Identification

The hole patterns can be separated into the groups of either a known and interpretable patterns or a group of unknown patterns. The pattern identification process works in analogy to the human expert who classifies the boards into structural parts such as cupboard units, side and middle walls, bottom boards and doors. The group of known and interpretable patterns used for the automated drilling system consists of a set of holes in rows, for metal fittings and for construction holes. The holes in rows are later called x-rows because they are a set of holes in x-direction and the construction holes are called y-rows. An x-row pattern can be defined as at least three holes in combination that fulfil the conditions: a) placed on the same work piece, b) equal diameter, c) equal mode, d) equal y-coordinate, and e) distance between holes = grid or 2 times grid. The holes of an identified pattern are stored in a fuzzy way to the generated tables [12]. Further processing can identify that there is the same x-row on different boards. The metal-fitting pattern has a hole with diameter ≥ 12mm and the location is close to the edge of the board and most time they have further holes, e.g. for mounting near the main hole. Algorithms identify these patterns and save them to a fitting-pattern database with the diameter of the main hole, the diameter of the accessory holes for screws, the distances between the accessory holes and the main hole, and the location of the pattern on the board, e.g. the bottom or top surface. With this the pattern is independent from the real x- and y-coordinates on the board. This enables a later check to determine if the same drill-combination planned for one board can be used on a different board.

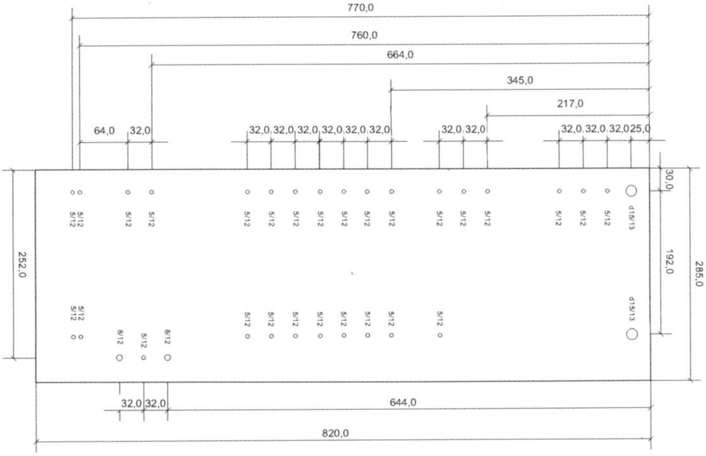

Fig. 2. Schematic of a board.

Unknown patterns are an arrangement of at least three holes that are in a modulo grid distance in the x- and y-direction from each other. They cannot be interpreted as a previous pattern but they can be used to find similar patterns on different boards. They are saved in a pattern-database independent from the real coordinates of each hole. Only the number of the first hole – with this it is possible to get all information about the hole like coordinates and other parameters – and the distances between the holes and the parameters of the holes like the diameter and the mode are saved. Furthermore the gear and the spindle place number are saved in relation to the gear and the spindle place number of the corresponding drill.

3.2. Initial Number of Supports, Gears and Spindles

Before the design process can be started an initial configuration defining the number of supports and the number of spindles on each gear is entered by the operator. The operator has to decide if special gears, that are mainly used to drill the metal-fitting holes, are required. The program supports the operator by giving a visual prompt that shows the number of fitting holes compared to all holes of all boards and the number of required supports to drill the fitting pattern in least number of cycles. The human expert has to decide if he wants the program to include and plan special supports for metal-fittings because this decision is dependent upon a value judgement made by the customer. The customer has to consider the impact of additional process cycles as against the additional costs for special supports. Furthermore the operator is given information if x-row drills are useful. If the amount of x-rows is large and the operator decided to use special rows for x-row drills, the operator has to initially define the number of spindle places to be planned for x-row drilling on each gear. These values and customer decisions are used to allocate spindle places in the database that are planned in a grid distance in columns and rows for all non-fitting gears. Each spindle place is related to the gears and has the coordinate on the gears as well as a data field for the tool number. The tool number characterises a drill by its diameter and drill mode. The initial tool number is 0 and indicates a free spindle place.

3.3. Design Process

The automated design process depends on heuristic knowledge from the human expert and on the knowledge derived from the different pattern databases. One heuristic is, e.g. that the mapping of holes to the drills or free spindles starts from a right support and runs from this start-support for the board to next support on the left side of the start-support and then support by support to the most left support required for the board. If it is not possible to use the planned supports because there is not enough drills or free spindles than the next left support of the initial start-support is chosen as a start-support and the whole process starts again until it is possible to produce the board. This heuristic was transferred to the program in the way that in the first step an optimised start-support is identified by checking all possible start-supports regarding their suitability. This step is done by selecting a set of holes from the right part of the board dependent on the width of the actual support. If the best start-support is found the holes for this support are processed and then the next set of holes is chosen dependent on the width of the next left support and the left edge of the support before. These holes are processed and so on until the end of the board is reached. If all holes cannot be processed in the first cycle, a second cycle is required. Dependent on the complexity of the board and on the suitable drills and free spindles on the gears further cycles are required until all holes of the board can be drilled. Between each cycle the supports can only be driven in a maximal driving distance between two cycles. This is covered in such a way that for a second or further cycle maximal and minimal x-positions of each support are calculated dependent on the position of the supports in the previous cycle.

```
DO until all boards are processed
   DO until all holes of actual board can be drilled
      select holes for work area of actual support
      IF first cycle AND first support of board THEN
         use search- and pattern-DB algorithm to find
            optimised start-support
         process holes for the start-support
      END IF
      FOR all supports
         select holes for work area of actual support
         use pattern-DB to identify similar pattern
         IF similar pattern on gears THEN
            search for possible gears/supports
            IF possible gears/supports exist THEN
               process holes for the best support/gears
            ELSE
               use search algorithm to find best gears
            END IF
         END IF
      NEXT support
      new cycle
   LOOP 'all holes
LOOP 'all boards
```

Fig. 3. Pseudo-code algorithm for the board-by-board processing.

The board-by-board processing (Fig. 3) is characterised by the two major algorithms "pattern-DB algorithm" and "search algorithm". The first algorithm is based on a pattern database where all hole patterns ever found on the boards before are saved. A pattern identification algorithm returns the gears that contain suitable drill pattern for the hole pattern or a part of the hole pattern which is actual processed. All gears containing the required drill combinations are checked regarding the constraints of the machine and the maximal driving distances. Only gears that fulfil the constraints are considered for the search for additional drills or free spindles to produces all holes of the actual hole pattern. The gear that can produce the most holes in one cycle is used. If no gear is found by the pattern-DB algorithm the actual hole pattern is processed by the search algorithm. This algorithm is searching for the best gear by variation of x- and y- coordinates of the supports and gears under the mentioned restrictions and is searching for the best gears for the actual hole pattern. The best gear is that which can produce the most holes in one cycle.

The pattern-DB algorithm and the search algorithm are used together to find an optimised start-support. The support is chosen as start-support that can produce the maximal number of holes. This has the advantage that free spindle places on more left supports are considered, e.g. on a right support there is 6 suitable drills and 2 suitable free spindle places but a support more right has only 3 suitable drills but 8 free suitable spindle places. In this case the more right support is chosen as the start support because with that one 11 holes can be produced in one cycle compared to the more left support which can produce only 8 holes in one cycle. This proceeding is similar to the heuristic knowledge of the expert and enables the uniform distribution of drills on the gears.

4 Results

The tool was successful tested using a real-world dataset containing 68 boards. The design of the machine found by the software-tool allows to manufacture all boards. The positions of supports and gears as well as the used tools are saved for each board in a project database. The whole design and optimisation process takes about 8 hours. The time strongly depends on the similarity of the boards because the pattern detection algorithm is very fast compared to the search algorithm. Special supports for fittings where planned to minimise the number of cycles required for the fitting-boards.

Moreover the tool was tested successful using different machine parameters e.g. different minimal distances between two supports or two gears or different initial sizes of the gears.

5 Summary and Conclusions

After introducing the target machine important design goals for multi-spindle drilling gear configuration where shown and a technical concept for automated multi-drilling gear design dependent on a heuristic knowledge and pattern detection and -identification was described. The generalised pre-placement of the automated concept is based on the detection of characteristic interpretable pattern and a pre-configuration of the machine depended on generalised interpretable pattern. Moreover the iterative board-by-board process is characterised by a pattern-DB algorithm and a search algorithm. The pattern-DB algorithm uses the information of already equipped drill pattern to find optimised gears while the search algorithm is used if the first algorithm is not successful. Both algorithms are used in competition to define an optimised start-support for each board. The results of the design tool show that the automated design is very fast compared to the design of the human expert. Different solutions depending on variation of the parameters of the target machine could be found for a real-world test dataset. A further advantage is the documentation of the expert's design procedure and its verification in this application software tool. The provision of a database providing knowledge as to which drill is used for each single hole provides a simple interface to the CNC-programs that will control the manufacturing machine.

References

1. Berthold, M., Hand, D. J., (eds.): Intelligent data analysis: an introduction. Springer Verlag Berlin Heidelberg New York (1999).
2. Whitelock, P., Kilby, K.; Linguistic and Computational Techniques in Machine Translation System Design, UCL Press (1995), London.
3. Karamjit, S. G., (ed.): Human Machine Symbiosis: the foundations of human centred system design, Springer Verlag Berlin Heidelberg New York (1996).
4. Steger, A.: Diskrete Strukturen: Kombinatorik, Graphentheorie, Algebra, Springer Verlag Berlin Heidelberg New York (2001).
5. Klene, G., Grauel, A., Convey, H. J., Hartley, A. J.: Intelligent Data Analysis for Design of Multi-Drilling Gear Machines, Proc. European Symposium on Intelligent Techniques, Fotodruck Mainz GmbH, Aachen (2000) 257–262.
6. Zimmermann, H.-J., Gutsche, L.,: Multi-Criteria Analyse: Einführung in die Theorie der Entscheidungen bei Mehrfachzielsetzungen, Springer-Verlag, Berlin (1991).
7. Ludwig, L.-A.: Computational Intelligence: Nutzenpotentiale der Methodenklasse CI für ausgewählte Problemstellungen der Produktionswirtschaft unter Berücksichtigung konfliktionärer multikriterieller Anforderungen. HNI Verlag, Paderborn (2000).
8. Klene, G., Grauel, A., Convey, H. J., Hartley, A. J.: Design of Multi-drilling Gear Machines by Knowledge Processing and Machine Simulation, Proc. Intelligent Data Engineering and Automated Learning, Springer-Verlag, Wien New York (2002) 93–98.
9. Klene, G., Grauel, A., Convey, H. J., Hartley, A. J.: A Heuristic Approach for Automated Configuration of Multi-Drilling Gear Machines, Proc. International Conference on Recent Advantages in Soft Computing, Nottingham, UK, 2002, pp. 368–373.
10. Klene, G., Grauel, A., Convey, H. J., Hartley, A. J.: Cost Oriented Automation by Automated Expert Decision Process, Proc. 6th IFAC Symposium on Cost Oriented Automation, IFAC Publications, Elsevier Science Ltd (2001), 121-126.
11. Klene, G., Grauel, A., Convey, H. J., Hartley, A. J.: Automated Experts Decision Process for Customised Design of Multi-Spindle Drilling Machines. Proc. 8th IFAC/IFIP/IFORS/IEA Symposium on Human-Machine Systems, IFAC Preprints, Elsevier Science Ltd. (2001) 683-687.
12. Klene, G., Grauel, A., Convey, H. J., Hartley, A. J.: Data Mining and Automation of Experts Decision Process Applied to Machine Design for Furniture Production, Proc. Intern. Conference on Artificial Neural Networks and Genetic Algorithms, Springer-Verlag, Wien New York (2001) 453–456.

Construction of Shape-preserving Affine Takagi-Sugeno Systems via Linear-rational Spline Fuzzy Partitions

Felipe Fernández, Julio Gutiérrez, Juan Carlos Crespo, and Gracián Triviño [1]

Dep. Tecnología Fotónica, Facultad de Informática Universidad Politécnica de Madrid, 28660 Madrid, Spain.
felipe.fernandez@es.bosch.com, jgr@dtf.fi.upm.es, crespozj@dtf.fi.upm.es, gtrivino@dtf.fi.upm.es

Abstract. The output function of standard affine Takagi-Sugeno systems presents shape failures and does not have continuous derivatives in the corresponding output function. These characteristics give an undesirable behaviour in many practical applications such us fuzzy control or fuzzy modelling. In order to avoid these problems, this paper presents a new method for deriving a shape-preserving affine Takagi-Sugeno model, using a standard trapezoidal fuzzy partition specification. Each univariate trapezoidal fuzzy partition is primarily transformed into an equivalent triangular one. An adaptive first-order symmetrical B-spline filter is later applied on each triangular partition in order to derive a mixed linear-rational spline fuzzy partition. The obtained model retains monotony, convexity and positivity of the corresponding control points defined on the cores of the original fuzzy partition. These properties are due to the characteristics of the linear-rational spline partitions obtained. Finally, a simple example is given to practically show the usefulness of the approach.

Keywords. Takagi-Sugeno models, Shape-preserving splines, rational splines, adaptive filters.

1 Introduction

First order or affine Takagi-Sugeno (ATS) fuzzy systems combine a global rule-based specification with local affine descriptions. These models are easy to identify, because each rule defines a different fuzzy region where the output depends on the inputs in an affine form. The construction of these systems implies the design of a set of fuzzy rules that approximates a system function with a certain accuracy level.

---

[1] This research has been supported by CICYT TIC2000-1420

Independently, splines functions are today widely used to fit a smooth continuous function through discrete data in such theoretical fields as data interpolation and approximation or in practical fields as computer graphics and signal processing. Some aspects of these techniques are analogous to fuzzy modelling and therefore we can borrow many mature tools and concepts from these related fields.

Some fuzzy software tools and research papers [4,5,2] have considered basic spline partitions, but in our opinion the analogy and possibility of deep synergy among the referred areas do not have been exploited enough. This paper presents a contribution in this hopeful way of creating new effective bridges among fuzzy systems, spline approximation and signal processing.

The main goal of this paper is to obtain a new smooth shape-preserving function from a standard trapezoidal ATS specification, using an adaptive B-spline filter on the derived triangular partitions. In order to avoid the suppression of local details specified by narrow fuzzy linguistic labels, the support length of the univariate filters applied changes in a piecewise affine form through each univariate input domain, according to the cores and borders lengths of the corresponding antecedent trapeziums.

An outline of the rest of the paper is as follows: Section 2 reviews the basic characteristics of standard ATS models; Section 3 briefly discusses the undesired slope and shape failures of these models; Section 4 introduces the basic structure of the shape preserving ATS model considered; Section 5 – which presents the main results of the paper – describes the main characteristics of the nonuniform B-spline adaptive filter applied; Section 6 gives the analytical form and main properties of the mixed linear-rational fuzzy partition obtained; Finally, Section 7 shows an example and concludes the work.

2 Standard ATS model

Standard conjunctive ATS fuzzy systems consist of rules $\mathrm{R_r}$ with the following structure

$$\{\mathrm{R_r}\text{: If } x_1 \text{ is } A_{\mathrm{r},1} \text{ and ... and } x_\mathrm{n} \text{ is } A_{\mathrm{r,n}} \text{ then } z=\mathbf{a}_\mathrm{r}^\mathrm{T}\mathbf{x}+b_r\}_{\mathrm{r}=1..\mathrm{R}}$$

where $\mathbf{x} = (x_1, \ldots, \mathbf{x}_\mathrm{n})^\mathrm{T}$ is a crisp vector of the multivariate input domain $\mathrm{U^n}$, $(A_{\mathrm{r}1}, \ldots, A_{\mathrm{rn}})$ are trapezoidal membership functions that belong to the corresponding univariate fuzzy partition [1] and $z_\mathrm{r}=\mathbf{a}_\mathrm{r}^\mathrm{T}\mathbf{x}+b_r$ is the affine output function of the r-th rule. This set of rules can also be described by the multivariate-closed form

$$\{\mathrm{R_r}\text{: If } \mathbf{x} \text{ is } \mathbf{A}_\mathrm{r} \text{ then } z_\mathrm{r}=\mathbf{a}_\mathrm{r}^\mathrm{T}\mathbf{x}+b_r\}$$

where $\mathbf{A_r} = \mathbf{A}_\mathrm{r}(\mathbf{x}) = A_{\mathrm{r}1}(\mathrm{x}_1) \times \ldots \times A_{\mathrm{rn}}(\mathrm{x_n})$ is a trapezoidal tensor product that belongs to the corresponding multivariate fuzzy partition. This multivariate partition of unity holds the following constraint on the entire input domain:

$$\sum_{\mathrm{r}=1}^{\mathrm{R}} \mathbf{A_r}(\mathbf{x}) = 1$$

By considering this condition and assuming that the rule base is complete in the sense that it covers the whole input domain U^n, it immediately follows that the global output function of the standard product-sum model simplifies to

$$z = \sum_{r=1}^{R} \mathbf{A_r}(\mathbf{x}) \times (\mathbf{a_r}^T \mathbf{x} + b_r)$$

Taking into account the trapezoidal fuzzy partitions defined, the multivariate input domain is partitioned into a grid of disjoint intervals of two types: *multidimensional interpolation intervals* (or multidimensional antecedent term cores) where $\mathbf{A_r}(\mathbf{x})=1$ and *multidimensional approximation intervals* where $0<\mathbf{A_r}(\mathbf{x})<1$. The corresponding univariate intervals are called respectively univariate interpolation intervals $\{IP_j\}$ and univariate approximation intervals $\{AP_j\}$.

3 Limitations of standard ATS models

Unfortunately, the output of a standard ATS model does not have continuous derivatives and is not shape preserving since it exhibits wild wiggles that are not inherent in the core data [1]. In this section we briefly reviewed these undesirable properties of standard ATS systems. For this purpose, we consider the following SISO ATS example (see Figure 1):

$$\{R_1\text{: If } x \text{ is } A_1 \text{ then } z_1 = 1.5; \quad R_2\text{: If } x \text{ is } A_2 \text{ then } z_2 = a\,(x\text{-}1) + 2.5\}$$

with slope $a \geq 0$ and where the trapezoidal fuzzy partition $\{A_1, A_2\}$ is defined by the knot sequence (-0.3, 0, 1, 1.3). The standard output function z of this system in the approximation interval AP=(0, 1) is given by

$$z\,(x) = (1\text{-}x)\,1.5 + x\,(a(x\text{-}1)\text{+}2.5) = ax^2\text{+}(1\text{-}a)x\text{+}1.5$$

Its output derivative $z'(x) = 2\,a\,x + (1\text{-}a) = 2\,a\,(x - (a\text{-}1)/2a)$ is not equal in the corners of the approximation interval ($z'(0)= (1\text{-}a)$, $z'(1)=(1\text{+}a)$) to the corresponding derivatives of the specified affine functions ($z'_1=0$, $z'_2=a$). Moreover, this derivative becomes negative for slopes $a>1$, in the interval $[0, (a\text{-}1)/2a]$, even though the gain or slopes of the local affine models considered are nonnegative.

Therefore, the global output function of an ATS system does not have continuous derivative and it is not shape preserving [3] since it does not conserve the positivity, monotony and convexity of the defined data in the corresponding multidimensional antecedent term cores. A general reason of this objectionable behaviour is that the same membership functions are used to blend the constant part b_i and linear part $a_i x$ of the affine function $a_i x+b_i$. Notice that classical Hermite spline interpolation uses different types of blending functions for these two components.

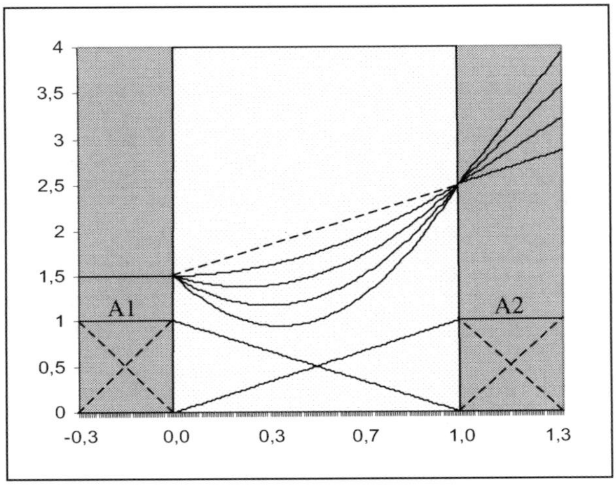

Fig. 1. Shape-failures of a standard ATS system

4 Shape-preserving ATS model

In this paper, in order to avoid this problem, each original univariate trapezoidal antecedent fuzzy partitions $\{A_{ji}\}$ of a domain U_j, defined by the knots sequence $T_j=(a_j=x_{jo}<x_{j1}<..<x_{jm}=b_j)$, is transformed into an equivalent univariate triangular one $\{B_{jk}\}$ using the same knots sequence, in order to obtain a shape-preserving piecewise output function. See example in Figure 1, where T=(-0.3, 0, 1, 1.3). The derived zero-order TS model has a rule for each corner of the original multidimensional antecedents core. The new equivalent zero-order TS system gives the following non-smooth C^0 output function

$$z = \sum_{r=1}^{R'} \mathbf{B}_r(\mathbf{x}) \times c_r$$

where $R' = R \times 2^n$ is the number of new derived rules and c_r are the constant output functions of the corresponding rules. This output function is a shape-preserving piecewise multilinear interpolator.

For the previous example (Figure 1), the outputs of their triangular rules are respectively c_0=1.5 c_1=1.5, c_2=2.5 c_3=2.5+0.3·a, and the global output is a shape-preserving piecewise affine function also shown by a dashed line in Figure 1.

5 Adaptive B-spline filter

An additional nonuniform adaptive first-order B-spline filter is applied to each univariate triangular fuzzy partition, in order to get a smooth C^1 output function. The filtered output function obtained is a piecewise linear-rational approximator that preserves the affine functions on the reduced cores of the initial multivariate antecedent terms.

The nonuniform triangular fuzzy partition $\{B_i\}$ of each univariate input domain U is transformed into a linear-rational spline fuzzy partition $\{Q_i\}$, called in this paper LR-spline partition (more deeply described in Section 6). This way, each triangular term B_i is transformed into a linear-rational one $Q_{i,w}$ using the following nonuniform first-order B-spline filter:

$$Q_{i,w}(u) = \frac{1}{2w(u)} \int_{u-w(u)}^{u+w(u)} B_i(v)\, dv$$

where $2w(u)$ is the kernel support length function of the nonuniform symmetrical first-order B-spline filter applied: $N^1_w(v) = \{|v| \le w \rightarrow 1/2w\ ;\ 0\}$.

The function $w=w(u)$ has a piecewise affine form. In each univariate *filter interval* $IW_i=[x_i-w_{i0}, x_i+w_{i1}]$, relative to an intermediate knot x_i, the corresponding affine function is given by

$$w(u) = w_{i0} + (w_{i1} - w_{i0})\, u, \qquad u \in [0,1]$$

where $u = (x - (x_i - w_{i0}))/(w_{i1}+w_{i0})$ is the considered local parameter of the filter interval IW_i, and $2w_{i0}$ and $2w_{i1}$ are respectively, the kernel support lengths in the left and right boundaries of IW_i.

To provide an adaptive behaviour, the lengths of kernel support $2w_{i0}$ and $2w_{i1}$ are in this paper defined as proportional to the lengths of the corresponding intervals:

$$w_{i0} = s_{i0} \cdot (x_i - x_{i-1}) \quad (0 \le s_{i0} \le 1/2); \qquad w_{i1} = s_{i1} \cdot (x_{i+1} - x_i) \quad (0 \le s_{i1} \le 1/2)$$

where s_{i0} and s_{i1} are locally specified parameters of the corresponding spline. Figure 2 shows an example of a univariate LR-spline fuzzy partition derived from the corresponding triangular one, defined by the knots sequence T =(0, 4, 5), and with

$$w_{10} = s_{10}(x_1 - x_0) = 0.2 \times 4 \ \text{ and } \ w_{11} = s_{11}(x_2 - x_1) = 0.3 \times 1.$$

The mixed linear-rational nature of the splines obtained is a consequence of the affine invariance property of nonuniform symmetrical filters. Outside the filter intervals $\{IW_i\}$, the filter applied does not modify the corresponding triangular partition $\{B_i\}$ (affine preservation property):

$$Q_{i,w}(u) = \frac{1}{2w(u)} \int_{u-w(u)}^{u+w(u)} B_i(v)\, dv = \frac{1}{2w(u)} \int_{u-w(u)}^{u+w(u)} (a \cdot v + b)\, dv = a \cdot u + b = B_i(u)$$

The nonuniform filter considered is a generalisation of the uniform one. This nonuniform filter increases the smoothness and continuity order of the corresponding shape-preserving output function and transforms the intermediate C^0 piecewise multilinear output function into a C^1 piecewise linear-rational one. Moreover, the adaptive nature of this filter avoids removing the effect of the narrow antecedent labels on the implicated output function.

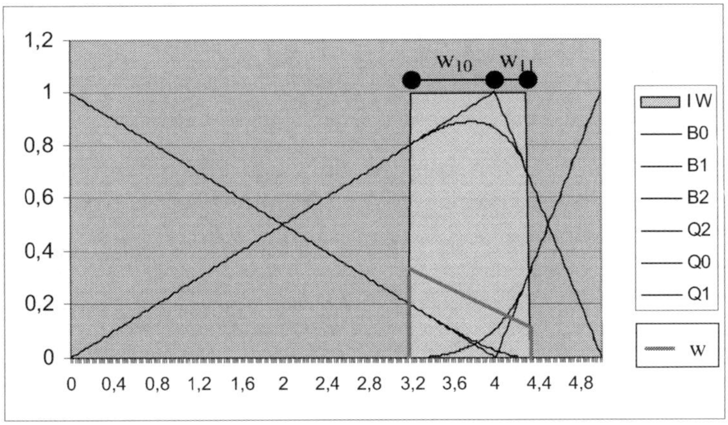

Fig. 2. Univariate LR-spline fuzzy partition

6 Linear-rational spline partitions

Univariate LR-spline fuzzy partitions are obtained by means of the referred adaptive first-order B-spline filter N^1_w applied to a triangular fuzzy partition $\{B_i\}$. The local tension parameter t_i, in each filter interval IW_i, is defined in this paper by the complementary armonic mean: $t_i=1-4/((1/s_{i0})+(1/s_{i1}))$ where $(0\le t\le 1)$.

In each filter interval $IW_i = [x_i-w_{i0}, x_i+w_{i1}]$ relative to each intermediate knot x_i, the rational blending function Q_{i-1}, Q_i, Q_{i+1} are easily derived. For $k = i-1$, i and $i+1$ we have:

$$Q_{k,w}(u) = \frac{1}{2w(u)} \int_{u-w(u)}^{u+w(u)} B_k(v)\, dv$$

Outside these filter intervals, the resultant blending functions $\{Q_i\}$ match the original triangular blending functions $\{B_i\}$.

Within each filter interval $IW_i = [x_i-w_{i0}, x_i+w_{i1}]$ of an intermediate knot x_i, the corresponding rational splines $\{Q_{i-1}(u), Q_i(u), Q_{i+1}(u)\}$ for $u\in[0,1]$, have the following easily deduced expressions:

$$Q_{i-1}(u) = \frac{1}{2w(u)} \int_0^{2w_{i1}\cdot u} \frac{v}{(x_{i+1}-x_i)}\, dv = \frac{w_{i1}{}^2}{(x_{i+1}-x_i)} \times \frac{u^2}{w_{i0}+u(w_{i1}-w_{i0})}$$

$$Q_{i+1}(u) = \frac{1}{2w(1-u)} \int_0^{2w_{i0}\cdot(1-u)} \frac{v}{(x_i - x_{i-1})} dv = \frac{w_{i0}^{\;2}}{(x_i - x_{i-1})} \times \frac{(1-u)^2}{w_{i1} + u(w_{i0} - w_{i1})}$$

$$Q_i(u) = 1 - Q_{i-1}(u) - Q_{i+1}(u)$$

where the parameter $u = (x - (x_i - w_{i0}))/(w_{i1} + w_{i0})$ represents the referred local variable defined in each filter interval IW_i. It is important to note that these formulas only give a single rational segment of the corresponding LR-splines over the corresponding filter interval $IW_i = [x_i - w_{i0}, x_i + w_{i1}]$. Notice that for $(w_{i0} = w_{i1})$ the corresponding rational blending functions degenerate into quadratic ones.

Theorem 1. Main properties of LR-splines:

1. **Mixed form**: LR-splines $\{Q_i\}$ are formed by C^1 piecewise rational and linear (affine) functions.
2. **Affine preservation**: LR-splines $\{Q_i\}$ are affine functions outside the filter intervals $\{IW_i\}$: $Q_i(x) = B_i(x)$ for $x \notin \{IW_i\}$.
3. **Variation diminishing property**: it means that there are no more local extreme points in the linear combination of LR-splines than there are in the data set of the corresponding cores control points.
4. **Shape-preserving**: LR-splines conserve the positivity, monotony or convexity of the corresponding control data set.
5. **Multivariate extension**: univariate LR-splines are directly extended to an arbitrary number of dimensions, while preserving the above properties, by means of the corresponding tensor product.

Sketch of proof. These properties are directly deduced from the nonuniform box filter properties applied to triangular fuzzy partitions, and from the described blending functions. □

7 Application example and conclusions

Figure 3 and its embedded table describes a SISO ATS system defined by means of the following rules:

$$\{R_r\text{: If } x \text{ is } A_r \text{ then } z_2 = a_r(x - x_r) + b_r\}_{r=1..3,}$$

where the trapezoidal fuzzy partition $\{A_0, A_1, A_2\}$ is defined by the knot sequence T=(0, 8, 12, 17, 22, 25.4). The support length $w(u)$ of the nonuniform first-order B-spline filter applied on the triangular fuzzy partition is also shown in this figure, where it is also possible to observe that the presented shape-preserving model allows to adjust easily the appropriate smoothing characteristics of the corresponding output function. Moreover, the involved blending functions are easily computed using the analytical expressions obtained.

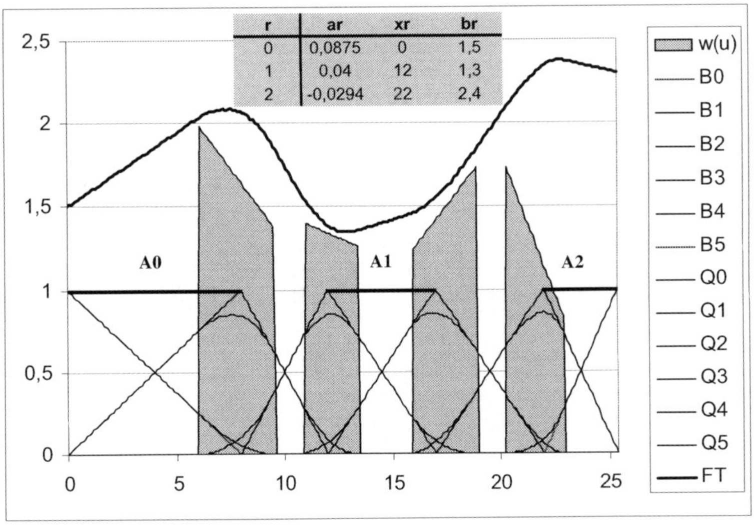

Fig. 3. Example of a SISO shape-preserving ATS model via LR-splines.

References

1. Babuska R, Fantuzzi C, Kaymak U, and Verbrunggen H B (1996) Improved inference for Takagi-Sugeno models, In *Proc. Fifth IEEE International Conference on Fuzzy Systems*, New Orleans, USA, pp 701-706
2. Babuska R (1998), Fuzzy Modeling for Control, Kluwer Academic Publisher
3. Fernández F and Gutiérrez J (2000), "Structured Design of an Extended TS controller Using Global Fuzzy Parameters and Fuzzification Transform", *Proc. of 8th International Conference IPMU 2000*,Vol.II, pp 724-731
4. Kvasov B (2000), Methods of Shape-Preserving spline approximation, Word Scientific
5. Zhang J and Knoll A (1996) Constructing Fuzzy Controllers with B-spline Models, *Proc. of the IEEE International Conference on Fuzzy Systems 1996,* pp 416-421
6. Zhang J and Knoll A (1997) Unsupervised Learning of Control Surfaces Based on B-spline Models, *Proc. of the IEEE International Conference on Fuzzy Systems 1997*, pp 1725-1730

Identification of Dynamic Cause-Effect Relations for Systems Performance Evaluation

Gancho Vachkov, Yuhiko Kiyota and Koji Komatsu

Faculty of Engineering, Kagawa University, Hayashi-cho 2217-20,
Takamatsu-Shi, 761-0396 Kagawa-ken, Japan
E-mail: {vachkov, kiyota, komatsu}@eng.kagawa-u.ac.jp

Abstract. Incremental cause-effect relations are used in this paper in order to create respective CER dynamic models that are further used to reveal the existing relationships between the pairs of parameters in the observed set of system parameters.
Two Identification schemes are presented, namely for off-line and for real-time identification, based on the Least Squares Estimation and on the *Widrow-Hoff*'s learning rule. They are used to create the so called normal operation model and the current operation models of the system. The degree of discrepancy between these models during the operation of the system is used as a measure for detecting the abnormality of its operation.
Computer simulations demonstrate the main idea of identification and its applicability for real-time system monitoring and fault detection.

Keywords. Dynamic Models, Cause-Effect Relations, Real Time Identification, Widrow-Hoff's Learning Rule, System Evaluation, Fault Detection

1 Introduction

Contemporary industrial plants, systems and machinery have become more complex and advanced from a technological viewpoint. Therefore a reliable solution to the problem of *real-time* evaluation of the system performance and early fault detection is of big importance for the smooth and faultless operation of the system.

Different methods for data acquisition, pre-processing and data analysis are applied for such purpose. An appropriate computation scheme for revealing the system dynamic behavior, its current trend as well as the sudden change in behavior is necessary for the real time system performance evaluation. In this paper, the identification of the cause-effect relations between selected pairs of observed parameters is used as a tool for *real-time* system monitoring. The cause-effect approach to dynamic modeling and two identification schemes are proposed that can be successfully used for real time evaluation of the system behavior.

2 Incremental Cause-Effect Relations (CER) Dynamic Models

The dynamics of a process with control input $u(t)$ and a process output $y(t)$ can be represented in many different formats and structures [2-6]. When a discrete dynamics is taken into account, the cause-effect relation scheme could be a suitable way of modeling the dynamic behavior.
Let k be the current time sampling and n be the pre-determined *memory length* (buffer) of the dynamic model. Then the basic idea of the cause-effect relations dynamic model is to predict the future process output $y(k+1)$ of the plant by taking into the current output $y(k)$, as well as a sequence of n inputs, namely the current and the previous controls: $u(k), u(k-1), u(k-2),..., u(k-n+1)$.
The proposed model in [5,6] is an *incremental* type of cause-effect relations (CER) model that uses increments rather than absolute values of the inputs and outputs, as follows: *changes-of-the-current and past pa t inputs:* $\Delta u(k), \Delta u(k-1), ..., \Delta u(k-n+1)$ and the resulted *change-of-the-future -output:* $\Delta y(k+1)$. These increments are calculated as follows:

$$\Delta u(k-i+1) = u(k-i+1) - u(k-i), \quad i=1,2,...,n \tag{1}$$

Then the increment of the output can be generally described as:

$$\Delta y(k+1) = F[\Delta u(k),..., \Delta u(k-i+1),..., \Delta u(k-n+1)] \tag{2}$$

where ***F*** is a Cause-Effect Relation function of the input increments., which could be nonlinear in the general case. In this paper, as in [5] and [6], we assume the following structure of the CER Model:

$$\Delta y(k+1) = \sum_{i=1}^{n} \Delta u(k-i+1) f(i) = \sum_{i=1}^{n} S_i \tag{3}$$

Finally, the predicted output of the model is incremented as follows:

$$y(k+1) = y(k) + \Delta y(k+1) \tag{4}$$

It is worth to note that if the dynamic process is linear and time invariant, the elements : $f(1), f(2),... f(i),..., f(n)$ are constants and each of them shows the *strength of the relation* between the observed change-of-the output $\Delta y(k+1)$ and the respective change-of-the past input $\Delta u(k-i)$. In the nonlinear case all these strength levels are functions of the amplitude of the inputs. The following Fig. 1. illustrates the idea and the computational structure of the incremental CER Dynamic Model.

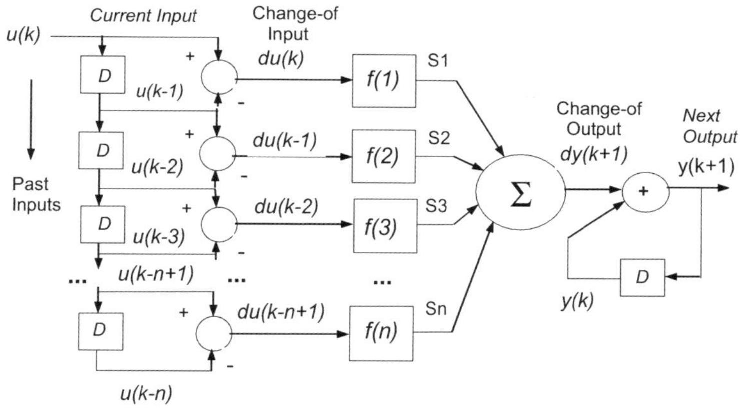

Fig.1. Computational Structure of the One-Dimensional CER Dynamic Model

The strength levels sequence $f(1), f(2),...f(i),..., f(n)$ can be plotted as a Cause-Effect Relation function of the dynamic model. It is graphically shown for two different CER Models in Fig. 2.

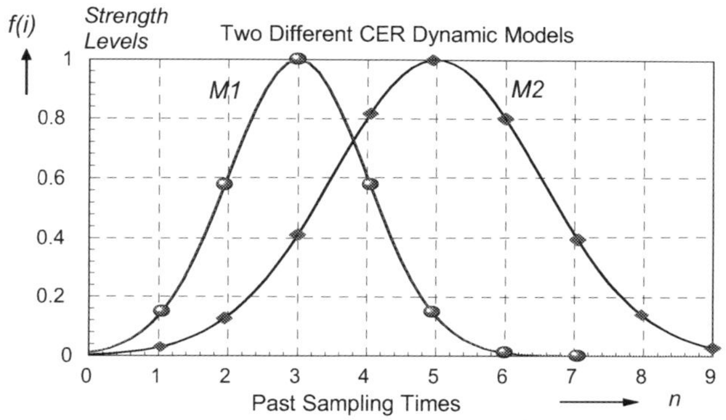

Fig.2. Example of two Different CER Models

Constructed in such incremental way, the CER dynamic model can be regarded as a kind of "rule-based model" that shows the strength levels between a specified past change-of-input and the current change-of-output. Since most real systems have a time delay, the general shape of the CER-functions is smooth and usually resembles a non-symmetrical membership function that first monotonically increases and then decreases, as seen in Fig. 2. Such shape has a clear physical meaning and is quite understandable for the human operator.

In the next Fig. 3. two dynamic responses produced by the above two CER-models (*M1* and *M2*) are displayed, based on the same input signal. One practical problem is to recognize in real time whether and when the system behavior has been changed from M1 to M2. Further on two Identification algorithms for CER dynamic models are described that can be useful for solving the above problem.

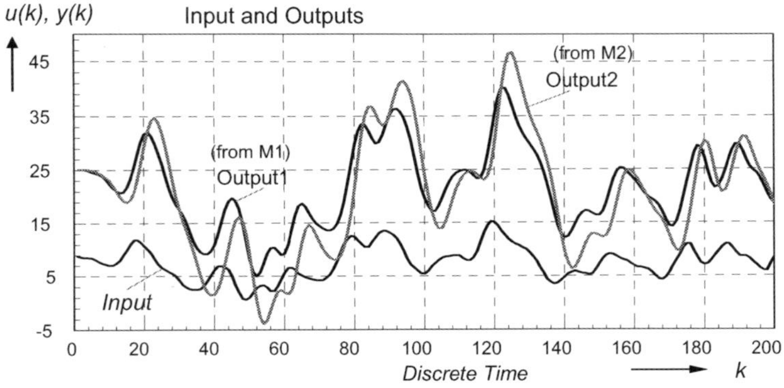

Fig. 3. Dynamic Responses from the two CER Models in Fig. 2.

3 Off-Line Identification of the CER Dynamic Models

The off-line identification problem arises where a preliminary defined number of n strength degrees $f(i),\ i=1,2,...,n$ of the CER-function have to be identified based on a preliminary saved collection of M input-output experimental data: $\{u(k),d(k)\},\ k=1,2,...,M$. During the Identification, the mean squared error (MSE) between the measured output $d(k)$ and the estimated output $y(k)$ is minimized, as follows:

$$MSE = \frac{1}{m}\sum_{k=n+1}^{M}\{d(k)-y(k)\}^2 = \sum_{k=n+1}^{M}\{d(k)-[y(k-1)+\sum_{i=1}^{n}\Delta u(k-i)f(i)]\}^2 \rightarrow \min \tag{5}$$

The actual number of data used for the evaluation is: $m = M - n$. Then the implementation of the Least-Squares Estimation Algorithm, as in [5,6] leads to the following linear system of equations:

$$\mathbf{A}_{m\times n}\,\mathbf{f}_{n\times 1} = \mathbf{d}_{m\times 1} \tag{6}$$

where $\mathbf{d}$ is a vector-column $\mathbf{d}_{m\times 1} = [\Delta d(n+1), \Delta d(n+2), ... , \Delta d(M)]^T$ of the measured output increments: and the vector-column $\mathbf{f}$ with size n contains the unknown strength levels of the CER-function: $\mathbf{f}_{n\times 1} = [\ f(1)\ \ f(2)\ \ ...\ \ f(i) ...\ f(n)]^T$.

Then the solution of the Identification solution, based on the Least-square estimation, can be written as:

$$\mathbf{f} = (\mathbf{A}^{\mathrm{T}}\mathbf{A})^{-1}\mathbf{A}^{\mathrm{T}}\mathbf{d}\ , \tag{7}$$

where the rectangular matrix $\mathbf{A}$ has the following structure:

$$\mathbf{A}_{m\times n} = \begin{bmatrix} \Delta u_{n+1}(k-1) & \Delta u_{n+1}(k-2) & \ldots & \Delta u_{n+1}(k-n) \\ \Delta u_{n+2}(k-1) & \Delta u_{n+2}(k-2) & \ldots & \Delta u_{n+2}(k-n) \\ \ldots & \ldots & \ldots & \ldots \\ \Delta u_M(k-1) & \Delta u_M(k-2) & \ldots & \Delta u_M(k-n) \end{bmatrix} \quad (8)$$

The above off-line Identification scheme could be useful for creating a *normal operation model* for a pair of parameters, based on a collection of reliable data about the normal operation of a system.

4 Jacobian Based Computational scheme of the CER Dynamic Model

From a computational viewpoint, the above described incremental CER model can be considered as a numerically calculated *linearized Jacobian* [1,2], as follows:

$$\Delta y(k+1) = \mathbf{J}(k+1)\Delta\mathbf{u}(k) \quad (9)$$

where:

$$\mathbf{J}(k+1) = [J_1(k+1), J_2(k+1), \ldots J_i(k+1), \ldots, J_n(k+1)] \quad (10)$$

is the ***Jacobian*** in the form of a $(1\times n)$ row-vector consisting of the following partial derivatives:

$$J_i(k+1) = \partial y(k+1)/du_i(k) = f(i),\ i=1,2,\ldots,n \quad (11)$$

The $(n\times 1)$ column-vector $\Delta\mathbf{u}(k) = [\Delta u_1(k), \ldots, \Delta u_i(k), \ldots, \Delta u_n(k)]^T$ containing all the *change-of-inputs* has the following elements:

$$\Delta u_i(k) = \Delta u(k-i+1) = u(k-i+1) - u(k-i),\ \ i = 1,2,\ldots,n \quad (12)$$

From (3), (9) and (11) it is seen that the elements of the CER-function: $f(1), f(2), \ldots f(i), \ldots, f(n)$ are represented now as linearized elements of the ***Jacobian***. The computational scheme of the CER model based on the ***Jacobian*** calculation is shown in Fig. 4.

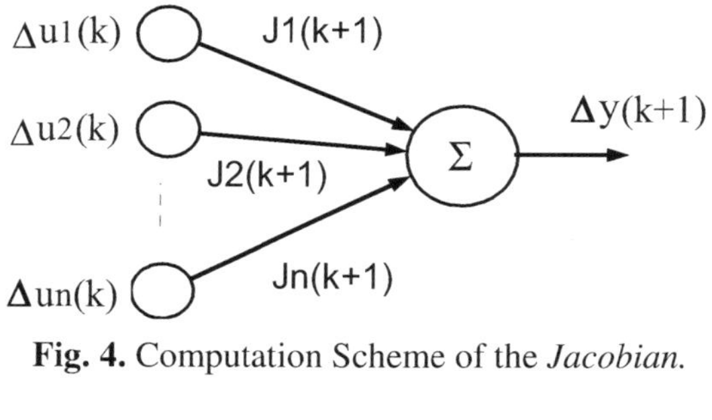

Fig. 4. Computation Scheme of the *Jacobian*.

It is clear that if the computation (9) is carried out at each sampling period, then $\mathbf{J}(k+1)$ will serve as a real time approximation of the actual non-linear *Jacobian*.

5 Algorithm for Real Time Identification of the CER Dynamic Model

During the operation of the system, at each sampling period T_s the observation data for the current measurements are taken and saved into a buffer data base with usually fixed length. This allows us to create *real-time* CER dynamic models for all (or selected) pairs of parameters. These models will be updated at each sampling and will serve as *current operation models* for the monitored system. Obviously each *current operation model* is a kind of variable model that reflects the current system condition in the presence of disturbances and/or faults to the system. In such way it will be an indicator for the current system performance, when compared to the *normal operation model* (obtained by the off-line identification) for the same relation.

In order to update the CER dynamic model at each sampling period, a *real-time* updating algorithm should be used, such as *Widrow-Hoff's* learning rule (Least Mean Squares *LMS* Method) [1,2] or weighted recurrent least-squares algorithm (WRLSA, Kalman Filter) [2,3] or possible others.

In this paper we use the *Widrow-Hoff's* learning rule in order to recursively update the evaluation $\hat{\mathbf{J}}(k+1)$ of the ***Jacobian*** (10).

The recursive computation scheme is explained in the following steps:

Initialization Step.

Calculations start when have measured the set of the first $n+2$ input-output data, as follows:

$\{u(i), d(i)\}$, $i = 1,2,\dots, n+2$ and set: $k = n+1$. Then the initialization step includes the following calculations:

- Calculate the initial n increments (*change-in-inputs*) from (12).
- Calculate the first *change in-output* as follows:

$$\Delta y(k+1) = d(k+1) - d(k) \tag{13}$$

- Set the first old evaluation of the ***Jacobian*** as follows:

$$\hat{\mathbf{J}}(k+1)_{old} = \mathbf{J}_0 \tag{14}$$

where the initial ***Jacobian*** $\mathbf{J}_0$ usually contains random elements or is defined by using some kind of preliminary human operator knowledge.Then, the updating of the ***Jacobian*** is performed for each time sampling: $k > n+2$.

Step 1. Compute the new evaluation of the ***Jacobian***. It is done by the *Widrow-Hoff's* updating rule that is written in our particular case as follows:

$$\hat{\mathbf{J}}(k+1)_{new} = \hat{\mathbf{J}}(k+1)_{old} + \alpha \frac{[\Delta y((k+1) - \hat{\mathbf{J}}(k)\Delta \mathbf{u}(k)]\Delta \mathbf{u}^T(k)}{\Delta \mathbf{u}^T(k)\Delta \mathbf{u}(k)} \tag{15}$$

where $0.0 < \alpha < 2.0$ is the *learning rate*. It is well known that if α is large, the learning process and the convergence are faster, but is too large α may lead to instability in the learning and subsequent inaccurate identification.

Step. 2. Update the old evaluation of the ***Jacobian*** with the new one, i.e.

$$\hat{\mathbf{J}}(k+1)_{old} \leftarrow \hat{\mathbf{J}}(k+1)_{new} \tag{16}$$

Step. 3. Perform a new sampling: $k \leftarrow k+1$ and measure the new *input-output* pair of data: $\{u(k), y(k)\}$ Then go to Step 1. for the new evaluation.

Normally there is no stop criterion in this *real-time* model evaluation procedure since it is used for the purpose of *real-time* monitoring of the operation status of the system (to extract a sequence of *current operation models*). However when a case of "improper learning" occurs (i.e. big oscillations with increasing amplitude) the learning could be stopped and resumed with different learning rate α.

6 Simulations Results for System Performance Evaluation

The following simulations by using the above algorithm for *real-time* Identification of a test dynamical system demonstrate the detection of a sudden change in the dynamics of the system. This could be an indicator that a sudden change-of-load or a malfunction in the system operation has occurred.

First, we suppose that the CER Model *M1*in Fig. 2. represents the normal operation of the system the CER model *M1* in te same figure corresponds to some abnormality in the system operation (slower dynamics with a bigger gain).

Then we simulate a case in which a sudden change in the system dynamics (from model *M1* to model *M2*) occurs at discrete time *100*, as shown in Fig. 5.

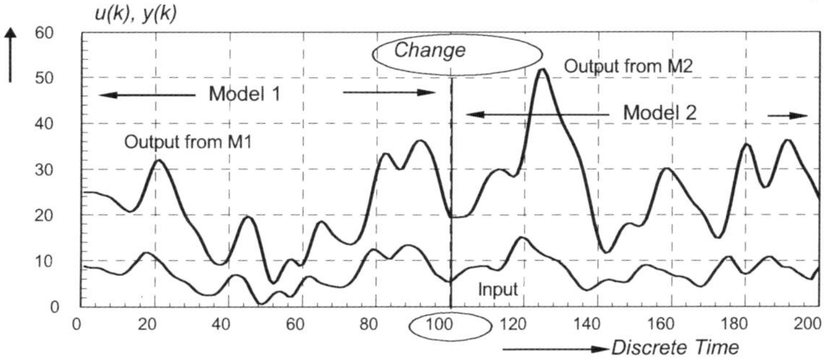

Fig. 5. Dynamic System Behavior after Sudden Change in the Operating Conditions

As seen in Fig. 5., it is impossible to discover directly the change in the system behavior. However, the *real-time* Identification by *Widrow-Hoff's* algorithm, as shown in Fig. 6. clearly indicates that system parameters suddenly change at around discrete time *100*, because of some system malfunction.

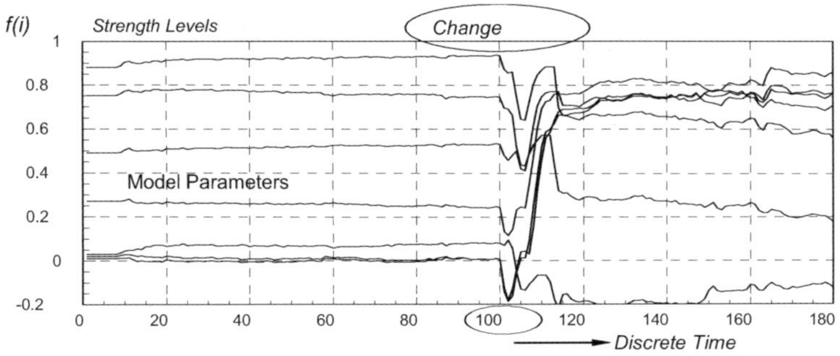

Fig. 6. Real Time Identification that Discovers the abrupt Change in System Dynamics.

7 Conclusions

Two computation schemes: for *off-line* and *real-time* identification of cause-effect relation models are described in the paper. They are used to create the *normal* and the *current operation models* of the system. A repetitive comparison of these models (rather than direct comparison of the measured data) can be used for real time evaluation of the system operation, thus discovering the system abnormality. Further research is aimed at improving the robustness of the identification against errors that would lead to more plausible fault detection.

References

[1] K.J.Astrom and B. Wittenmark, Adaptive Control, Reading, MA: Addison-Wesley, 1989.

[2] D. Filev, Applied Intelligent Control – Control of Automotive Paint Process, CD-ROM Proc. of the FUZZ-IEEE 2002 in the *World Congress of Computational Intelligence* (WCCI-2002), Honolulu, Hawaii, May 12-17, 2002.

[3] M. Hojati and S. Gazor, Hybrid Adaptive Fuzzy Identification and Control of Nonlinear Systems, *IEEE Transactions on Fuzzy Systems*, Vol. 10, No. 2, April, 2002, pp.198-210.

[4] C.-H. Lee and C.-C. Teng, Identification and Control of Dynamic Systems Using Recurrent Fuzzy Neural Networks, *IEEE Transactions on Fuzzy Systems*, vol. 8, no. 4, 2000, pp.349-366.

[5] G. Vachkov and T. Fukuda, Cause-Effect Relations Based Dynamic Modeling and Its Application to Control, *Communications in Information and Systems* (CIS), Vol. 1, No. 4, Dec. 2001, pp. 407-432.

[6] G. Vachkov and N. Christova, Cellular Dynamic Models with Built-in Causal Relation Functions, Cellular Dynamic Models with Built-in Causal Relation Functions, CD ROM Proc. of the *IEEE Int. Symposium on Intelligent Systems*, IS-2002, Sep. 10-12, 2002, Varna, Bulgaria, pp. 33-38.

A Multi-layered Immune Inspired Machine Learning Algorithm

Thomas Knight and Jon Timmis

Computing Laboratory, University of Kent, Canterbury, Kent, CT2 7NF, United Kingdom.
Email {tpk1, jt6}@kent.ac.uk

Abstract. Artificial Immune Systems (AIS) have recently been proposed as an additional soft computing paradigm. This paper proposes a new multi-layered unsupervised machine learning algorithm inspired by the vertebrate immune system. The algorithm has been tested on benchmark data and has shown a great deal of potential for data reduction and clustering tasks. This paper presents an overview of the algorithm, drawing analogies to the vertebrae immune system where appropriate. Results are presented for three data sets and observations are made about the potential for adapting the algorithm for a continuous learning paradigm.

1 Introduction

Soft computing has been described as computational systems that exploit tolerance for imprecision, uncertainty, partial truth and approximation [1]. Such systems include *artificial neural networks, fuzzy systems, evolutionary algorithms* and *probabilistic reasoning. Artificial Immune Systems* (AIS) have recently been proposed as an additional soft computing paradigm [2]. It has been argued that AIS exhibit similar characteristics to other soft computing paradigms and therefore can be used to complement and augment existing soft computing techniques. AIS can be defined as *adaptive systems, inspired by theoretical immunology and observed immune functions, principles and models which are applied to problem solving* [3]. Preliminary results on three numerical data sets are given and future directions are discussed. This paper argues that this new algorithm adheres to the soft computing philosophy and therefore proposes this as a novel algorithm with regard to both soft computing and immune inspired algorithms.

2 Context of Work

Building on work in [4], a motivation for the development of an algorithm has been drawn which is *"To develop an immune inspired machine learning algorithm capable of continuous learning"*. In addition to being able to operate in a continuous environment, the algorithm is also required to perform data reduction, and cluster identification

The algorithm described in this paper was originally devised with the goal of continuous learning in mind. However, after initial development it was discovered that the algorithm was performing well for static learning (i.e. learning on data that doesn't change with time). It is these results that are presented in this paper, as part of ongoing work to create a truly continuous learning algorithm.

Previous work

The natural immune system is interesting from a computational perspective as it is *distributed, diverse, self-organizing, dynamic* and capable of *recognition, learning* and *memory* [5]. Based on these perspectives, an immune inspired a data mining algorithm, AINE (Artificial Immune Network) was proposed by [6]. The motivation behind this work was to abstract processes and characteristics from the immune system, more specifically the immune network theory [7] to develop a novel *unsupervised machine learning* algorithm. The core mechanism captured by AINE is the co-stimulation and suppression interactions between B-cells, inspired by Jerne's Immune Network theory [7], and the resultant response of these B-cells to antigenic stimulation. AINE was applied to the Fisher Iris data set and was reported to perform well on this benchmark task [6]. It was anticipated by the authors that this algorithm would be suitable for continuous learning with initial work showing that stable populations within the networks could be achieved [6]. However, more recent work has shown that the networks suffer strong evolutionary pressure and converge to the strongest class represented in the data [4]. Whilst this is an interesting development that could potentially be applied to optimization, with regard to continuous learning viewpoint, it would not be a preferential outcome: it is clearly more desirable if all patterns persist over time rather than the strongest.

Parallel to the work in this paper a form of the original algorithm that is capable of finding stable clusters even if the input patterns are removed has been developed [8]. Here, a different population control mechanism based on exponential decay of stimulation levels has been used. The work employs an improved resource allocation mechanism and different stimulation level calculations for B-cell interaction. However, one drawback of this work is that there is poor data compression in the final networks. These observations prompted a step back from the existing work to re-evaluate the approaches taken. It was noted that a more holistic approach might provide a better solution in the search for an immune inspired data mining algorithm capable of continuous learning.

Immunology

Before discussion on the proposed algorithm can be undertaken, it is necessary to provide a brief overview of the relevant immune system mechanisms that have served as inspiration to this work. Our bodies are under constant attack by viruses and bacteria that are foreign to our bodies, these are known as *antigens*. As part of the adaptive immune system a B-cell is capable of recognizing free antigens without the assistance of MHC (Major Histocompatibility Complex) molecules. B-cells have surface receptors known as *antibodies* that are antigen specific (will only respond to specific antigen). On encountering an antigen the B-cell will become stimulated and will proliferate and differentiate into plasma cells that secrete antibody molecules in high volumes. These neutralize the antigen and lead to the removal of the antigen from the system. Some of these activated B-cell and T-cells differentiate into memory cells which remain in the immune system for long periods of time and help produce a more rapid response in the future. Previous work [6] has concentrated on the interactions of B-cells only as described in the immune network theory, however this paper proposes a new algorithm that draws its inspiration from not only B-cells, but the antibodies and memory cells that they produce. Conceptually it is possible to divide the immune system into processes, with each process having its own specific interactions. Thus, a multi-layered immune system is proposed in the following section.

In this section, a brief description of the processes in the *adaptive immune system* have been described and therefore if the reader wishes to know more about the immune system, then they are directed to [3], which provides a more detailed description.

3 A Multi-layered Artificial Immune System

A great deal of work in the field of AIS has been in developing systems based on the interactions of B-cells. Both clonal selection and the immune network theory have been tried with some success [11, 12, 6]. As an addition to these existing techniques, this paper proposes a multi-layered algorithm that is inspired by the clonal selection theory and incorporates a feedback mechanism much like the co-stimulation that is seen in the immune network theory model. The proposed algorithm also incorporates the idea of a *primary immune response* to handle the event of unknown data being presented to the system. The primary immune response is the initial response provided by the immune system to previously unknown antigenic patters and helps build a B-cell repertoire that can recognize these patterns.

The proposed algorithm is divided into three layers; A *free-antibody layer,* a *B-cell layer* and a *memory cell layer*. In order to simplify the algorithm T-cell, MHC and APC (Antigen Presenting Cell) interactions are not directly considered within the algorithm, however the signaling action that they employ has been incorporated in part to the feedback mechanism between the free-antibody layer and the

B-cell layer. The first level of interactions an antigen has is with the free-antibody layer. At the next level we consider the antigen interacting with the B-cells which occurs in the B-cell layer. Finally the memory cells that are produced by the B-cells interact in a memory cell layer. A more detailed description of each layer is given in the following sections. Throughout the description of the new algorithm training data will be referred to an antigen, and the free antibodies, B-cell and memory cells all have data items which exist in the same shape space. Both antigenic data and antibodies can be considered as the representation in the system, i.e. feature vectors.

Free-antibody Layer

The free-antibody layer is the first layer that the antigenic data is presented to. This layer performs a more generalized search functions than that of the B-cell layer. This layer takes a single antigen as its input, which is then presented to *p* percent of the free-antibody population where the free-antibodies are given the chance to bind to the antigen. Binding occurs if the inverse of the Euclidean distance (known as the *affinity*) is less than the free-antibody affinity threshold (σ_{fab}). The free-antibodies that bind to the antigen are then removed from the free-antibody population. The free-antibodies that are attached to the antigen are key to the stimulation of B-cells in the B-cell layer.

B-cell Layer

The function of the B-cell layer is to perform a more specific search of the antigenic data. Its role is also to learn and generalize the patterns found in the antigenic data. The antigen is presented to each B-cell in the B-cell population until the affinity binding value is below the pre-defined affinity threshold for binding (σ_{bcell}). The antigen then stimulates the B-cell according to equation 1.

$$s_t = \left(\sum_{t=0}^{t-1} a_t\right) + a_t \qquad (1)$$

where s_t is the stimulation level of the B-cell at time *t*, and a_t is the stimulation received (the number of attached free-antibodies) from the antigen at time *t*. Once stimulated, if the stimulation level is above the stimulation threshold (ζ) the B-cell will produce a mutated clone, a memory cell (which is an exact copy of the data vector) and a number of free antibodies (determined by equation 2).

$$n = |(d_{\max} - d_{(a,b)}).K| \qquad (2)$$

where *n* is the number of free-antibodies produced, d_{max} is the maximum possible distance in the data space between two cells, $d_{(a,b)}$ is the affinity between the antigen (*a*) and the B-cell (*b*) and *K* is a constant. The cloned B-cell is mutated using *affinity proportional mutation*. The size of the mutation is proportional to the affinity between the B-cell and the antigen that caused the mutation. The data vector is mutated so that the point in space that the cloned B-cell represents moves towards the space that the antigen occupies (equation 3).

$$V_e = V_b + (d_{(a,b)} \cdot (V_a - V_b)) \tag{3}$$

where V_e is the new data vector for the clone, V_a is the data vector for the antigen that caused the response, V_b is the data vector of the B-cell that created the clone and $d_{(a,b)}$ is the affinity between the B-cell and the antigen. The memory cell is then added to the memory-cell layer.

Once a B-cell has responded to the current antigen in the layer, the antigen is no longer presented to the rest of the population. The reasoning for this is that the immune system does something similar when an antigenic substance is detected. The reaction is local to the point of infection and does not incur a reaction by the whole system. From a computational perspective, this is also a valid reaction as this potentially reduces the time complexity of the algorithm.

If the antigen is presented to the whole B-cell population without eliciting a reaction then a *primary response* occurs. The data vector contained in the antigen is incorporated into a new B-cell and added to the B-cell population. This response allows new data to be incorporated into the algorithm and then gives it a chance to survive if the data is a repeating pattern. This allows learning to occur without an initial population of B-cells.

Memory Layer

The memory layer maintains the generalised patterns that the B-cell and free-antibody layers have produced. The cells that are stored in this layer can have no effect on the learning process and the layer itself has no outputs. For each memory cell presented to the memory layer, the following processes take place. The new memory cell is presented to all of the current cells in the memory population and the Euclidean distance, $d_{(n, mi)}$ between the new cell and the existing cells is measured. Where n is the new memory cell and m_i is the ith memory cell ($I = 0, \ldots, M$) and M is the population of memory cells in the memory layer. If the memory cell with the smallest affinity $d_{(n,mi)min}$ to the new cell, has an affinity to the new cell of less than $\square_{mem}$ then the new memory cell is considered to represent data that is already known, the existing memory cell is stimulated (see Population control section) and then the new cell is deleted. However, if $d_{(n,mi)min}$ is greater the $\square_{mem}$ then it is considered the new memory cells represents previously unknown data, and is therefore added to the memory population, M.

If the memory cell is inserted into the memory cell population, it is then presented to all memory cells in that population and connections are made if the affinity $d_{(n,mi)} \leq \square$, where $\square$ is the memory cell connection threshold. These connections have no effect on the learning process and are there simply to help understand the structure of the memory layer.

Population Control

The population control mechanism acts on all layers of the algorithm at the end of each iteration, to remove cells that have not had any form of *interaction* within a given time-frame defined by □. The term interaction is defined for each type of cell as the following; *free-antibody* – a free antibody is considered to have interacted when it attaches to an antigen, *B-cell* – a B-cell is considered to have interacted when it has received stimulation from an antigen, *memory cell* – a memory cell is considered to have interacted when it has received stimulation from a new memory cell. When an interaction occurs in a cell, the iteration number is recorded, t_i.

The population control mechanism operates only once on each iteration, for the iteration based algorithm, at the end of the iteration. An iteration is considered to have been a completed when each individual of the entire training population has been presented to the algorithm once.

4 Results

Initial results have been determined from benchmark data. The algorithm has been tested on 3 numeric data sets; Simulated Data (3 classes, 2 dimensions, 30 items), 3Circle data (3 classes, 2 dimensions, 600 items) and the Doughnuts data (2 classes, 3 dimensions, 221 items). A plot of the 3Cirlce data can be seen in Figure 1 and the 3D data set known as the doughnut data originates from [11]. The algorithm was run for 500 iterations using a default set of parameters. The parameters used were empirically defined during the initial development of the algorithm. The experiments were repeated 50 times and the population sizes for each layer were recorded on each iteration. Earlier, a set of criteria that the algorithm should meet were proposed, they were that it should be; capable of identifying clusters within data, data simplification, adaptation to new data and continuous learning. The following experiments assess the performance of the algorithm in respect to these criteria

Table 1. Average Mean ($\overline{x}$) and Standard Deviation (σ) of populations' sizes within the three layers. Values have been rounded to the nearest integer.

Layer	Simulated Data		3Circle Data		Doughnuts Data	
	□	$\overline{x}$	□	$\overline{x}$	□	$\overline{x}$
Free-antibody	14	162	77	1143	77	1441
B-cell	4	52	43	112	18	388
Memory	1	5	4	50	5	75

The mean population size for each data set was calculated and the standard deviation recorded. These results are presented in Table 1. Figure 2 shows the resulting pattern in the memory layer after 5 iterations of the algorithm. It can be seen that after a short period of time the pattern is visually representative of the original data. By looking at the mean number of cells we can see that the algorithm achieves good compression rations (92% for the 3 circle data and 66% for the doughnut data). The tests have also shown that over 500 iterations the system does not suffer from loss of information as seen in the AINE system.

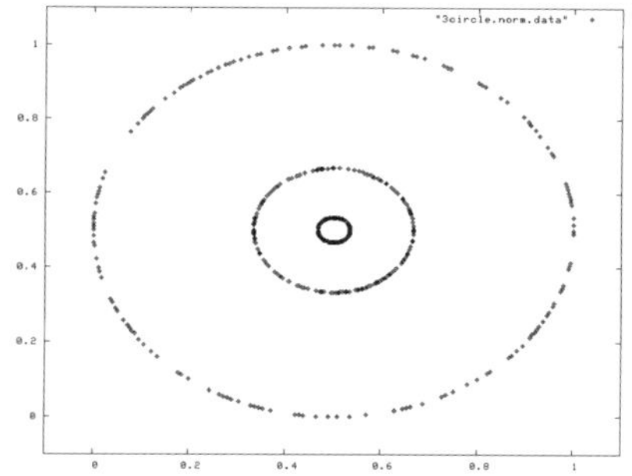

Figure 2: The 3Circle training data set (600 items) in normalized form.

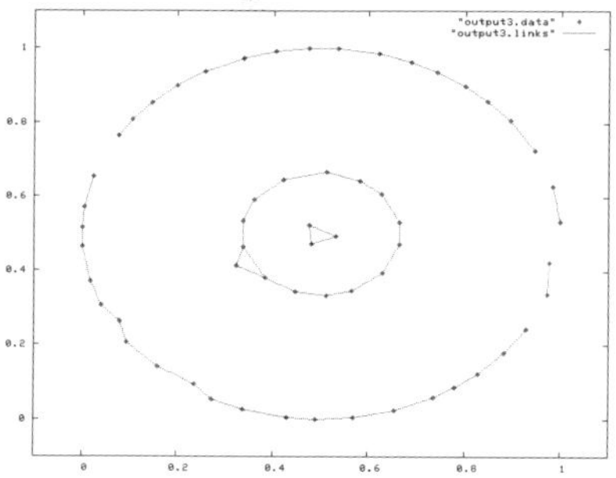

Figure 2: Patterns in the memory layer evolved by the new algorithm after 5 iterations (58 items).

With regard to data simplification, the algorithm can be deemed to perform well, with suitable compression ratios in the memory layer. The algorithm has also shown, through visual inspection of the memory layer output, that it is capable of identifying clusters within data. As the algorithm can be run using a starting population of zero B-cells, it can be said that it is capable of adaptation to new data. Continuous learning is the only criteria not met; however, as explained earlier, the algorithm has shown to be capable of learning on static data sets, and therefore the task of applying it to continuous data has been reserved for future work.

5 Conclusions

Previous work has shown that aspects of the human immune system lend themselves to solving computational problems such as data mining. Previously, an artificial immune system for data mining (AINE) [6] was created and was used with some success on benchmark data. However, more recent work has shown that AINE does not perform as the authors intended it to. It was identified that there were problems concerning the stability of networks produced and how an AIS that produces stable networks is desirable. Parallel to this work [8] has re-designed AINE and succeeded in producing the performance originally intended. The relative youth of the field also has identified the there may well be benefits from trying different solutions to those that already exist. To this end, a new algorithm has been devised that seeks to produce reproducible results and explores new ideas. It can be said to adhere to the soft computing philosophy as the algorithm exploits the imprecision, approximation and uncertainty of the natural immune system.

6 Future Work

Future work will comprise of further evaluation and testing of the algorithm on benchmark data sets, such as the Fisher Iris data and Wisconsin Breast Cancer data. Comparisons with other techniques will be made and a full investigation of the parameters undertaken. Analysis of the clusters generated will also be undertaken to establish accuracy of the results.

Acknowledgements

Thomas Knight would like to thank SUN Microsystems for their continual financial support for his PhD studies.

References

1. Jin, Y., *What is Soft Computing*. 2002.
2. deCastro, L.N. and J.I. Timmis. *Artificial Immune Systems as a Novel Soft Computing Paradigm*. in *Soft Computing*. 7(7). 2003.
3. deCastro, L.N. and J.I. Timmis, *Artificial Immune Systems: A New Computational Intelligence Approach*. 2002a.
4. Knight, T.P. and J.I. Timmis, *AINE: An immunological approach to data mining*, in *IEEE International Conference on Data Mining*. 2001: San Jose, CA. p. 297-304.
5. Dasgupta, D., *Artificial Immune Systems and their Applications*. 1998: p. 3-21.
6. Timmis, J. and M. Neal, *A resource limited artificial immune system for data mining*. Knowledge Based Systems, 2001. **125**(C): p. 73-389.
7. Jerne, N.K., *Towards a Network theory of the Immune System*. Annals of Immunology, 1974. **125c**: p. 373-389.
8. Neal, M., *An artificial immune system for continuous analysis of time-varying data*, in *1st International Conference on Artificial Immune Systems (ICARIS)*. 2002: Canterbury, UK. p. 0-2.
9. Tizzard, I.R., *Immunology: An Introduction*. 4th Edition ed. 1995: Saunders College Publishing.
10. Burnet, F.M., *The Clonal Selection Theory of Immunity*. 1959.
11. deCastro, L.N. and F.J.V. Zuben, *Data Mining: A Heuristic Approach*. 2001: p. 231-259.
12. Hunt, J.E. and D.E. Cooke, *Learning using an artificial immune system*. Journal of Network and Computer Applications, 1996. **19**: p. 189-212.

Using Plausible Reasoning for Developing Intelligent Systems

Bita Khandzad*, Elham Ashoori*, Farhad Oroumchian**, Babak Nadjar Araabi*

* Faculty of Engineering, University of Tehran,
email: {khandzad, ashoori, araabi}@ece.ut.ac.ir
** University of Wollongong-Dubai,
email: foroumchian@acm.org

Abstract. In this paper, the application of plausible reasoning in different domains has been investigated. The theory of plausible reasoning has been introduced on early 1990s. However after that there has not been a significant work to show its applicability to real world problems. After several years of experimentation in different fields such as Information Retrieval, Information Filtering and Intelligent Tutoring systems, we feel confident that this theory could be applied to real world problems instead of or in conjunction with other theories such as Fuzzy logic, Dempster-Shaffer theory of evidence etc. Here we describe several of our implementations and experiments with this theory.

1 Introduction

For approximately 15 years, Collins and his colleagues have been collecting and organizing a wide variety of human plausible inferences made from incomplete and inconsistent information. These observations led to the development of a descriptive theory of human plausible inferences that categorizes plausible inferences in terms of a set of frequently recurring inference patterns and a set of transformations on those patterns. According to the theory, a specific inference combines an inference pattern with a transformation that relates the available knowledge to the questions based on some relationship (i.e. generalization, specialization, similarity or dissimilarity) between them. The primitives of the theory consist of basic expressions, operators and certainty parameters. In the formal notation of the theory, the statement "coffee grows in the Llanos" might be written:

$$\text{GROWS-IN (Llanos) = Coffee, } \gamma\text{=1.0}$$

This statement has the *descriptor* GROWS-IN applied to the argument Llanos and the *referent* Coffee. The certainty of the statement (γ) is 1.0 since it declares a fact about the Llanos. The pair descriptor and argument is called a *term*. Expressions are terms associated with one or more referents. All descriptors, arguments and referents are nodes in (several) semantic hierarchies. Any node in the semantic network can be used as a descriptor, argument or referent when appropriate. There are many parameters for handling uncertainty in this theory. However there is no complete agreement on their computational definitions and different computer models of the theory have implemented them in different ways. The complete definitions of the most important ones are reported in [1].

The theory has a rich set of transforms and inferences, which provides a means of converting one statement to another and inferring unknown concepts from the known ones. Interested readers are referred to the references [1] and [2] for an in depth explanation of the theory.

We believe this theory has many practical applications in developing intelligent systems. We have successfully applied it to several domains. Section 2 describes our experiments in Information Retrieval, which is a challenging field to develop knowledge-based systems. Section 3 discusses the application of plausible reasoning on developing filtering agents. At the end, the architecture and experiments with an ITS system based on this theory are reported.

2 Document Retrieval by Plausible Reasoning

There are four elements in a logic based IR system. Those are the description of documents, the representation of queries, a knowledge base containing domain knowledge and a set of inference rules. Here, a document is retrieved only if its partial description can be inferred from a query description. Thus the retrieval process is expanding a query description by applying a set of inference rules continuously and inferring other related concepts, logical terms and statements until locating a document or documents which are described partially by those concepts or logical terms or statements.

2.1 Representing a Query as an Incomplete Statement

A query can be represented as an incomplete logical statement in which the descriptor is the keyword REF (reference) and its argument is the subject in which the user is interested. The referents of this statement i.e. the desired documents, are unknown. A typical query in logical notation will have the form:

$$\text{REF (A-Subject)} = \{?\}$$

Queries with more than one concept or term can be represented as a set of simple queries and the system can retrieve a set of references for each one separately and then reexamine the sets by combining the confidence on references, which are members of more than one set. Then the sets can be joined and the elements of the resulting set can be sorted according to their confidence values.

2.2 Document Retrieval through Plausible Inferences

Many of the inferences of the core theory such as argument-based transforms (generalization, specialization, similarity, dissimilarity) have been implemented and experimented with in a system called PLIR5][7].

For reasons of brevity only one of them, the specialization (SPEC-) based referent transform is described in detail here.

The SPEC-based referent transform in the core theory is a strategy to utilize the part-of and kind-of relationships to find other referents for a given statement. In the IR situation this strategy could be applied to the concepts found relevant in earlier stages of retrieval in order to find other relevant concepts and their associated references.

The example in Figure 1 is based on the query number 7 from the CACM collection, where a user requests references for "distributed algorithms" and expresses interest in "synchronization by using message passing" among other things. Document number 3128 is about "synchronization of concurrent processes".

The inference starts by identifying "concurrent programs" as an example of distributed algorithms. Concurrent processes are specialization of "concurrent programs" in the context (CX) of "synchronization", which is of interest to the user. The dependency between "synchronization algorithms" and "distributed algorithms" can be established by looking at their co-occurrence in the collection. In this case they have appeared in the same documents in 10% of cases. Line 5 concludes that concurrent process is also an example of distributed algorithm. Line 6 identifies document number 3128 as a reference for concurrent process, and therefore it is a reference for distributed algorithm.

Query: REF (algorithm (distributed))= {?}

1- Algorithm (distributed) = {**concurrent_program**}

: $\gamma 1 = 0.30$

2- **Concurrent_process** SPEC **concurrent_program**
in **CX** (**algorithm**, synchroronization (**algorithm**))

: δ 1= 0.40, A 1 = 0.5

3- Syncbronization (algorithm) <----> distributed (**algorithm**)

: $\alpha 1 = 0.10$, γ **2**= 0.6

4 - **Distributed** SPEC **distributed**

: δ 2 =l. 0, A2 = 0

5 - Algorithm (distributed) = (**concurrent_process**)

: γ **3** = 0.03

Fig.1 Finding References by SPEC-based Referent Transform

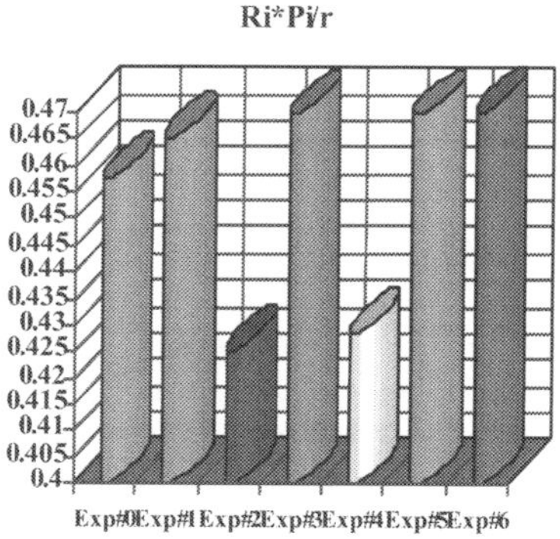

Fig 2 A Comparison of overall performance on all queries for the experiments with both global and local combination approach

2.3 Experiments

Several experiments have been conducted using the CACM collection to investigate the effectiveness of plausible inferences. 48 queries out of 64 standard CACM queries were used in these experiments. The vector space model with tf.idf weighting also was implemented and used as benchmark. For global combination approach, two different methods were used. In global-pessimistic approach, the highest value of different certainty values was considered. On the other hand, the global-optimistic approach considered the inferences to be independent of each other and combined the confidences as independent probabilities. The local combination approaches combined the confidence values in three stages. The Dempster-Shafer (DS) theory of evidence [3], the weighted addition and the Smets rule [11][12]have been investigated in combining the evidences provided by plausible inferences in the first stage. In the second level, averaging all the different degrees of relevance has been used to estimate the overall degree of relevance to the whole query. Table 1 shows different experiments for the local approach. Interested readers are referred to8for the complete details of the local approach experiments.

The experiments demonstrated that PLIR outperforms the Vector Space retrieval and the Dominance weight is better than the tf.idf weights. But there is no difference between the performance of global approaches [5][7]. Other experiments focused on comparing local & global approaches. In table 1, the experiment Exp#0 is the global approach.

Figure 2 shows a comparison of overall performance on all queries. In all experiments with local approaches the ranking has improved over the global approaches. Each one of experiments produced more ranks with lesser number of documents in each rank. In these experiments Exp#6 has produced better results.

In general, the local approaches for combining confidence values produce more qualitative ranking than global approaches. Since PLIR is a high precision retrieval system that retrieves few documents, it seems reasonable to believe the combination of PLIR with DS theory and a good query-based threshold could be useful for information filtering. Since PLIR allows adaptation to user needs therefore threshold can be learned from user relevance judgments. Then this approach could be more useful for adaptive information filtering.

3 Document Filtering by Plausible Reasoning

Following the definition of retrieval process in 5], filtering process is defined as expanding a query description by applying a set of inference rules continuously on the description of the query and inferring other related concepts, logical terms and statement while considering the user's preferences and feedback until locating a document that is described partially by those concepts or logical terms or statements. An example of a modified inference rule that can be used in information filtering is the "document filtering using a GEN-based argument transform" is depicted in figure 3.

REF (d (a)) = {?}

1- d (a') = r $\gamma 1$

2 - a GEN a' in CX (a , D (a)) $\delta 1, A1$

3 - D (a) < ---------------> d (a) α, $\gamma 2$

4 – d (a) = r

$\gamma 3$ = F1 ($\gamma 1$, $\delta 1$, A1, α, $\gamma 2$) from 1,2,3

by Gen-based Argument transform ,B1

5 – REF (r) = {doc#x} $\delta 2, A2$

Fig 3. Document Filtering By Argument Based Transform

Table 1. Different Experiment for Local Approach

Exp#	$m_i(doc_j)$	m_i (T)	$m_i(\phi)$	m (doc_j)
Exp#1	c_i	$1-c_i$	0*	$m1 \otimes m2 \otimes m3 \otimes ...m5$
Exp#2	$w_i * c_i$	$w_i * (1-c_i)$	< >0*	$m1 \otimes m2 \otimes m3 \otimes ...m$
Exp#3	$w_i * c_i$	1- $w_i * c_i$	0*	$m1 \otimes m2 \otimes m3 \otimes ...m$
Exp#4	$w_i * c_i$	$w_i * (1-w_i * c$	< >0*	$m1 \otimes m2 \otimes m3 \otimes ...m$
Exp#5	$w_i * c_i$			m1+m2+m3+..+m5
Exp#6	$w_i * c_i$	$w_i * (1-c_i)$	< >0**	$m1 \otimes m2 \otimes m3 \otimes ...m$

*B*1 can be computed individually for each user. It will represent the point of view of a particular user over time about different types of the argument "a" retrieved by the system. Details on the description of the above parameters can be found in [5]. In information filtering we try to investigate the dominant inference rules and the weight of each inference type of plausible reasoning for individual users.

4 Intelligent Tutoring Systems by Plausible Reasoning

Intelligent tutoring systems (ITS) are software programs that support the learning activities. They use new means for education that changes the role of the human tutor or teacher, and may optimize it[9]. The Intelligent Tutoring systems usually start by collecting information about a student to determine the depth of his/her background knowledge, preferences and interests. Normally this can be achieved through a set of initial tests and communications with student [14]. ITSs save all the gathered information in a comprehensive student model. This model is consulted and sometimes updated through the tutoring process.

An important characteristic of an ITS that contributes to its effectiveness in teaching is the ability to adapt to individual students. It seems reasonable to assume that, an ITS that could do this task more precisely, could better satisfy students by minimizing the learning time and choosing the best curriculum and tutoring strategy. Human tutors and instructors use uncertain inferences to assess individual student's progress and best tutoring strategy to follow.

4.1 The Intelligent Tutoring Model

Here we present a unified solution for the ITS problem based on plausible reasoning. Our model consists of 4 sub-models: Course Knowledge Base Model, Student Model, Evaluation Model and Tutoring Model. Below each sub-model is described.

The Course knowledge Base Model

The course materials are represented in this model by a semantic network. The hierarchical nature of the course material (Skills and the Problems that produce the skills) is captured with part-of relations in this network. Each skill is composed of some problems and each problem covers some percentage of a skill. Each node contains many fields such as title, author, date, or text. It is reasonable to assume that all the problems that are assigned by the tutor, do not contribute equally to acquire the skill level designed for that problem. Different types of relations exist in the semantic network in order to support the above relations.

The Student Model

The student model is divided into two parts. One part represents the state of knowledge, which is based on the domain or course knowledge base. The other part represents the student's personal variables such as learning style and other information collected from student [6].

This division of the model into two subparts makes it possible to adapt the system based on two different sets of parameters. In the beginning, the user model will be filled with personal information about background and anything else that can be collected from the student. Then the course begins, as the student progresses and learns the skills, parts of the structure of the domain/course knowledge are copied into the student model. Therefore, system maintains a history of student progress [4].

The two different parts of the user model can be used in different ways by the inference engine. The course knowledge part mostly defines what to study next and the personal section mostly controls how the tutoring should proceed.

The Evaluation Model

The tutor needs to evaluate the apprehension level of a student before making a decision about how to proceed with the instruction. This decision could be letting a student to proceed to the next skill level, or to review previous skills or to work more on the current skill level. The quality of this decision is influenced by the quality of the evaluation of the situation by the tutor. Some information, such as the actual time a student has spent in a section, needs to be gathered and evaluated before guessing the apprehension level of the student.

Figure 4 depicts an argument-based transform that estimates the system's confidence on the actual time (T_a) spent on a problem by a student to be *enough* for comprehending that problem. The first line of the reasoning expresses the confidence of the system on the amount of $T_{i,j}$ time for the problem Pr_i by student St_j to be *Enough* (γ_1,ϕ). This estimation can be done from information gathered in the student model. The system's confidence on the time T_a being enough is influenced by the below two factors:

- The similarity of the *actual* time T_a to the *normal* time $T_{i,j}$ (second line)
- The relationship between the time needed for learning a problem and the time a *normal* student uses to study a *normal* problem (line4)

Therefore γ (the confidence) is a function of all certainty parameters involved as expressed in the figure 4.

The interesting thing is that the fourth line can be rewritten in a number of ways. In each case it represents a different method of estimating the certainty of the relationship between the time needed for the learning a problem and the time a *normal* student uses to study a *normal* problem. Then the tutoring system can utilize the values of these parameters to estimate the overall apprehension of the student on the current skill level.

The Tutoring Model

After the tutor has assigned a confidence to the apprehension of a problem by the student, this confidence can be translated into a confidence on the apprehension of a skill level (superset of the problem) and the confidence on the apprehension of a skill is translated into the confidence on the next best action [13].

Study_Time (Pr_i, St_j, $T_{i,j}$) = Enough	γ_1
T_a SIM $T_{i,j}$ in CX (Time, Learning (Time))	σ
T_a , $T_{i,j}$ SPEC Time	$\gamma_3 \rightarrow 1, \gamma_4 \rightarrow 1$
Learning (Time)$\leftrightarrow$ Study_Time (Problems, Students, Time)	
$\alpha \rightarrow 1, \gamma_5 \rightarrow 1$	

Fig. 4. The Inference for Estimating How Enough is The Actual Time Spent by Student.

In this system, the Tutoring model makes its decision on what could be the next best action on estimates that the evaluation model provides about how much the student has comprehended a concept, skill or section.

In our implementation of ITS all these decisions are carried through plausible inferences. Interested readers are referred to [6].

4.2 Experimental Results with PACT

For comparing the decisions of our ITS system with an actual tutor's actions, we have used the PACT Geometry Tutor system dataset The PACT corpus contains data about students' use of the PACT Geometry Tutor; a cognitive tutor for high school geometry problem solving. This data set describes in detail the actions of 40 high school students as they worked through the 3rd unit of the PACT Geometry tutor.

In these experiments, several methods of estimating learning parameters such as time, number of errors and number of hints has been investigated. For example, a learning speed rate of between –1 and 1 is assigned to each student that suggests how this student compares to an average student with respect to the time students spend on each problem.

Also different methods of calculating the certainty (%) of our estimation have been studied. As a system based on plausible reasoning, our ITS system has to assign a confidence degree to each of its actions or conclusions.

In general, the actions and judgments of our ITS system has been understandably close to that of the human tutor, except that it takes the students through the course much faster than the human tutor. This could be related to the way we estimate our certainty parameters. Figure 5 depicts the comparison of the ITS recommendations in two different modes with those of the human tutor. First ITS is used without its learning capabilities. In this mode it only uses the estimations done in the beginning of the course and what happens during the sessions does not affect the values of these parameters.

In the other mode, ITS uses it's learning capabilities and modifies its estimations based on the experience it gains from the students. As you can see in the fig-

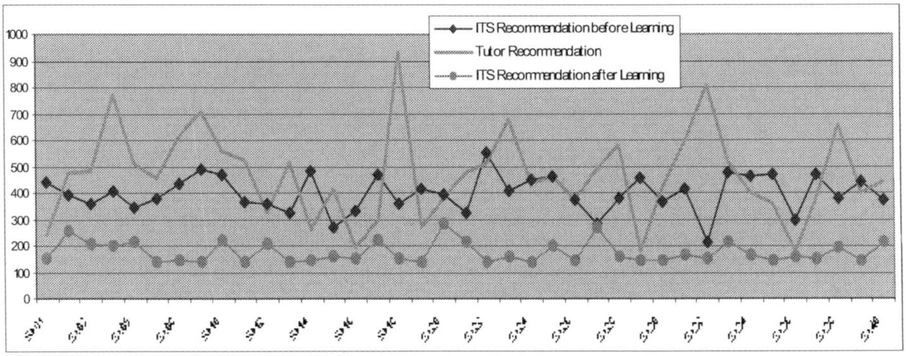

Fig. 5. Comparing Tutor Recommendation with ITS Recommendations

ure 7, learning causes ITS to help the students to go with fewer steps through the course material. Approximately for all of the students, the ITS completes the course in fewer steps.

5 Final Remarks

In this chapter, we have presented several applications based on the theory of plausible reasoning. In general it seems that plausible reasoning provides a viable vehicle for building intelligent systems in different domains. The theory of plausible reasoning is more general than many other well known theories such as Fuzzy logic, Dempster-Shaffer, etc.

In this article, briefly the PLIR system, our implementation of an IR system based on plausible reasoning and related experiments have been described. In PLIR experiments, the local and global approaches for combining evidences have been compared. It is demonstrated that local approaches based on DS theory of evidence produce more qualitative ranking than the global approaches.

Other systems such as a Filtering system based on PLIR and an ITS system based on plausible reasoning also introduced. In experiments with ITS system, the learning ability of the system and its impact on tutoring decisions has been demonstrated. Our ITS system generally mimics the actual tutor's decisions however it tutors the students with fewer steps than the human tutor. The relationship between the number of steps and our methods of estimating the confidence values are currently under investigation.

In future, we will expand our experiments in different domains and different methods of calculating certainty parameters.

References

1 A. Collins, and R. Michalski. The logic of Plausible Reasoning A core theory, *Cognitive Science*, vol. 13, pp. 1-49, 1989.

2 A. Collins, M. H. Burstejn, Modeling A Theory Of Human Plausible Reasoning, *Artificial Intelligence III*, 1988.

3 A. Dempster, Upper and lower probabilities induced by a multivalued mapping, Ann. Math. Stat., vol. 38, no. 2, pp. 325–339, 1967.

4 A. B Quinn, C. M. Mitchell, and A. R. Champell. The Importance of Student Models in Intelligent Tutoring for Complex Dynamic Systems, IEA/HFES 2000 Congress Proceedings. 2000.

5 F. Oroumchian. Information Retrieval By Plausible Inferences: An Application Of The Theory Of Plausible Reasoning Of Collins And Michalski, Ph.D. Dissertation, School Of Computer And Information Science, Syracuse University, 1995.

6 F. Oroumchian, Bita Khandzad. Modeling An Intelligent Tutoring System by Plausible Inferences, 1st International Conference on Fuzzy systems and knowledge discovery, Volume 1, pp529, 2002.

7 F. Oroumchian, and R. N. Oddy. An Application Of Plausible Reasoning To Information Retrieval, In *Procs. The 19th ACM SIGIR Conference on Research and Development in Information Retrieval*, pp. 244-252, 1996.

8 F. Oroumchian, B. N. Araabi, E. Ashoori. Using Plausible Inferences and Dempster shafer Theory Of Evidence For Filtering, In *Procs. 4th International Conference on Recent Advances in Soft Computing (RASC2002),* Nottingham, United Kingdom, Dec 12-13. 2002.

9 H. Gamboa, A. Fred. Designing Intelligent Tutoring Systems: a Bayesian Approach, *3rd International Conference on Enterprise Information Systems*, ICEIS'2001.

10 J. Rocchio. Relevance feedback in Information Retrieval, In *Procs of the SMART Retrieval System -Experiments in Automatic Document process*, pp 313-322, 1971.

11 P. Smets. The nature of the unnormalized beliefs model, D. Dubois, M. P. Wellman, B. D'Ambrosio, and Ph. Smets, editors, *Uncertainty in Artificial Intelligence 92*, Morgan Kaufman, pp. 292–297, San Mateo, CA, 1992.

12 P. Smets, Belief Functions and The Transferable Belief Model, at the web site of the Imprecise Probabilities Project: http://ippserv.rug.ac.be, 31/5/2003.

13 M. N Ainsworth, S. E. Developing. ITSs using a Psychologically Motivated Authoring Environment. *AAAI Fall symposium*, AAAI Press: Menlo Park, CA. ISBN 1-57735-038-3, pp. 53-59, 1997.

14 N. Matsuda, K. VanLehn. Decision Theoretic Instructional Planner for Intelligent Tutoring Systems, in Procs. *Workshop on Modeling Human Teaching Tactics and Strategies*, ITS2000, pp.72-83, 2000.

Towards Model-Independent Mode Detection and Characterisation of Very Long Biomedical Time Series

Karl Pauwels[1], Temujin Gautama[2], Danilo P. Mandic[3] and Marc M. Van Hulle[4]

[1,2,4] Laboratorium voor Neuro- en Psychofysiologie, K.U.Leuven, Campus Gasthuisberg, Herestraat 49, B-3000 Leuven, BELGIUM,
{karl, temu, marc}@neuro.kuleuven.ac.be
[3] Department of Electrical and Electronic Engineering, Imperial College of Science, Technology and Medicine, London, SW7 2BT, U.K.,
d.mandic@ic.ac.uk

Abstract. A novel technique, the Delay Vector Variance method, which provides model-independent characterisation of time series in terms of their predictability is introduced and applied in a biomedical context. The merits of the procedure are demonstrated in a mode segmentation context on a set of long nonstationary physiological signals, obtained from subjects undergoing different sleep and wake stages. It is shown that the features extracted remain consistent within and across subjects. Next, the presence of nonlinearity associated with the different modes is investigated. A comparison with other measures supports the obtained results, namely that the signals show a higher degree of nonlinearity during wake than during sleep stages.

Keywords. time series characterisation, mode detection, nonlinearity

1 Introduction

The recent progress in mathematical modelling and the increasing availability of computational resources have initiated the resolution of problems of high complexity, involving high degrees of nonlinearity and nonstationarity. Too often, overly complex models are used to address these problems, resulting in weak generalisation ability and poor training performance. For this reason, the assessment of the signal's multimodality and the nature of these modes in terms of required model order and model nonlinearity, should precede the development of data processing or prediction systems [4].

The method proposed here draws upon the 'Delay Vector Variance' (DVV) method [2]. It characterises the signal in terms of its predictability, and achieves this in a model-independent manner which is closely related to the dynamical structure of the signal. The method is computationally and conceptually appealing. Its simplicity enables an incremental calculation, allowing for on-line signal processing applications.

The application discussed in this paper is concerned with the analysis of nonstationary physiological signals recorded from subjects undergoing different sleep stages. Currently, the most successful approaches towards this problem employ Hidden Markov Models [1, 3]. Since the sleeping patterns (i.e., the transitions from one stage to another) are very similar across subjects, these approaches can achieve good results without the need to directly grasp the dynamics of the signal modes. The proposed DVV method on the other hand, is strictly data-driven and yields excellent results, without exploiting temporal information such as state transition probabilities. This is a strong indication that the features extracted are closely related to the underlying dynamics of the system.

Biomedical Time Series Used

To illustrate the method, a set of biomedical signals obtained from subjects during different wake and sleep stages is used. The set contains recordings of three successive naps for five different subjects. For each nap a manual labelling assigned by a medical expert is available. The labels segment the data into different wake and sleep classes. These data sets are publicly available[1] and have been used before in the context of mode segmentation [3]. The analyses are limited to the respiration (RES) and electro-encephalogram (EEG) signals of the first two subjects.

2 The Delay Vector Variance Method

A time series is first transformed into a set of delay vectors (DVs) of a given embedding dimension m, $\mathbf{x}(k) = [x_{k-m1}, \ldots, x_{k-1}]$. The next sample, x_k, is referred to as the target of the DV. The proposed DVV method examines the mean target variance of all sets Ω_k, generated by grouping the DVs that are within a certain distance to $\mathbf{x}(k)$. To make the results comparable over different embedding dimensions as well as across different dynamical ranges of the time series, the distance threshold is varied in a standardized manner. For a given embedding dimension, the proposed 'Delay Vector Variance' method can be summarised as follows [2]:

- The mean, μ_d, and standard deviation, σ_d, are computed over a discrete set of all pairwise distances between DVs, $\|\mathbf{x}(i) - \mathbf{x}(j)\|$ $(i \neq j)$;
- The sets Ω_k are generated such that $\Omega_k = \{\mathbf{x}(i) \mid \|\mathbf{x}(k) - \mathbf{x}(i)\| \leq d\}$, i.e., sets which consist of all DVs that are within a certain uniformly spaced distance d from $\mathbf{x}(k)$, taken from the interval $[\mu_d - n_d\sigma_d \,;\, \mu_d + n_d\sigma_d]$, where n_d is a parameter controlling the span over which to perform the DVV analysis;

[1] http://www.first.gmd.de/persons/Kohlmorgen.Jens/publications.html

- For every[2] set Ω_k, the variance of the corresponding targets, σ_k^2, is computed[3]. The average over all sets Ω_k, normalised by the variance of the time series, σ_x^2, yields the measure of unpredictability, σ^{*2}:

$$\sigma^{*2} = \frac{\frac{1}{N}\sum_{k=1}^{N} \sigma_k^2}{\sigma_x^2} \tag{1}$$

An example of a typical 'DVV plot' is shown in Figure 1. Small target variances for small spans (left hand side of Figure 1) are an indication of a strong deterministic component (high predictability). For large spans, the curve smoothly converges to unity, since all DVs belong to the same universal set and therefore the variance of the corresponding targets will be equal to the variance of the time series.

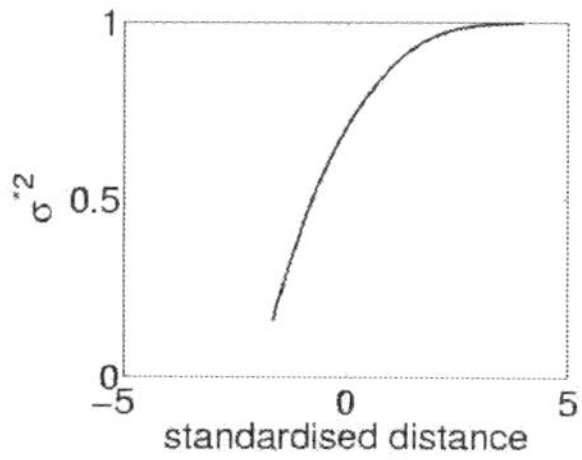

Fig. 1. Example DVV plot

To cope with nonstationarity, DVVs are calculated on short, overlapping windows that are slided over the signal one sample at a time.

3 Mode Segmentation

The goal of the proposed mode segmentation procedure is to identify the sleep stages, labelled by a medical expert. In the following, the analysis is limited to the three respiration signals of patients 1 and 2. Since these signals are very noisy and contain many outliers, a pre-processing step is first performed. Categories labelled as artifacts are removed and DVVs are calculated in such a way that the windows contain only a single mode. Next, the on-line version of the algorithm is used to extract DVV plots from the resulting signals[4]. Since the extracted features are high-dimensional and still contain redundancy, a dimensionality reduction using the Principal Component Analysis (PCA) is applied before proceeding to the segmentation.

---

[2] To improve robustness, only sets containing at least 30 DVs are taken into consideration.

[3] The reasoning is that, for a stationary signal, similar DVs should have similar targets.

[4] A window length of 400 samples, embedding dimension 4 and a span of 3 standard deviations were used in the DVV analyses.

Table 1. Classification accuracies obtained while testing within-subject generalisation

signal		accuracy (%)	
train	test	subject 1	subject 2
1	2	86	82
1	3	93	69
2	1	76	68
2	3	87	74
3	1	79	70
3	2	92	87
mean		86	75

Using a standard multi-layer perceptron[5], the data is segmented from the extracted features into a wake and a sleep mode. A first set of analyses examines the subject-dependent generalization performance of the network. One signal of a particular subject is used to train the network that is subsequently used to segment a signal, recorded during a different nap of the same subject. To prevent the network from overfitting the data, cross-validation (early stopping) is used. Table 1 contains the classification accuracies for the three signals of both subjects. The high accuracies clearly indicate that DVV extracts features that remain consistent over different sleeps of the same subject. A subsequent analysis investigates whether these features remain consistent over different subjects as well by using all DVV plots from one subject to classify the DVV plots of the other subject. This procedure results in classification accuracies of 64 % when training on subject 1 and testing on subject 2, and 74 % when training on subject 2 and testing on subject 1. Such a high accuracy is remarkable given the fact that medical experts need different signals (EEG, EOG, etc.) to correctly identify sleep stages. In the following, the DVV method is used to further examine the nature of the signals during sleep and wake stages.

4 Nonlinearity Analysis

To help clarify the differences observed during sleep and wake stages, the proposed characterisation method is applied to investigate the degree of nonlinearity present in the signals. Although the different procedures explained in this section consistently indicated strong nonlinearity in the respiration signals during both wake and sleep stages, no significant differences were found between the stages. For this reason, the analyses in the remainder of this section focus on the EEG signals only. For subject 1, three EEG signals are available. In order not to introduce any artifacts or spurious correlations, the windows belong strictly to one class and do not overlap. This results in a data set containing 133 and 166 signal windows belonging to respectively the sleep and wake class.

[5] In all simulations, a network containing one hidden layer (consisting of 2 neurons) and initialised with different random seeds was used.

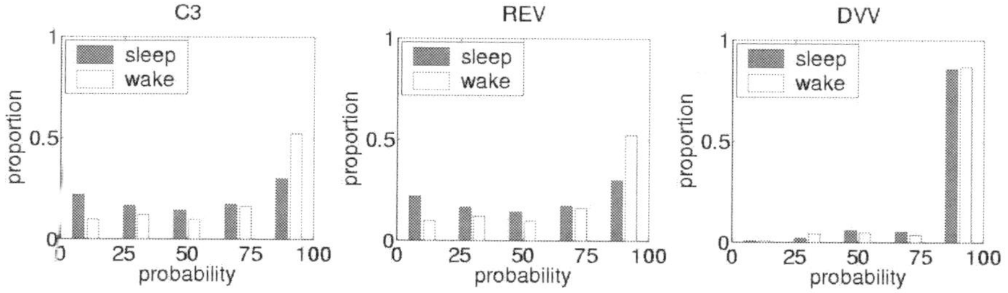

Fig. 2. Histograms of the rankings obtained with C3, REV and DVV on the sets of wake and sleep signals. All rank indices have been translated into probabilities

It is plausible to assume that a higher cognitive awareness and the larger number of processes that exert an influence on the signal during awakeness manifests itself in a higher degree of signal nonlinearity during wake than during sleep stages. This hypothesis is investigated using DVV and validated with two well-known measures of nonlinearity: the third-order autocovariance (C3) and the asymmetry due to time reversal (REV) [7]. All three measures provide a characterisation of the time series. By comparing these measures to those calculated on linearised versions, 'surrogates', of the original signals, an assessment of the degree of nonlinearity can be made [6]. A rank test is used to establish whether or not the null hypothesis of linearity can be rejected. Unlike C3 and REV, DVV provides a range of values. Single measures are obtained by computing the distances, for all surrogates and the original signal, between their respective DVV plot and an average DVV plot (averaged over all surrogates). Figure 2 contains the binned rank indices for all windows and for all three measures[6]. The larger amount of rejections on wake signals as compared to sleep signals in C3 and REV, is already a strong indication that the signals contain a higher degree of nonlinearity during wake stages. Using the DVV analysis, nearly all windows reject the null hypothesis of linearity. This result is in line with previous nonlinearity analyses, performed on EEG signals [5].

To obtain more quantitative results, a statistical test is performed on the measures themselves. The measures calculated on the original signal provide a direct indication of nonlinearity. To allow for inter-signal comparisons, these measures are first normalised by the spread of the values obtained on the surrogates. Next, a non-parametric statistical test, the Wilcoxon rank sum test, is used to investigate the null hypothesis of identical distributions. The following p-values are obtained: $4.5 \cdot 10^{-6}$ for C3, 0.025 for REV and 0.0058 for DVV. All three measures thus consistently reject the null hypothesis at the 0.05 level of significance. For this subject, the working hypothesis that the EEG signals exhibit higher nonlinearity during wake than sleep is validated.

[6] The rankings of the two-sided tests (C3 and REV) have been rearranged so that the rightmost bin represents the rejection of the null hypothesis.

5 Conclusions

It has been demonstrated that the proposed method for time series characterisation, the 'Delay Vector Variance' method, is able to extract features, related to the predictability of the respiration signals, that allow for a subject-independent segmentation into wake and sleep stages. In a subsequent study, these features were used to examine the difference in degree of nonlinearity of EEG signals. The results obtained (higher degree of nonlinearity during wake as opposed to sleep) are in line with those obtained using established nonlinearity measures.

Acknowledgements

Karl Pauwels is supported by a scholarship from the Flemish Ministry for Science and Technology (VIS/98/012) and from the European Commission, 5th framework programme (QLG3-CT-2000-30161). The work of Temujin Gautama is supported by a scholarship from the Flemish Regional Ministry of Education (GOA 2000/11) and a research grant from the Fund for Scientific Research (G.0248.03). Danilo P. Mandic was supported by QLG3-CT-2000-30161 and a visiting fellowship from the K.U.Leuven (F/01/079) while at the Laboratorium voor Neuro- & Psychofysiologie, K.U.Leuven, Belgium. Marc M. Van Hulle is supported by research grants received from the Fund for Scientic Research (G.0185.96N and G.0248.03), the National Lottery (Belgium) (9.0185.96), the Flemish Regional Ministry of Education (Belgium) (GOA 95/99-06; 2000/11), the Flemish Ministry for Science and Technology (VIS/98/012), and from the European Commission, 5th framework programme (QLG3-CT-2000-30161 and IST-2001-32114).

References

1. Flexer A, Dorner G, Sykacek P., Rezek I (2002) An automatic, continuous and probabilistic sleep stager based on a hidden Markov model. Applied Articial Intelligence 16(3):199–207
2. Gautama T, Mandic DP, Van Hulle MM (2003) Indications of nonlinear structures in brain electrical activity. Phys. Rev. E 67:046204
3. Kohlmorgen J, Müller K-R, Rittweger J, Pawelzik K (2000) Identication of nonstationary dynamics in physiological recordings. Biol. Cybern. 83:73–84
4. Mandic DP, Chambers JA (2001) Recurrent neural networks for prediction: learning algorithms, architectures and stability. John Wiley & Sons, U.K., 1st edn.
5. Paluš M (1996) Nonlinearity in normal human EEG: cycles, temporal asymmetry, nonstationarity and randomness, not chaos. Biol. Cybern. 75:389–396
6. Schreiber T, Schmitz A (1996) Improved surrogate data for nonlinearity tests. Phys. Rev. Lett. 77:635–638
7. Schreiber T, Schmitz A (1997) On the discrimination power of measures for nonlinearity in a time series. Phys. Rev. E 55:5443–5447

Analysis of Rules Discovered by the Data Mining Process

Kenneth McGarry and James Malone

School of Computing and Technology, University of Sunderland, St Peters Campus
St Peters Way, Sunderland, SR6 ODD, United Kingdom
ken.mcgarry@sunderland.ac.uk
james.malone@sunderland.ac.uk

Abstract. This paper describes how symbolic rules may be extracted from Radial Basis Function neural networks and shows how they can be used by the data mining and knowledge discovery process. Rule extraction overcomes a major disadvantage of neural networks which is concerned with making the comprehensibilty of the learned internal model more open to scrutiny. Having extracted the symbolic rules we show how they are assessed and ranked for interesting or novel features. Two such techniques are presented here, the first is a data driven approach that uses objective mathematical measures to identify interesting patterns or features. The second is a goal driven method that uses subjective measures obtained from the user. The measures are applied to rules extracted from RBF neural networks trained on several data sets including benchmark sets from the UCI repository and a large real-world industrial data set.

Keywords. Rule extraction, neural networks, data mining, interestingness measures.

1 Introduction

Data mining and knowledge discovery from data bases has received much attention in recent years. However, the majority of this work has concentrated on producing accurate models without considering the potential gains to be had from understanding the details of these models. Recent work has to some extent addressed this problem but much work has still to be done [3, 5]. In this paper we outline our own methods of assessing the novelty,usefulness and comprehensibility of rules extracted from the data mining process. In fact one of the most insightful definitions of data mining states that to be truly successful data mining should be *"the process of identifying valid, novel, potentially useful, and ultimately comprehensible knowledge from databases"* that is used to make crucial business decisions, [2]. Currently, there are two main techniques of assessing the interestingness of discovered patterns. The first uses objective mathematical measures to assess the degree of interestingness, many such measures exist. The second method is to incorporate the users subjective knowledge into the assessment strategy. We consider how the objective approach may highlight regularities from the discovered patterns.

The particular system under investigation is comprised of symbolic rules extracted from Radial Basis Function Neural Networks. Neural networks are are accurate classifiers for low level pattern recognition tasks. Unfortunately, they are effectively "black boxes" because their internal representation is difficult if not impossible to decode by humans. Previous work performed by the author has seen this problem partly solved by extracting $< IF..THEN >$ type rules [6, 7]. However, what this work has been lacking is the automatic analysis of these rules using "interestingness" measures to asssess their novelty,usefulness and comprehensibility.

The remainder of this paper is structured as follows. Section two descibes the architecture of the radial basis function network and how symbolic rules may be extracted from its internal parameters. Section three introduces data mining and knowledge discovery and reviews the interestingness measures used for ranking the rules/patterns identified by the rule extraction algorithm. Section four shows the experimental results from several data sets. Section five dicusses the conclusions and areas for further work.

2 Rule Extraction from RBF Neural Networks

The RBF network consists of feedforward architecture with an input layer, a hidden layer of RBF "pattern" units and an output layer of linear units. The input layer simply transfers the input vector to the hidden units, which form a localized response to the input pattern. The activation levels of the output units provide an indication of the nearness of the input vector to the classes. Learning is normally undertaken as a two-stage process. The objective of hidden layer learning is to locate the radial basis function centres and to determine the optimum field widths in relation to the training samples. The right half of figure 1 shows the RBF architecture. Several schemes for locating hidden unit parameters have been suggested. Broomhead and Lowe used a uniformly distributed lattice of hidden units but this proved to be impractical for high input dimensionality as the number of hidden units grew exponentially [1]. The simplest procedure is to randomly set them as *prototypes* of a subset of the training data. The left half of figure 1 highlights the local nature of each hidden unit as it maps into only a limited part of the input space, thus enabling a single rule to be formed from each hidden unit.

2.1 LREX: Local Rule Extraction Algorithm

The development of the LREX algorithm was motivated by the local architecture of RBF networks which suggested that rules with unique characteristics

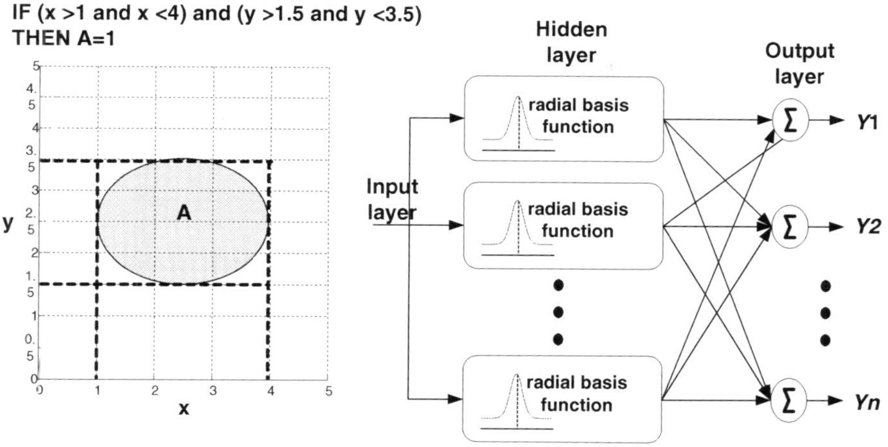

Fig. 1. Symbolic approximation of RBF centres

could be extracted. The LREX algorithm extracts IF..THEN type rules based on the premise that a hidden unit can be uniquely assigned to a specific output class. Therefore, by using the centre locations of the hidden units an input vector could be directly mapped to an output class. Experimental work performed on the simpler data sets tended to reinforce this belief. However, hidden unit sharing occurs within networks trained on non-linear or complex data. This phenomena reduces rule accuracy as several hidden units may be shared amongst several classes [6]. The functionality of LREX algorithm is shown in figure 2.

A single rule from the Iris domain is presented in figure 3. The antecedent is formed by calculating the lower and upper bounds for each hidden units 'spread'.

The use of information theory was made to reduce the number of antecedents within a rule. This involved implementing the ChiMerge algorithm and discretizing the data sets to simplify the process of calculating the information measure. The information measure or information gain about an input feature represents the importance of that feature to resolve class identity. Features with very low values may be excluded from the extracted rules and thus simplied the rule and aids comprehensibility.

3 Interestingness Measures for Knowledge Discovery

The term 'data mining and knowledge discovery' came into use around the late 1980's and has been an intense area of research activity since then. The reason for the surge in interest was due the vast quantities of data generated

Input:
Hidden weights μ (centre positions)
Gaussian radius spread σ
Output weights $W2$
Statistical measure S
Training patterns
Output:
One rule per hidden unit
Procedure:
Train RBF network on data set
Collate training pattern *"hits"* for each hidden unit
For each hidden unit
  Use $W2$ correlation to determine Class label
  Use *"hits"* to determine S
  Select S format $\{min, max, std, mean, med\}$
  For each μ_i
    $X_lower = \mu_i - \sigma_i * S$
    $X_upper = \mu_i + \sigma_i * S$
Build rule by:
  antecedent $= [X_lower; X_upper]$
  Join antecedents with AND
  Add Class label
Write rule to file

Fig. 2. LREX rule-extraction algorithm

Rule 1 :
IF (SepalLength $\geq$ 4.1674 AND $\leq$ 5.8326) AND
IF (SepalWidth $\geq$ 2.6674 AND $\leq$ 4.3326) AND
IF (PetalLength $\geq$ 0.46745 AND $\leq$ 2.1326) AND
IF (PetalWidth $\geq$ 0.53255 AND $\leq$ 1.1326)
THEN..Setosa

Fig. 3. Extracted rule from Iris domain

by modern companies as part of their daily activity. The means to analyse this data far outstretched the conventional statistical techniques. Clearly an automated approached was needed and techniques from machine learning have been applied to understand, describe and predict future patterns. The analysis and ranking of the discovered patterns may present the greatest challenge of all.

The majority of the data mining work concentrates on the discovery of accurate and comprehensible patterns. Whilst this approach will provide the user with a degree of confidence regarding the discoveries it falls far short of the notion of "knowledge discovery". The essential task for data mining algorithms is to search for patterns that are "suprising" in addition to being accurate and comprehensible. Criteria such as rule coverage, rule complexity, rule confidence and rule completeness are becoming used as a measure of the interestingness of the discovered patterns [3].

- Small disjuncts. Inductive systems are generally designed, where possible to produce models with large disjuncts i.e. a rules that will cover as many instances of the training set as possible. This aids the understandibility of the model through producing fewer rules. Unfortunately, for some data sets small disjuncts can make up perhaps 20% of the total number of rules in a classifier [4]. The presence of small disjuncts can indicate a noisy data set and/or some interesting exceptions.
- Class imbalance. Should one or two classes predominate the original data set in terms of numbers then the rules predicting the smaller, less well represented class may represent some interesting exception.
- Complexity. The complexity of a rule was assessed by the number of average number of antecedents within the rule body.

4 Experimental Results

The rule sets extracted from the RBF networks were analysed by of the measures discussed in the previous section. Most of the data sets will be familiar to the AI community apart from the Vibration data sets which are concerned with fault diagnosis on industrial machinery.

Table 1. Extracted rule set size and accuracy

Data set	No Rules	Attributes	Classes	Accuracy %
Monks1	20	6	2	83
Diabetes	65	8	2	76
Credit	50	15	9	93
Vibration 1	30	9	3	73
Vibration 2	100	20	8	94

In terms of new or suprising knowledge, the vibration rule set provided the best results. The rule extraction algorithm in conjunction with the interestingness measures was able to identify two input parameters not generally thought of as being important for identifing machine faults. The RBF training algorithm had used the most discriminating features for detecting the fault classes which was made explicit by the rule extraction algorithm. The rules were then ranked accordingly. The rules using these parameters are valid, accurate and account for 15-20% of the total number of extracted rules.

Table 2. Objective interestingness measures applied to extracted RBF rules

Interestingness Measure	**RBF Rule Sets**				
	Monks	Diabetes	Credit	Vibration 1	Vibration 2
Small Disjuncts	No	No	Yes	Yes	Yes
Class Imbalance	No	No	Yes	No	Yes
Size/accuracy	4.1	1.1	1.8	2.4	0.94
Complexity	6	8	9	9	20

5 Conclusions

This paper has presented the results of ranking and the analysis of rules extracted from RBF neural networks using both objective and subjective measures. The interestingness of a rule can be assessed by a data driven approach. Unfortunately, objective measures may still allow uninteresting patterns to emerge from the data mining process. As suspected there is no overall optimum measure that can be applied to all data mining patterns and by their nature are highly domain specific. Future work will explore the issue of integrating objective with subjective driven measures.

References

1. D. S. Broomhead and D. Lowe. Multivariable functional interpolation and adaptive networks. *Complex Systems*, pages 321–355, 1988.
2. U. Fayyad, G. Piatetsky-Shapiro, and P. Smyth. From data mining to knowledge discovery: an overview. In U. Fayyad, G. Piatetsky-Shapiro, P. Smyth, and R. Uthursamy, editors, *Advances in Knowledge Discovery and Data Mining*. AAAI-Press, 1996.
3. A. Freitas. On rule interestingness measures. *Knowledge Based Systems*, 12(5-6):309–315, 1999.
4. R. C. Holte, L. E. Acker, and B. W. Porter. Concept learning and the problem of small disjuncts. In *International Joint Conference on Artificial Intelligence*, pages 813–818, Detroit, Michigan, USA, 1989.
5. B. Liu, W. Hsu, L. Mun, and H. Y. Lee. Finding interesting patterns using user expectations. *IEEE Transactions on Knowledge and Data Engineering*, 11(6):817–832, 1999.
6. K. McGarry, S. Wermter, and J. MacIntyre. The extraction and comparison of knowledge from local function networks. *International Journal of Computational Intelligence and Applications*, 1(4):369–382, 2001.
7. K. McGarry, S. Wermter, and J. MacIntyre. Knowledge extraction from local function networks. In *Seventeenth International Joint Conference on Artificial Intelligence*, volume 2, pages 765–770, Seattle, USA, August 4th-10th 2001.

A Study of Query Generation Strategies for Interactive Constraint Acquisition *

Sarah O'Connell, Barry O'Sullivan, and Eugene C. Freuder

Cork Constraint Computation Centre, Department of Computer Science,
University College Cork, Ireland
{s.oconnell|b.osullivan|e.freuder}@4c.ucc.ie

1 Introduction

In many practical applications users find it difficult to articulate their constraints. While users can recognize examples of where a constraint should be satisfied or violated, they cannot articulate the constraint itself. For example, a customer may be trying to specify a constraint to an engineer, without being able to use the correct engineering terms for the relevant concepts.

Recently, researchers have become more interested in techniques for solving problems where users have difficulties articulating constraints. Freuder and Wallace have considered suggestion strategies for applications where a user cannot articulate all constraints in advance, but can articulate additional constraints when confronted with something which is unacceptable [2]. However, that work assumes that the user can actually articulate each of the constraints when required. In essence their approach is model focussed, in the sense that they are concerned with acquiring a partial model of the user's CSP which is sufficient to be able to find a solution that the user would find acceptable. Padmanabhuni *et al.* have proposed a framework for learning constraints [6]. However, they do not show how negative (unacceptable) examples can be handled by their approach. Rossi and Sperduti have studied the utility of using inductive learning, through the application of gradient descent techniques, for acquiring global and local preferences in the context of interactive soft constraint solving [7]. Their approach assumes that the constraints of the problem are known but that the local and global preferences are not. Freuder and O'Sullivan have begun studying the interactive acquisition of tradeoff constraints in interactive constraint-based configuration [1]. Their approach focuses on generating constraints which model tradeoffs between variables in problems which have become over-constrained during a interactive configuration session. Again, while some constraints in the problem are being acquired,

* This work has been supported by Science Foundation Ireland under Grant 00/PI.1/C075.

it is assumed that the user can accurately articulate constraints, and that the issue is how to remove over-constrainedness in a model. Finally, a preliminary report on the research reported here has also been presented [5]. Therefore, while constraint acquisition, in one sense or another, is receiving some attention, there is a lack of work on techniques which can support the acquisition of individual constraints necessary to model the user's problem. It is this gap in the research which the work reported here is beginning to address.

The remainder of the paper is organized as follows. In Section 2 the model of interaction which we base our experiments is presented. Section 3 presents the set of query generation strategies that we consider here. Section 4 presents the results of our experiments. In Section 5 some brief concluding remarks are made.

2 A Model for Interactive Constraint Acquisition

In this paper we present an approach to interactive constraint acquisition based on techniques from the field of machine learning. In our system, constraint acquisition is modelled as search through an "hypothesis space" of constraints over which a general-to-specific ordering is explicitly known, or is implicit in its representation. Examples provided by the user can be used to identify a *version space* ([3]) of constraints that the user could be attempting to articulate. The system then attempts to generalize the user's examples by choosing an hypothesis from the current version space and attempting to validate it. This attempt at generalization involves proposing a *qualifying example* (a query) to the user based on the the candidate hypothesis. If the qualifying example is accepted by the user, the system attempts to generalize again. An accepted qualifying example is implicitly positive and is therefore used to further refine the version space for the constraint. Generalization continues until either the version space for the user's constraint has been reduced to a single hypothesis, or an example has been proposed to the user which he rejects. A rejected example is implicitly negative. Therefore, these negative examples are also used to further refine the version space for the constraint. If the user rejects an example he is invited to provide another positive example if one is available. The process terminates once the version space has converged upon a single hypothesis.

3 Generalization Strategies

In the work presented here, generalization is performed interactively through the computer's generation of a qualifying question (a query) and presenting it to the user. Mitchell [4] suggests that the optimal query generation strategy to adopt when learning using version spaces is one that involves generating instances that satisfy exactly half of the hypotheses in the version space. For the purposes of this paper we will refer to this strategy as *50/50*. Given that we are concerned with the interactive acquisition of constraints, one of the key metrics that is relevant to us is the total number of interactions with the user (the dialog length). Obviously, we would wish this number to be minimal. Therefore, query generation strategies which have the potential to minimize the length of the interaction (dialog) with the human are of significant value. One could imagine that in the real-world the difference between a user using an interactive constraint acquisition system and abandoning it maybe their lack of tolerance of one more interaction.

Therefore, it is worthwhile posing the question: does the combination of acquisition problem, user profile and version space topology have an effect on the performance of different query generation strategies? In particular, can we tailor our query generation strategy so that the total number of interactions with the human user is minimized? The query generation strategies that we have studied are: (a) *50/50* – our baseline case, (b) *Least Generalization (Random)*, (c) *Least Generalization (Maximum Connectivity)*, (d) *Least Generalization (50/50)* and (e) *Random Choice*. We outline the differences below:

50/50 (Baseline case) – This heuristic represents the conventional wisdom in the machine learning field and generates queries which satisfy half of the hypotheses in the current version space. This is the best known general-purpose query generation strategy for version space learning.

Least Generalization (Random) – One disadvantage with the previous strategy is that while it is based on a sound principle, it can give rise to the generation of queries that are difficult for the user to explain from the perspective of perspicuity. Therefore, it may be worthwhile to use strategies that generate queries which the user would regard as "logical" generalizations. The *Least Generalization (Random)* strategy generates a query at random from the set of examples that could be used to qualify a minimally generalized hypothesis.

Least Generalization (Maximum Connectivity) – This strategy selects a query whose set of generalizations is maximal. This is a risky strategy since it relies on a negative response to the query to significantly reduce the size of the version space.

Least Generalization (50/50) – This strategy attempts a least generalization step based on the query which reduces the version space by a similar amount regardless of whether it is accepted or rejected.

Random Choice – This strategy selects a query at random.

4 Evaluation

The objective of our evaluation was to compare the length of dialog (number of interactions) between the user and computer required to acquire a constraint.

The evaluation was performed by varying (a) the topology of the original hypothesis space and (b) the extent to which the user was trying to be helpful.

We studied two alternative hypothesis space topologies. Each topology was parameterized with a single integer value representing the number of atomic examples that formed the vocabulary of the interaction. Figure 1 presents the hypothesis-space topologies used for the purposes of our evaluation. We will refer to the topology illustrated in Figure 1(a) as the *tree topology*, while we will refer to the topology illustrated in Figure 1(b) as the *lattice topology*.

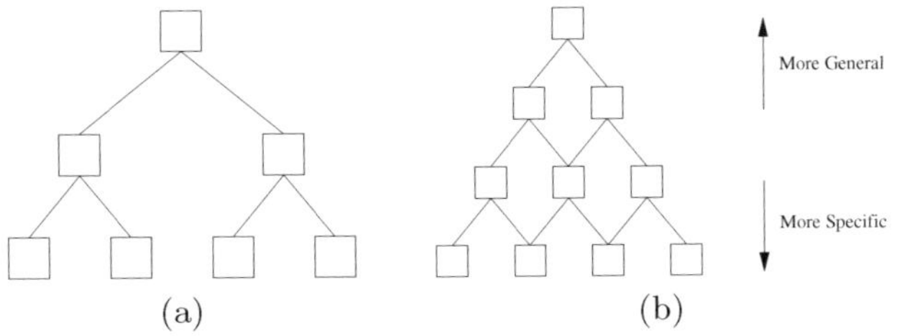

Fig. 1. The hypothesis space topologies used in our evaluation: (a) the tree topology and (b) the lattice topology. Note that we do not show the trivially most specific case, where nothing is acceptable.

For the purposes of our experiments, we developed an hypothesis-space based on the *tree topology* having 64 example nodes, giving a total of 127 possible constraints. We also generated an hypothesis space based on the *lattice topology* having 15 example nodes, giving a total of 120 possible constraints.

The extent to which the user was trying to be helpful was modelled as the number of examples that he would give to the computer when user-specified examples were required. An *unhelpful user* gave a single example to the computer, while a *helpful user* gave a randomly chosen subset of the examples that exemplify the constraint being articulated. This is a reasonable approach to adopt since the user must be able to present at least one example of the constraint, but may be able to provide more than one in attempting to be helpful.

In the context of the hypothesis space based on the *tree topology* we compared the *50/50* and *Least Generalization (Random)* strategies. The other two variants of the *Least Generalization* strategy, namely the *Least Generalization (Maximum Connectivity)* and *Least Generalization (50/50)* would behave in exactly way as *Least Generalization (Random)* the context of this hypothesis space topology, since there is never a situation where a principled choice could be made between alternative queries. In the context of the *lattice topology*, all strategies were compared.

The model of interaction used in the evaluation was based on that described in Section 2. Each experiment involved counting the number of interactions, between the simulated user and our constraint acquisition system, required to acquire constraints from each of the hypothesis spaces described above. The number of interactions (dialog length) was computed as the sum of the number of interactions used for user-specified examples and the number of interactions for system-generated queries. Since we are concerned with reducing the length of the dialog between the user and computer, we acquired each constraint in each hypothesis space, recording the total number of interactions required to acquire it and its depth in the hypothesis space. Depth, in this context, is a measure of the generality of the constraint. The more general the constraint, the deeper it is in the hypothesis space. The comparison presented below is based on averages over 50 runs of each experiment. In all the graphs, the x-axis represents the depth in the hypothesis space of the constraint being acquired.

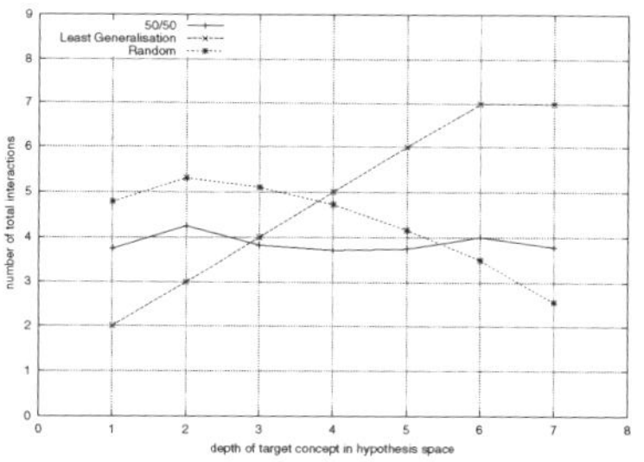

Fig. 2. Comparison of strategies on the tree topology when the user is not being helpful.

Figure 2 illustrates the average number of interactions required to acquire constraints at various depths in the hypothesis space. In this scenario, the hypothesis space is a (binary) tree and the user is not being helpful, only giving a single example at each point that a user-specified example is sought. It can clearly be seen that by far the best strategy to adopt is *50/50*. However, the *Least Generalization* strategy performs well for less general constraints.

Figure 3(a) again illustrates the average number of interactions required to acquire constraints at various depths in the hypothesis space. In this scenario, the only change over that illustrated in Figure 2 is that we have a *lattice* topology. Again, from Figure 3(a), it can be seen that the best strategy across the entire version space is *50/50*. However, for constraints which are less general, the *Least Generalization* strategies perform best.

Figure 3(b) presents a comparison of the the dialog lengths of each of the strategies in Figure 3(a), which shows the relative saving of the various strategies over *50/50*. On this graph, all points with a value greater than 1 represent an improvement over the *50/50* baseline strategy. It is clear to see that the *Least Generalization* strategies perform very well for less general constraints.

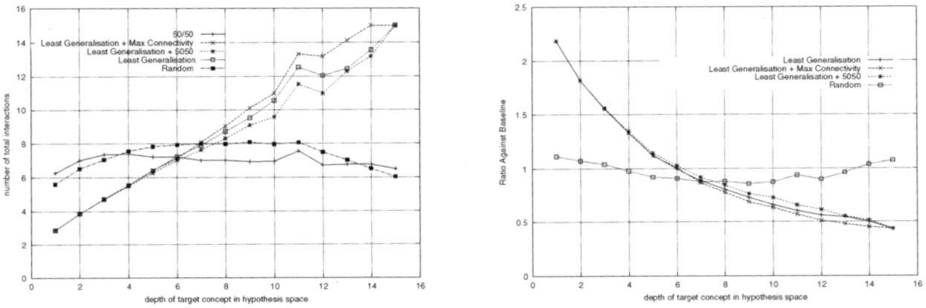

(a) Dialog Length (b) Comparison against the *50/50* strategy

Fig. 3. Comparison of strategies on the lattice topology when the user is not being helpful.

Figure 4 again illustrates the average number of interactions required to acquire constraints at various depths in the hypothesis space for a *tree* topology. The user in this scenario is more helpful than in the previous two scenarios. Here the user presents a random set of examples. It can be seen clearly that the best strategy across the entire hypothesis space is *Least Generalization.* This is a significant improvement over the scenario illustrated in Figure 2. It is clear evidence that the choice of generalization strategy can be dramatically informed by knowledge of the nature of the interaction with the user, in particular the degree to which he is helpful.

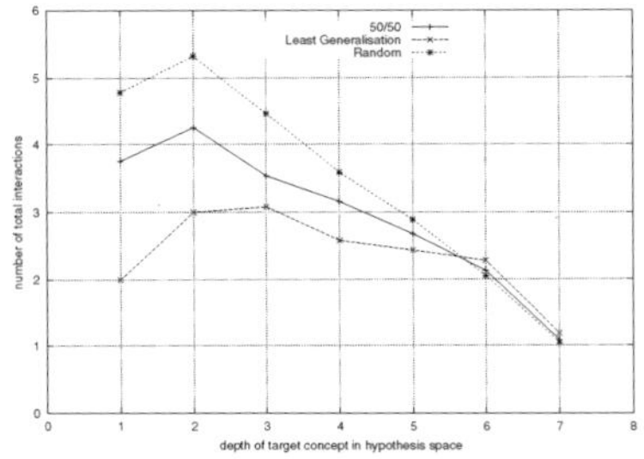

Fig. 4. Comparison of strategies on the tree topology when the user is being helpful.

Finally, Figure 5(a) illustrates the average number of interactions required to acquire constraints at various depths in the hypothesis space for a *lattice* topology. Again the user in this scenario is helpful. It can be seen clearly that while *50/50* performs consistently well across the entire hypothesis space, the best strategy is again based on *Least Generalization.* The improvement that these heuristics offer over *50/50* can be seen more clearly in Figure 5(b), which shows the relative saving of the various strategies over *50/50.* At almost every depth, the *Least Generalization (50/50)* strategies either save on interactions over the *50/50* approach, or closely follow it. This, again, is evidence that the choice of generalization strategy can be dramatically informed by knowledge of the nature of the interaction with the user. In general it can be seen that strategies based on *Least Generalization* are more powerful than *50/50* on constraints which are more specific in the version space and when the user is being helpful, perform competitively across the version space.

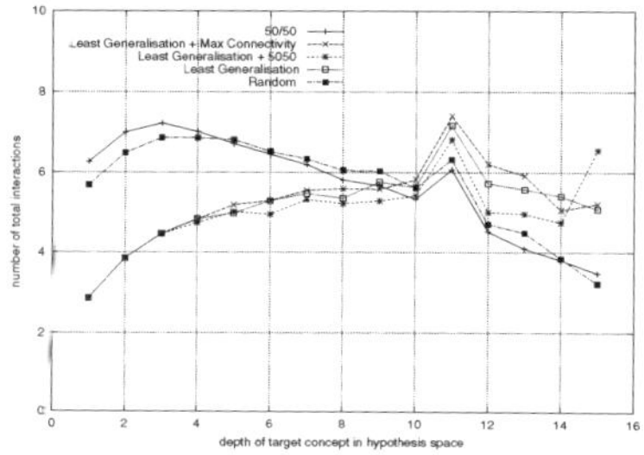

(a) Dialog Length

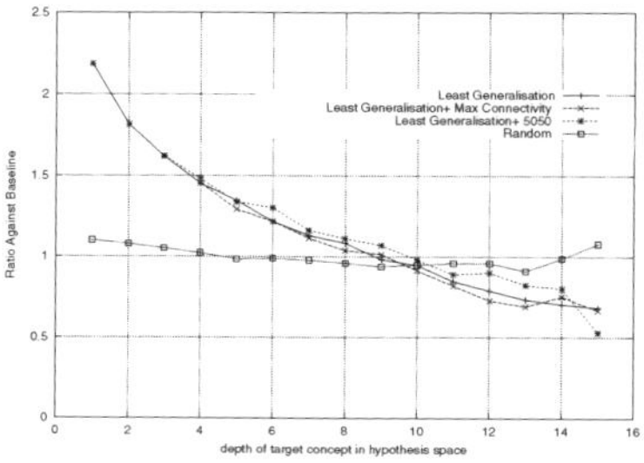

(b) Comparison against the *50/50* strategy

Fig. 5. Comparison of strategies on the lattice topology when the user is being helpful.

In order to get an indication of the magnitude of these savings, Table 1 presents a comparison of the average savings in the learning rates for all constraints for each of the generalization strategies over the *50/50* baseline. It can be seen that on the average the *Random* strategy never beats *50/50*, regardless of the type of user from whom we are acquiring constraints. However, the performance of the variants of the *Least Generalization* strategy are interesting.

Table 1. Comparison of learning rates against baseline

Heuristic	*Tree One Ex.*	*Lattice One Ex.*	*Tree Multiple Ex.*	*Lattice Multiple Ex.*
Least Generalization	-26%	-27%	21%	10%
Least Generalization + 50-50	-	-22%	-	12%
Least Generalization + Max Conn.	-	-33%	-	8%
Random	-11%	-5%	-19%	0%

While there was no benefit, on the average, from using a *Least Generalization* strategy when the user is not being helpful, once the user became helpful, there was a marked improvement in the performance of these strategies. While, on the average we can see that the benefit is not large, we have seen above that *Least Generation* strategies are best for more specific constraints.

We can also see from Table 1, that on tree-structured hypothesis spaces, a helpful user coupled with a *Least Generalisation* query strategy can give good reductions in dialog length. While acknowledging the general purpose power of the *50/50* strategy, we have some evidence to hypothesise that there are circumstances when it is worthwhile tailoring the query generation strategy of the acquisition system based on the degree to which the user (teacher) is being helpful, the topology of the hypothesis space and the generality of the target constraint.

5 Conclusions

The work presented in this paper is concerned with interactive constraint acquisition. In this paper, we have compared a number of different heuristics for query generation for interactive constraint acquisition. We have demonstrated that the choice of generalization strategy is closely related to the degree to which the user is being helpful and version space topology. Future work will address the utility of hybrid strategies for query generation for constraint acquisition. In addition, we will focus on learning disjunctive constraints and also, more importantly, sets of constraints. While in this paper we have compared a helpful teacher with one which is more helpful, a further investigation focused on how the teachers ability to provide good examples can affect the acquisition process needs to be undertaken. Such a study would more formally study the interaction between an intelligent learner with teachers of varying abilities and, in particular, study whether skill is more important than helpfulness.

References

1. Eugene C. Freuder and Barry O'Sullivan. Generating Tradeoffs for Interative Constraint-Based Configuration. In Toby Walsh, editor, *Proceedings of CP-2001*, pages 590–594, November 2001.
2. Eugene C. Freuder and Richard J. Wallace. Suggestion Strategies for Constraint-Based Matchmaker Agents. In *Proc. of CP-98*, pages 192–204, Oct. 1998.
3. Tom Mitchell. Generalization as Search. *AI*, 18(2):203–226, 1982.
4. Tom Mitchell. Concept Learning and the General-to-Specific Ordering. In *Machine Learning*, Chapter 2, pages 20–51. McGraw Hill, 1997.
5. Barry O'Sullivan, Eugene C. Freuder, and Sarah O'Connell. Interactive Constraint Acquisition – Position Paper. In *Proc. of the CP-2001 Workshop on User-Interaction in Constraint Satisfaction*, pages 73–81, Dec. 2001.
6. S. Padmanabhuni, et-al, A Framework for Learning Constraints. In *Proc. of the PRICAI-96 Workshop on Induction of Complex Representations*, Aug. 1996.
7. F. Rossi and A. Sperduti. Acquiring both Global and Local Preferences in Interactive Constraint Solving via Machine Learning. In *Proc. of the CP-2001 Workshop on User-Interaction in Constraint Satisfaction*, pages 83–95, Dec. 2001.

A Multi-agent System as Framework for Learning Algorithms Benchmarking

Luis Miramontes Hercog and Terence C. Fogarty

School of Computing, South Bank University
103 Borough Rd. London SE1 0AA, U.K.
email:{miramol, fogarttc}@sbu.ac.uk

Abstract. From a biological point of view a multi-agent system (MAS) can be seen as an ecological niche where different species are competing. A MAS that complies with the basic requirement of autonomous decision mechanism for each agent [3] is an ideal benchmark tool for different learning algorithms. In these experiments two learning classifier systems (LCS), the zero-th level classifier system (ZCS) [9] and the extended classifier system (XCS) are compared in the MAS framework. The comparison is based on the reward accumulated by the agents using either classifier system. It has been found that ZCS outperforms XCS in some cases, but most of the time is the latter that wins. The use of a MAS to compare XCS and ZCS, besides being novel, has yielded unexpected results, considering previous work of accuracy-based vs. strength-based fitness comparisons, where the former LCS always outperformed the latter kind. Further work is planned to exploit the MAS benchmark framework with a larger number of agents and different learning algorithms.

Keywords. Multi-agent systems, benchmark, learning algorithms.

1 Introduction

A multi-agent system (MAS) that complies with the basic requirement of autonomous decision mechanism for each agent [3] is an ideal benchmark tool for different learning algorithms. In these experiments two learning classifier systems (LCS), the zero-th level classifier system (ZCS) [9] and the extended classifier system (XCS) [2] – the former a strength-based fitness and the latter an accuracy based fitness LCS – are compared in the MAS framework. The problem the agents learn to solve is a bounded rationality problem: "El Farol" Problem [1]. From a biological point of view a MAS can be seen as an ecological niche where different species are competing.

The experiments reported here prove that a MAS solving collectively a problem without communication between the agents is ideal for benchmarking the learning algorithms. Furthermore, the MAS leaves open the option to benchmark as many learning algorithms as there are agents in the system, allowing the use of a different learning algorithm for each agent.

The multi-agent system offers a direct comparison for a competition, in contrast with previous comparisons of learning algorithms that are tested separately in a given problem and then compared the performance of each separately [8].

In this experiment, the learning algorithms co-evolve, without any communication, to solve the given problem. Moreover, each agent has an indicator of its accumulated reward to calculate its performance. The multi-agent system also provides a sociological tool to determine which set-up is fairer in the sense of the distribution of the reward among the agents, i.e. the computation of the welfare of the agents.

The idea comes from previous experiments, where XCS [5] and ZCS [6] have been used on their own to solve the same problem. In the next section the benchmark problem is described, then a concise but comprehensive description of XCS and ZCS is done. Thus the experimental configuration is presented followed by the results, analysis and discussion, drawing the conclusions.

2 The "El Farol" Problem

The "El Farol" Problem [1] is an ill-defined problem that is based on a public bar. The place is small and uncomfortable if crowded above a comfort threshold (Λ_{cm}). Each member of a group of agents has to decide simultaneously whether to go to the bar or not, each week, based on attendance information of the previous weeks. The reward given to the agents depends upon if they attend the bar and it is not overcrowded or if they stay at home and the bar is overcrowded. For this experiment the reward function works as displayed in Eq. 1 as (reward, attendance) pairs.

$$Reward = \{(1000, \Lambda_{cm}), (679, \Lambda_{cm} - 0.1), (679, \Lambda_{cm} + 0.1), (0, other)\}. \quad (1)$$

Using Eq. 1 as the reward function gives a reward only if the bar is full enough to be amusing. For example, if the bar is overcrowded by 0.1, those agents that stayed at home receive a reward of 679 and those that attended the bar 0; on the other hand, if the bar is full to $\Lambda_{cm} - 0.1$ of its capacity, then those that attended receive a reward of 679 and those that stayed at home receive 0.

The agents start with a set of random strategies, each with an indicator of how good the strategy, that are tested, is throughout the experiment. The indicator is updated according to the reward received. After several steps, the worst performing agent is replaced by a copy of the best agent, slightly mutated. The agents are not allowed to evolve their own strategies. For the experiments reported here, there is no deletion nor reproduction of agents and the agents are allowed to evolve their strategies using the ZCS and XCS learning algorithms. Furthermore, the problem is very difficult, as it is: a) non-Markovian, because the answer depends on all the actions of the agent, not on the agents individually, therefore the states perceived by the agents are not unique based on the information provided by the sensors; b) non-stationary, because for a specific state, the correct answer is not the same always.

3 Strength-based fitness and accuracy-based fitness classifier systems

Learning Classifier Systems (LCS) [7] are evolutionary reinforcement learning methods and as such, interact with an environment, trying to establish a model of it and obtaining the maximum reward possible. The model is represented by condition→action rules, represented generally by simple strings, which are recombined by a genetic algorithm [7] [4] , based on their fitness.

Due to their complexity, the first classifier systems were difficult to analyse in detail. It was unclear how one component of the system affected the general performance. The "zero-th" level classifier system (ZCS) [9] has the minimal components of a LCS to analyse each part of the system independently. The population [P] of the rule base is initialised randomly.

A learning cycle of ZCS in a single step environment is: the system perceives its environment and encodes it into an environmental message that is compared with the classifiers in the rule base, putting those that satisfy it in the match set [M]. If no classifiers match or the average strength of the rules in [M] is a fraction (generally 0.5) of the population's average fitness a new rule that satisfies the environmental message is generated with the average strength of the population. Then, a probabilistic (exploration) or deterministic (exploitation) action is chosen from the classifiers in [M] and those classifiers that have it are put in the action set [A]. The action is posted to the environment and a reward is received. Once the reward (R) is received, it is divided by the size of [A] and distributed among the members of [A] using $S = S + \beta(R - S)$, β is the learning rate and S the classifier's strength. Those classifiers that are in [M], but not in [A] are taxed by using $S = S - \tau S$, τ is the tax rate.

XCS interacts and functions in the same way that ZCS does. The fitness calculation in different classifier systems besides its use for selection, also determines the action to take as a conflict resolution method when [M] contains contradictory actions. For a detailed explanation of XCS, the reader is referred to more detailed XCS articles [2][8] and ZCS [9].

4 The experiment

Ten agents have been tested for all the experiments, because this number gives better control than a larger one and in terms of analysis is less complex than using a large number of agents [5] [6]. For better control, all the agents either exploit or explore at the same time; individual random exploration/exploitation has yielded the same results reported here.

Unlike the "El Farol" problem there are no agents reproduced or killed from the agent population and different thresholds have been tested, to see if the system adapts: all the possible thresholds were tested, and the one selected to represent the behaviour of the system is 0.6. The 0.6 threshold is chosen as it is the most commonly reported result in the literature. Each agent accumulates the reward it receives for its actions throughout the experiment, this accumulation is the point of performance comparison of both classifier systems used.

For the experiments reported here, each learning algorithm perceives whether it was good (1) or not (0) to attend the bar in each of the previous 5 weeks (M=5), i.e. a 0 if the bar was overcrowded and a 1 if it was good to attend in the corresponding week. The rule conditions represent what was good to do last week with the leftmost bit and what was good to do 5 weeks ago with the rightmost bit.

To do a consistent analysis, different configurations have been tested for 10 agents: 2 XCS and 8 ZCS; 5 XCS and 5 ZCS; 8 XCS and 2ZCS. The heterogeneous multi-agent system is composed of homogeneous ZCSs and XCSs, i.e. the parameters and environment perceived by each ZCS and XCS are the same. The XCS parameters used are: don't care probability ($P_{\#}$)=0.333, explore/exploit rate=0.5, crossover rate (χ)=0.8, mutation rate (μ) = 0.02, θ_{GA}=25, minimum error (ϵ_0)=0.01 and learning rate (β) =0.2 for each XCS in the system. Effector covering is enabled: when a match set is formed and all the possible actions are not present, a new classifier is inserted to cover the missing action. Action set and GA subsumption deletion are used, the experience for subsumption is set to 50, since a lower value yielded fully-general rules in the early steps of the experiment leading the system to random behaviour.

The ZCS parameters used are: a population of 64 rules, learning rate (β)=0.2, tax rate (τ)=0.1, χ=0.8, μ=0.04, $P_{\#}$=0.33, the covering trigger is 0.5 (if $\Sigma S_{[M]} < 0.5 \Sigma S_{[P]}$) for every ZCS in the system. The rules of all the agents are initialised with 4 different random seeds, 64 rules are selected because this number allows the system to have each possible state for both actions in case no generalisation would be possible.

5 Results and discussion

The results reported are representative for 20 runs, which are very similar to one another. In order to test if there is a significant difference between the different numbers of agents that use ZCS, different number of agents have been tested: 2, 5, and 8 ZCS out of 10 agents in the system.

XCS outperforms ZCS in most of the experiments tested, the difference in performance is significant (Figs. 1 c). The difference between the final accumulated reward by the agents can be seen in Fig. 1 c. Sometimes, XCS performs worse than ZCS, e.g. agent 2 outperforms agents 3, 6 and 10, in Fig. 1.

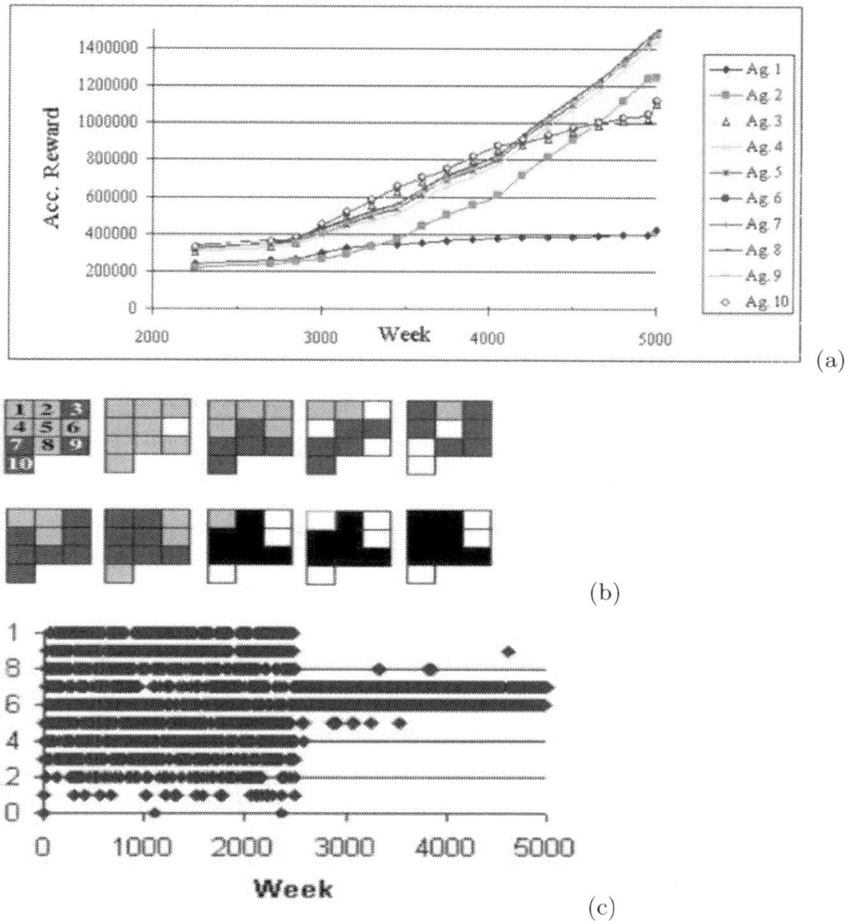

Fig. 1. The different aspects of a 2 ZCS 8 XCS configuration shows that agent 2 (ZCS) outperforms agents 3, 6 and 10 (XCS) agents (c) and (b) shows also that the attendance is not very far from the Λ_{cm} of 0.6. Individual actions (a) show the dynamics of the system.

An interesting feature in the system is that, the more ZCS are used, the lower the accumulated reward is and smaller the difference among the agents. The difference of accumulated reward between the best and worst agents is: when using 2 ZCS = 1052469 (Fig. 1c); when using 5 ZCS=557228; when using 8 ZCS= 278156.

Depending on the evaluation criterion that is used for the system, the interpretation can be different. If the evaluation criterion is the difference of accumulated reward among the agents, more ZCS are better for the system. On the other hand, if the evaluation criterion would be the total accumulated reward by the agents, more ZCS decrease the performance of the system. The populations of the agents show that no consistent generalisations are made, as the numerosity of all the rules is 1, in fact most rules are generated after exploration is switched off (time stamp>2500). Agent 1, the worst performer shows that all (not shown), but one, have been generated by covering after the exploration has finished. The case of the XCS agents is quite different regarding generalisation. The difference of the best (agent 9) and worst (agent 3) of the 8 XCS agents for the experiment with 2 ZCS are quite similar, the only significant difference between the populations is the prediction of each rule.

It can be observed for all the cases that most of the agents fix their behaviour and there are 1 or 2 ZCS that are always changing their behaviour, oscillating between going and not going (Fig. 1a). The accumulated reward during the exploration phase is quite even, until the system only exploits. When there are 8 ZCS and 2 XCS, both XCS are the top lines in the graph.

The final populations of the ZCS agents preserve very few rules from step 2500 until 5000. The final population of ZCS agents show how dynamic the ZCS population is, by having only 5 classifiers that are the same as at step 2500, and 2 of them have been generated by chance by covering. On the other hand, XCS agents kept most of the classifiers that it had at step 2500 because the fitness drops at the same time for all the classifiers in the same niche, covering is not triggered in XCS as often as in ZCS. It could seem that it is generalisation that allows the agents to achieve the best performance, in order to test this hypothesis, the analysis of the worst XCS performer is done for the 2 ZCSs and 8 XCSs experiment and then for the other experiments.

For most of the experiments and the ZCS agents that perform worse than the XCS agents, the number of rules taken from step 2500 are very low. It seems ZCS needs the genetic algorithm to keep good population control, as ZCS does not have non-genetic population reduction operators, it depends totally on the genetic algorithm to generalise.

An explanation for the XCS agents being consistently wrong could be the subsumption deletion itself, as it narrows the space search by locking the population in one specific point of the search space and then, since there is no mutation, there is no possibility of moving away from the bad spot where the XCS is stuck.

6 Conclusions

The flexibility and control of the multi-agent system for benchmarking learning algorithms is encouraging. The initial proposition of using MAS for benchmarking has proved to be effective as the research reported here illustrates. The observations and detail that can be obtained from a MAS have proved valuable for the current research.

The advantages that a MAS offers for benchmarking as many learning algorithms as agents are in the system and the direct interaction of the learning algorithms are precious and unique to a MAS.

The use of a MAS to compare XCS and ZCS, besides being novel, has yielded unexpected results, considering previous work of accuracy-based vs. strength-based [8] fitness comparisons, where the former LCSs always outperformed the latter kind.

References

1. W. B. Arthur. Complexity in Economic Theory: Inductive Reasoning and Bounded Rationality. *The American Economic Review*, 84(2):406–411, May 1994. http://www.santafe.edu/ wba/Papers/El_Farol.html.
2. M.V. Butz and S.W. Wilson. An algorithmic description of XCS. In P. L. Lanzi, W. Stolzmann, and S. W. Wilson, editors, *Advances in Learning Classifier Systems. Third International Workshop (IWLCS-2000)*, Berlin, 2001. Springer-Verlag.
3. J. Ferber. *Multi-Agent Systems: An introduction to distributed artificial intelligence.* Adisson Wesley, 1999.
4. D.E. Goldberg. *Genetic algorithms in Search, Optimization and Machine Learning.* Addison Wesley, 1989.
5. L. Miramontes Hercog and T.C. Fogarty. Co-evolutionary classifier systems for multi-agent simulation. In D. B. Fogel, Mohamed A. El-Sharkawi, X. Yao, G. Greenwood, H. Iba, P. Marrow, and M. Shackleton, editors, *Proceedings of the 2002 Congress on Evolutionary Computation CEC2002*, pages 1798–1803. IEEE Press, 2002.
6. L. Miramontes Hercog and T.C. Fogarty. Social simulation using a Multi-Agent Model based on Classifier Systems: The Emergence of Vacillating Behaviour in the "El Farol" Bar Problem. In P.L. Lanzi, W. Stolzman, and S. W. Wilson, editors, *Advances in Learning Classifier Systems. Proceedings of the Fourth International Workshop in Learning Classifier Systems 2001.* Springer-Verlag, 2002. LNAI series 2321. ISBN: 3-540-43793-2.
7. J.H. Holland. *Hidden Order.* Addison-Wesley, 1995.
8. T. Kovacs. Strength or Accuracy? A comparison of two approaches to fitness calculation in learning classifier systems. In *2nd. International Workshop in Classifier Systems*, 1999.
9. S.W. Wilson. ZCS: A Zeroth Level Classifier System. *Evolutionary Computation*, 2(1):1–18, 1994.

Soft Learning: A Conceptual Bridge between Data Mining and Machine Learning

Flaviu Adrian Mărginean

Department of Computer Science, The University of York
Heslington, York YO10 5DD, United Kingdom
flav@cs.york.ac.uk

1 Introduction

It has been felt for some time that, despite employing different formalisms, being served by their own dedicated research communities and addressing distinct issues of practical interest, problems in Data Mining and Machine Learning connect through deep relationships. The paper [5] has taken a first step towards linking Data Mining and Machine Learning via Combinatorics by showing a correspondence between the problem of finding maximally specific sentences that are interesting in a database, the model of exact learning of monotone boolean functions in computational learning theory and the hypergraph transversal problem in the combinatorics of finite sets. [5] summarises and concludes a series of valuable Data Mining research on fast discovery of association rules by the levelwise algorithm, series that includes [1, 4, 11]. Intuitively, a Data Mining task may consist of finding many weak predictors in a hypothesis space whereas in Machine Learning one strong predictor is sought.

Data Mining therefore appears as a form of *soft* Machine Learning. It is shown in [5] that:

> *"The field of data mining has recently expanded considerably: both applications and methods have been developed at a fast pace. However, the theory of the area is still at its infancy: there is virtually no information about the complexity of data mining problems, and there even is no generally accepted formal framework for the area."*

The authors of [5] therefore take preliminary steps towards laying a theoretical foundation for Data Mining. In this paper we take this a step further, pinpointing elements of unity between many apparently distinct problems in Data Mining and Machine Learning. In the same vein with the approach in [5], the connection is provided by Combinatorics.

The paper is organised as follows: in Section 2 we describe the initial motivation that led us to pursuing the development of the concept of Soft

Learning. Section 3 outlines a definition for Soft Learning and discusses its relevance to Soft Computing. Section 4 presents a PAC-learnability analysis for a simple but interesting instantiation of Soft Learning. Finally, Section 5 concludes.

2 The Roots of Soft Learning: Association Rule Data Mining, Learning in Databases, Clause Lattice Search, Version Spaces

The research on Association Rule Data Mining [1, 4, 11] has led to the development of the following formal framework [5]: we are given a database $\mathbf{r}$, a language $\mathcal{L}$ for expressing properties or defining subgroups of the data, and an interestingness predicate q for evaluating whether a sentence $\phi \in \mathcal{L}$ defines an interesting subclass of $\mathbf{r}$. The task will then be to find the theory of $\mathbf{r}$ with respect to $\mathcal{L}$ and q, i.e. the set $Th(\mathcal{L}, \mathbf{r}, \mathrm{q}) \stackrel{\mathrm{def}}{=} \{\phi \in \mathcal{L} \mid q(\mathbf{r}, \phi)\}$. It has generally been assumed that the predicate q is monotone with respect to some specialisation/generalisation ordering on $\mathcal{L}$, a property that is then used to search for $Th(\mathcal{L}, \mathbf{r}, \mathrm{q})$.

The study of relevance for learning in deductive databases [10] is about identifying, in a preprocessing stage, those literals in a Datalog language $\mathcal{L}$ that are relevant to learning. Relevance is defined as a maximality property with respect to a quasi-order derived from positive and negative coverage, and the costs of literals.

Cautious induction [2] is an approach in Inductive Logic Programming in which the space of first-order clauses defined by the training data and background knowledge is searched, indeed mined, for *candidate* clauses, all of which meet some quality criteria. A subset of those candidate clauses are then used to assemble a theory that covers all the positive examples and none of the negative ones, and which is consistent with background knowledge.

Version Spaces and Disjunctive Version Spaces [6, 8, 13, 14, 16] collect all the strong predictors consistent with training data. Once again, the underlying ordering of the hypothesis space is used to compute the Version Spaces.

There is no a priori reason that we should consider these distinct problems, except for the fact that they have evolved from different necessities, are cast in different terminologies and deal with representations of various descriptive power. Beyond these rather superficial aspects, however, deeper unity lies: all these problems require the search of large search spaces, whether they be transaction databases, sets of literals defined by the language or hypotheses spaces defined by the data and background knowledge available. In all these cases, the ordering endowing the search space is used to mine, learn or search for certain predictors that, by themselves, will not necessarily be strong enough to be satisfactory: in transaction databases, an association rule will be just one of the many mined items that *collectively* rather than individually will give an indication of the patterns in customer preferences; in learning relevant literals from a database, it will not be one literal or another that matters but the *collection* of all such literals from which clauses are then built; in lattices of clauses mined by cautious induction algorithms, it will not be the good individual clauses that are the final output, but the *collection* of such clauses from which theories are then built according to a certain procedure.

3 Definition of Soft Learning and Relevance to Soft Computing

Definition 1 (Soft Learning — First Definition). *Soft Learning is the Machine Learning of soft targets, in particular those representable as sets of weak predictors.*

Whereas in Machine Learning the accent is on inferring a unique strong predictor, in Soft Learning weak predictors are sought that congregate to collectively rather than individually communicate some useful information: for instance, relevant literals identified in the Soft Learning phase are then used to build Datalog clauses; or, Horn clauses identified in a cautious induction phase are then used to build theories. In Data Mining the set of mined association rules conveys information about customer preferences in transaction databases.

Soft Computing has been defined by L.A. Zadeh as covering those forms of computing that are tolerant of imprecision, uncertainty, partial truth, or approximation. Under this definition, Soft Learning will be that form of Machine Learning that is tolerant of partial predictability as well as multiplicity of predictors. There are many familiar intuitions, albeit loose analogies, for the relationship *(one-strong : many-weak)* predictors. For instance, the relationship between a set and its parts (*one-set : many-parts*); or, at a higher level of abstraction, the relationship between a formal topological space and its locales *(one-space : many-locales)*. More importantly for the framework we give, the relationship *(one-maximum : many-maximals)* in any quasi-order.

Soft Computing can also be said to be the computing that emerges from Hard Computing by interpolation between crisp choices, a process otherwise known as *fuzzification*. At the end of this process we might get not so much a semantic extension as the transformation of absolute qualifications in relative ones. For instance, the fuzzification of the known antinomy *(tall : short)* is often thought to produce intermediate characterisations of tallness by numerical assignments. Whilst this is formally correct according to fuzziness research, the true significance thereof is that a change of emphasis takes place, from *(tall : short)* to *(taller : shorter)*. A set of individuals can easily be separated into two categories with respect to the former antinomy, but arranged in a scale with respect to the latter — without any numerical assignments, as long as we have the ability to compare those individuals directly. It is reasonable to assume that the human skill can make such determinations directly, bypassing any numerical assignments. The moral of this example however is that we can marry fuzzification with higher abstraction, for instance using orderings to interpolate between crisp absolute values. We intend to exploit this insight to provide an abstract definition for Soft Learning. The essence of the insight is that we can change the antinomy *(strong : weak)* predictor to the antinomy *(stronger : weaker)* by introducing a quasi-order on the class of predictors (hypothesis space). Consistently with the desired level of abstraction, this will bypass any definition of strength for a predictor. We thus assume the ordering as given rather than describe any particular case of arriving at such an ordering, thereby ensuring a precise, yet abstract and free of clutter, definition.

Definition 2 (Soft Learning — Second Definition). *Soft Learning requires the determination of maximal elements according to a quasi-order describing the desirability of predictors in the hypothesis space (Optimal Soft Learning) or the determination of all predictors meeting some threshold requirements with respect to the said quasi-order (Threshold Soft Learning). A combination of the two is also possible by demanding the determination of those maximal predictors satisfying certain threshold requirements (Optimal-Thresholded Soft Learning).*

This definition can be shown to give us a common combinatorial (order-theoretic) denominator to many of the forms of learning described in Section 2. All these problems can be restated as finding maximal or good (in conformity with some criterion) elements in a given quasi-order. Notably, the use of negative information in relevance search and cautious induction will lead to this quasi-order being quite different from the generality relation on the hypothesis space. For Version Spaces, on the other hand, the emphasis on strong predictors might severely limit its applicability. Association Rule Data Mining and Cautious Induction appear as forms of Threshold or Optimal-Thresholded Soft Learning, whilst the Relevant Literal Learning and Version Spaces belong to Optimal or Optimal-Thresholded Soft Learning. We have no space to develop these ideas in here, except to say that the route to a more unified approach involves getting rid of tight boolean monotony assumptions tying the desirability quasi-order to the generality order on the hypothesis space. In the next section we describe one instantiation of Optimal Soft Learning for the well-known PAC-setting of learnability.

4 PAC-learnability Analysis for Optimal Soft Learning: Soft Version Spaces and Sample Complexity

We first present the PAC model of learning, generally following Mitchell [14].

In the PAC model of learning (whereof many variants exist), we are customarily given an instance space X and two subsets of the powerset of X, the target concept space $\mathcal{C}$ and the hypothesis concept space $\mathcal{H}$: $\mathcal{C}, \mathcal{H} \subseteq 2^X$; they can equally well be thought of as spaces of Boolean-valued functions over X. These two spaces are usually implicitly defined by some representational schemes, for instance DNF or CNF formulae, first-order representations etc. However, this will not be important in our analysis. It is assumed in the classical model that $\mathcal{C} \subseteq \mathcal{H}$, i.e. any possible target is representable in the hypothesis space of the learner. This entails the possibility of finding a unique strong predictor in the hypothesis space.

In Soft Learning, we no longer make this rather strong assumption. Models of learning that do not make this assumption have been described before under the name robust or agnostic learning [7, 14], however those models differ in a number of ways from the treatment that we propose here.

It is further assumed in the PAC model that a distribution $\mathcal{D}$ over X is given, unknown to the learner but fixed in advance. The purpose will be to probably approximately learn $c \in \mathcal{C}$ by querying an oracle that makes random independent draws from the distribution $\mathcal{D}$. Every time the oracle is queried by the learner, it draws an instance $x \in X$ at random according to $\mathcal{D}$ and returns the pair $\langle x, c(x) \rangle$ to the learner. An approximation $h \in H$ to $c \in \mathcal{C}$ is evaluated with respect to the distribution $\mathcal{D}$ over X: the error of the approximation is the probability that an instance from X randomly drawn according to $\mathcal{D}$ will be misclassified by h. It is required that a learner, using reasonable amounts of computational and informational resources (time, space, queries to the random oracle), output a hypothesis that with high confidence approximates the target well-enough. The use of resources such as time and space define the computational complexity of PAC learning while the number of queries to the random oracle needed to probably approximately infer the unknown target defines the sample complexity or information-theoretic complexity of the PAC learning. It is the latter we will be concerned with in this paper. Given that we no longer can rely on the assumption $\mathcal{C} \subseteq \mathcal{H}$ which would allow us to define the nonempty version space of the hypotheses consistent with the data, we have to find another way of defining interesting hypotheses from the point of view of consistency with the data. We do so by introducing two quasi-orders on $\mathcal{H}$, which are relative to the training data D and target concept c. The orders are subscripted $\mathcal{OD}$, indicating *desirability* of hypotheses in $\mathcal{H}$ (or Order of Desire). We shall then use the two quasi-orders as a bridge between the case treated in the classical theory of learning and our new setting.

The first quasi-order on $\mathcal{H}$ (classical case) is defined as follows: $h_1 \sqsupseteq_{\mathcal{OD}(c,D)} h_2$ *iff* $[\forall \langle x, c(x) \rangle \in D : h_2(x) = c(x)] \Rightarrow [\forall \langle x, c(x) \rangle \in D : h_1(x) = c(x)]$. Another way of writing this quasi-order is:

$$h_1 \sqsupseteq_{\mathcal{OD}(c,D)} h_2 \quad \textit{iff} \quad h_2|_D = c|_D \Rightarrow h_1|_D = c|_D$$

The second quasi-order on $\mathcal{H}$ (new setting) is defined as follows:

$$h_1 \succeq_{\mathcal{OD}(c,D)} h_2 \quad \textit{iff} \quad \forall \langle x, c(x) \rangle \in D : \; h_2(x) = c(x) \Rightarrow h_1(x) = c(x)$$

It is trivial to verify that the quasi-order axioms (transitivity, reflexivity) are satisfied in both cases. The relationship between the two orders is as follows:

$$\forall h_1, h_2 \in \mathcal{H}: \; h_1 \succeq_{\mathcal{OD}(c,D)} h_2 \;\;\Rightarrow\;\; h_1 \sqsupseteq_{\mathcal{OD}(c,D)} h_2$$

Another way of writing the definition of the new quasi-order $\succeq_{\mathcal{OD}(c,D)}$ is as follows:

$$h_1 \succeq_{\mathcal{OD}(c,D)} h_2 \;\; \textit{iff} \;\; E^+(h_1) \supseteq E^+(h_2) \; \textit{and} \; E^-(h_1) \subseteq E^-(h_2)$$

where:

$$E_+(h) \stackrel{\text{def}}{=} \{x \in X \mid h(x) = 1, \, c(x) = 1\} \; \textit{and}$$

$$E^-(h) \stackrel{\text{def}}{=} \{x \in X \mid h(x) = 1, \, c(x) = 0\}$$

Therefore, the distinction between the two quasi-orders is one of degree of fineness: the classical one is much coarser. Indeed, let $\sim_{\sqsupseteq_{\mathcal{OD}(c,D)}}$ be the equivalence relationship on $\mathcal{H}$ canonically induced by the quasi-order $\sqsupseteq_{\mathcal{OD}(c,D)}$:

$$h_1 \sim_{\sqsupseteq_{\mathcal{OD}(c,D)}} h_2 \;\; \textit{iff} \;\; h_1 \sqsupseteq_{\mathcal{OD}(c,D)} h_2 \; \textit{and} \; h_2 \sqsupseteq_{\mathcal{OD}(c,D)} h_1$$

The partial order $\mathcal{OD}(c,D)/\sim_{\sqsupseteq_{\mathcal{OD}(c,D)}}$ on $\mathcal{H}/\sim_{\sqsupseteq_{\mathcal{OD}(c,D)}}$ is then simply the boolean order $B_1 \stackrel{\text{def}}{=} (0 < 1)$, with the subclass of consistent hypotheses corresponding to 1 and the subclass of inconsistent hypotheses corresponding to 0. This is a pretty rough distinction. Using the two quasi-orders, the notion of Version Space $VS_{\mathcal{H},D}$ [14] generalises as follows:

$$VS_{\mathcal{H},D} \stackrel{\text{def}}{=} \{h \in \mathcal{H} \mid h|_D = c|_D\} = \{h \in \mathcal{H} \mid \nexists\, h' \in \mathcal{H} \; \textit{such that} \; h' \sqsupset_{\mathcal{OD}(c,D)} h\}$$

$$SVS_{\mathcal{H},D} \stackrel{\text{def}}{=} \{h \in \mathcal{H} \mid \nexists\, h' \in \mathcal{H} \; \textit{such that} \; h' \succ_{\mathcal{OD}(c,D)} h\}$$

We call the generalised Version Space, Soft Version Space $SVS_{\mathcal{H},D}$. The word *soft* we use to indicate the graceful degradation of hypotheses with respect to consistency. With classical Version Spaces, hypotheses are classified in a crisp black-and-white manner: hypotheses are either in (consistent) or out (inconsistent) of the Version Space. Soft Version Spaces $SVS_{\mathcal{H},D}$ retain the hypotheses maximally consistent with the data D, but the gap between them and the hypotheses left out of the version space is not as dramatic as in the classical case. Rather, consistency comes in degrees and there may be all sorts of shades, i.e. hypotheses that are more or less consistent with the data D according to the finer order $\succeq_{\mathcal{OD}(c,D)}$. The problem with the classical coarse order $\sqsupseteq_{\mathcal{OD}(c,D)}$ is that it needs $\mathcal{C} \subseteq \mathcal{H}$. If this assumption does not hold, the order $\sqsupseteq_{\mathcal{OD}(c,D)}$ may be subject to catastrophic failure: accumulation of data may lead to an empty version space, i.e. all hypotheses becoming inconsistent. In this case, the order $\sqsupseteq_{\mathcal{OD}(c,D)}$ no longer communicates any useful information: from the point of view of the order $\sqsupseteq_{\mathcal{OD}(c,D)}$, all hypotheses form just one equivalence class. On the contrary, the finer order $\succeq_{\mathcal{OD}(c,D)}$ handles the lack of the assumption $\mathcal{C} \subseteq \mathcal{H}$ pretty well: with accumulation of data, the Soft Version Space may come to consist of more than one equivalence class. Nevertheless, under reasonable assumptions, it is never empty.

When $D = X$ we denote the Soft Version Space by $SVS_{\mathcal{H},c}$, the set of hypotheses in $\mathcal{H}$ maximally consistent with c over the entire instance space X.

Definition 3 (($\mathcal{C}, \mathcal{H}$) is (η, d)—Well-Behaved). *Let $\mathcal{C}, \mathcal{H} \subseteq 2^X$ be a target concept space and a hypothesis concept space, respectively. The pair $(\mathcal{C}, \mathcal{H})$ is said to be* well-behaved *if the following two conditions are satisfied:*

1. *$\mathcal{C}$ is a family of Borel sets that does not have infinite ascending chains in the quasi-order $\succeq_{\mathcal{OD}(c,X)}$ and there exists a natural number η such that for all $c \in \mathcal{C}$, $1 \leq |SVS_{\mathcal{H},c}| \leq \eta < \infty$.*
2. *$\mathcal{H}$ is well-behaved in the sense of [3, A.1, p. 953][1] and has finite Vapnik-Chervonenkis dimension $d < \infty$.*

The following theorem establishes the sample complexity of Soft Learning for maximally consistent learners[2]. A proof can be found in [12].

Theorem 1 (Soft Version Spaces are $\epsilon, \delta, \eta, d$−Exhaustible). *Let $\mathcal{C}, \mathcal{H} \subseteq 2^X$ be a target concept space and a hypothesis concept space, respectively, such that $(\mathcal{C}, \mathcal{H})$ is well-behaved. For all $0 < \epsilon, \delta < 1$, for all $D \in (X \times \{0,1\})^m$ training data such that $m \geq \frac{2}{\epsilon^2} \max\left(\ln \frac{4\eta}{\delta}, 2\epsilon \log_2 \frac{4\eta}{\delta}, 4\epsilon d \log_2 \frac{13}{\epsilon}\right)$ and for all $h \in SVS_{\mathcal{H},c}$ there is $h' \in SVS_{\mathcal{H},c}$ and $h'' \in SVS_{\mathcal{H},D}$ such that, with probability at least $1 - \delta$, $error(h', c) \leq error(h, c) + \epsilon$ and $error(h', h'') < \epsilon$.*

5 Conclusion

This paper has introduced the concept of Soft Learning as a bridge between Data Mining and Machine Learning. For those who find such pseudo-equations appealing or insightful, a nice way of summarising things is:

$$\text{Data Mining} \subseteq \text{Soft Learning} \stackrel{\text{def}}{=} \text{Machine Learning} \cap \text{Soft Computing}$$

The benefits of seeing Data Mining as a form of learning is that this may contribute to a more solid foundation for the Data Mining enterprise. Much good work has been done in devising and experimentally testing algorithms for Data Mining tasks, but so far we largely miss the kind of theoretical analysis that, for Machine Learning, is provided by computational learning theory. Our learnability settings work well for unique strong predictors, but we need to extend this form of analysis to settings in which soft targets, consisting of many weak predictors, are to be sought.

---

[1] For the purposes of this paper, the exact definition of $\mathcal{H}$'s well-behavedness is not essential. "Virtually any $\mathcal{H}$ that one may consider in the context of machine-learning applications will be well-behaved [3, A.1, p. 953]."

[2] Compare with similar results in [14] regarding the complexity of PAC learning for consistent learners.

Dedication

In loving memory of my father, Dr. Emil Mărginean, on the 64th anniversary of his birth, 16th June 2003.

References

1. R. Agrawal, H. Mannila, R. Srikant, H. Toivonen, and A. Verkamo. Fast Discovery of Association Rules. In U. Fayyad, G. Piatetsky-Shapiro, P. Smyth, and R. Uthurusamy, editors, *Advance in Knowledge Discovery and Data Mining*, pages 307–328. AAAI Press/MIT Press, 1996.
2. S. Anthony and A. Frisch. Cautious Induction in Inductive Logic Programming. In Lavrač and Džeroski [9], pages 45–60.
3. A. Blumer, A. Ehrenfeucht, D. Haussler, and M.K. Warmuth. Learnability and the Vapnik-Chervonenkis Dimension. *Journal of the Association for Computing Machinery*, 36(4):929–965, 1989.
4. L. Dehaspe and L. De Raedt. Mining Association Rules in Multiple Relations. In Lavrač and Džeroski [9], pages 125–132.
5. D. Gunopulos, H. Mannila, R. Khardon, and H. Toivonen. Data Mining, Hypergraph Transversals, and Machine Learning (extended abstract). In *Proceedings of the* 16th *ACM SIGACT-SIGMOD-SIGART Symposium on Principles of Database Systems*, pages 209–216, Tucson, Arizona, USA, 1997. ACM Press.
6. D. Haussler. Quantifying inductive bias: AI learning algorithms and Valiant's learning framework. *Artificial Intelligence*, 36:177 – 221, 1988.
7. D. Haussler, S. Ben-David, N. Cesa-Bianchi, and P. Long. Characterizations of Learnability for Classes of {0,...,n}-valued Functions. *J. Comp. Sys. Sci.*, 50(1):74–86, 1995.
8. H. Hirsh. *Incremental Version Space Merging: A General Framework for Concept Learning*. Kluwer, 1990.
9. N. Lavrač and S. Džeroski, editors. *Inductive Logic Programming, 7th International Workshop, ILP-97*, volume 1297 of *Lecture Notes in Artificial Intelligence*. Springer, 1997.
10. N. Lavrač, D. Gamberger, and V. Jovanoski. A Study of Relevance for Learning in Deductive Databases. *Journal of Logic Programming*, 40(2/3):215–249, 1999.
11. H. Mannila and H. Toivonen. Levelwise Search and Borders of Theories in Knowledge Discovery. *Data Mining and Knowledge Discovery*, 1(3):241–258, 1997.
12. F.A. Mărginean. *Combinatorics of Fineness and Refinement*. PhD thesis, Department of Computer Science, The University of York, 2001.
13. T.M. Mitchell. *Version Spaces: An Approach to Concept Learning*. PhD thesis, Electrical Engineering Department, Stanford University, 1979.
14. T.M. Mitchell. *Machine Learning*. McGraw-Hill, New York, 1997.
15. L. Pitt and L. G. Valiant. Computational limitations on learning from examples. *Journal of the ACM*, 35(4):965–984, 1988.
16. M. Sebag. Delaying the Choice of Bias: A Disjunctive Version Space Approach. In R. Bajcsy, editor. *Proceedings of the* 13th *International Conference on Machine Learning, IJCAI 1993*. Morgan-Kaufmann, August–September 1993.

Heuristics for Kernels Adaptation in Support Vector Machines *

Ernesto Saavedra , Ingo Renners , Adolf Grauel[1] D. Morton and H.J. Convey[2]

[1] University of Applied Sciences - South Westphalia, Soest Campus,Center of Computational Intelligence and Cognitive Systems. Soest 59494, Lübecker Ring 2, Germany. [saavedra,renners,grauel]@fh-swf.de

[2] Bolton Institute of Higher Education, Faculty of Technology, Technology Development Unit, Deane Campus. Bolton, BL3 5AB, England [dm3,hc1]@bolton.ac

1 Introduction

Support Vector Machines are an algorithm introduced by Vapnik and co-workers [9], [10]. They are based on the idea that if input points are mapped to a high dimensional feature space then, a separating hyperplane can be easily found. **SVM** and kernel methods have been applied to a wide class of problems including approximation and classification and they have proven a remarkable performance on real world problems. An important step in their design is the setting of the kernels parameters which defines the structure of the high dimensional feature space where a maximal margin hyperplane will be found. Too rich feature space, e.g. small kernel parameters, will over-fit the data and hence result in a poor generalisation error, whereas if the kernel parameter is too big, the model will not be able to separate the data. In this paper we firstly propose a heuristic that permits the individual control of the growth in each kernel, which results in more sparse models with higher prediction accuracy. Secondly, a heuristic resulting from the combination of **SVM** trained by linear programming **(LP)** and **EC** for the optimisation of the kernels width is proposed.

2 Heuristics for SVM classifiers

This approach is based on Kernel Adatron algorithm **(KA)** [1], [2], which converges rapidly to the maximal margin solution and has proved to be robust and easy to implement. The decision function for classification is:

* This work has been supported by the Ministry of Education and Research, North-Rhine Westphalia - Germany.

$$f(x) = sign(\sum_{i \in SV} y_i \bar{\alpha}_i K(x, x_i) + \bar{\lambda}), \tag{1}$$

where $\bar{\alpha}$, $\bar{\lambda}$ are the last values of α_i^t and λ^t found in the last iteration. In order to adjust the kernels width, [1] proposes the following algorithm:

1. Initialise σ to a very small value.
2. Start the **KA** algorithm for classification.
3. If the margin is maximised, then
   - compute the generalisation or validation error,
   - increase the kernel parameter $\sigma \leftarrow \sigma + \delta\sigma$,
4. else go to step 2.
5. Stop when a predetermined value of σ is reached.

The kernels parameter σ is set to a small value and then slightly increased. Every time the **KA** algorithm has finished, a validation error correspondent to that value of σ is calculated. The algorithm stops when a certain value of σ has been reached. Finally a **SVM** network with the same kernel parameter for all support vectors and a small generalisation error is obtained.

2.1 Heuristic for individual setting of the kernel parameter

Using the **KA** algorithm, the support vectors coefficients α_i are initially set to zero and σ to a small value. The result of this will be a **SVM** network with all input samples as support vectors whose coefficients are the values that correspond to the output at that point if the maximum value for the support vectors (**C**) has not been set to a value smaller than the maximum output. If σ is incremented, then less input samples will be chosen as support vectors and the coefficients α_i will tend to **C**. However, it could also be possible to monitor the **SV** coefficients every time they are returned by the **KA** algorithm and check whether some of them are already equal or bigger than **C**. If that is the case, σ is decreased in one step and tagged so that its correspondent kernel parameter could not be any longer incremented. The algorithm finishes when the kernels correspondent to each sample cannot be modified or when a certain number of increments on the kernel parameter has been reached. The outcome of this heuristic will be a **SVM** network with a very small generalisation error and less support vectors as consequence of calculating σ individually for each support vector. The algorithm is listed below:

1. Initialise $\alpha_i^0 = 0$, $sv_{best} =$ number of training samples.
2. Set all kernel parameters to a initial value $\sigma_i \leftarrow \sigma_0$.
3. While ($p < p_{max}$ and $sv > sv_{max}$ and $svTag < svTag_{max}$ and $svC < svC_{max}$)
   - go to the **KA** algorithm
   - count support vectors (sv), tagged support vectors ($svTag$), support vectors with maximum coefficient value (svC)
   - if $svC > svC_{out}$
     – for $i = 0$ to $i < svC_{max}$

- get σ_x of support vector correspondent to i
- decrease σ_x by $\delta\sigma$
- tag support vector correspondent to i

- else if $sv < sv_{best}$ then store **SVM** network
- increase kernels of all not tagged support vectors $\sigma_x \leftarrow \sigma_x + \delta\sigma$
- $sv_{best} \leftarrow sv$
- $p = p + 1$

4. End while loop.

where :

- sv_{best} initially set to the number of training samples. It keeps track of the minimum number of **SV**.
- σ_0 is a initial value set to all possible kernels.
- p_{max} is the maximum number of iterations in the algorithm. For each iteration the **KA** algorithm returns a **SVM** network.
- sv_{max} is the stop criteria. The algorithm will stop if a network has a number of support vectors smaller than this value.
- $svTag_{max}$ is the number of support vectors that can be tagged. Its maximum possible value is the number of training samples.
- svC_{max} is the maximum number of support vectors in a network whose kernels can be modified at the same time.
- svC_{out} is the maximum number of support vectors whose coefficients can be bigger than **C**. This, in order to permit some error during the training process.

2.2 Testing the heuristic for SVM classifiers

The heuristic was tested on two well known classification problems: Sonar data set [3] which contains 208 samples and 60 features, and Wisconsin breast cancer data set [7] *WBC* with 683 samples and 9 features. For all cases the data was split in 2/3 part for training 1/3 part of the data set for testing selected randomly and proportionally to the number of elements in each class. For each case, the data was split 25 times yielding the same number of **SVM** networks. The best and worst results obtained by using the the heuristic proposed by [1] with same widths for all **SV**, and the heuristic we propose for kernels with variable size are presented in table 1. The columns CC_{train} and CC_{test} refer to the percentage of correctly classified patterns.

3 Heuristics for SVM function approximators

Motivated by the approaches proposed by [4], [8] we combined **EC** with **SVM** trained by **LP** [5],[6]. As a first aim, support vectors are found by considering all samples at the same time and optimising their correspondent kernels via **EC**. A second heuristic was implemented on which the **EC** chunks the data set and with each part it tries to find a network such that all outputs are within the tube. Also for this case the **EC** optimises the kernels shape. The proposed algorithm is:

1. Create a population *Pop* of N individuals *Indi*. An individual can represent a network with all samples as kernels or with only a set of them, depending on the minimum and maximum number of kernels. Each kernel is also restricted to a minimum and maximum width.
2. While $i <$ *max. generations*
   - If $i > 0$ then **crossover** $Pop(N)$
     - $loop = 0$
     - Do
       - If $j > 0$
         - If $loop > 0$ and $loop < 3$ then **mutate** *Indi(j)*
         - Else if $loop > 3$ then **initialise** *Indi(j), loop = 0*
       - *Indi(j)* to **SVM(j)**
       - Calculate activation matrix with samples chosen as possible **SV**
       - Calculate **SV** with **LP**. It returns the min. square error *(error)*, **SV**, samples out of tube *(sot)*
       - $loop = loop + 1$
     - While *error == NaN*
     - set fitness of *Indi(j)* in table *(sot, SV, error)*
     - Store coefficients of **SVM(j)** in *Indi(j)*
   - **tournament** selection in *Pop(N)*
3. End while loop.
4. Get the best individual *Indi(0)*.

Table 1. Classification results for the sonar and *WBC* data set. (CC: Correctly classified patterns)

KA parameters							Algorithm parameters					
Net.	t_{max}	ϵ	ϵ_0	μ	η	C	P_{max}	sv_{max}	svC_{max}	svC_{out}	σ_0	$\delta\sigma_0$
n_{1-4}	300	0.001	0.05	10.0	1.0	7.5	300	5	130	10	0.5	0.05
n_{5-8}	300	0.001	0.05	10.0	1.0	10.0	300	5	150	0	0.01	0.025

Results

Net.	*Algorithm*	*Data*	*SV*	σ_{min}	σ_{max}	CC_{train}	CC_{test}
n_1 (Best)	*Fixed Widths*	*Sonar*	73	1.32	1.32	100	94.12
n_2 (Worst)	*Fixed Widths*	*Sonar*	60	1.45	1.45	100	77.94
n_3 (Best)	*Diff. Widths*	*Sonar*	70	1.2	1.63	100	97.06
n_4 (Worst)	*Diff. Widths*	*Sonar*	57	1.22	2.22	100	73.53
n_5 (Best)	*Fixed Widths*	*WBC*	103	0.48	0.48	100	98.66
n_6 (Worst)	*Fixed Widths*	*WBC*	91	0.45	0.45	100	93.33
n_7 (Best)	*Diff. Widths*	*WBC*	33	0.41	1.97	100	99.11
n_8 (Worst)	*Diff. Widths*	*WBC*	25	0.5	1.7	100	94.2

3.1 Testing the heuristic for SVM approximators

We have used the approach for approximating the following one-dimensional and two-dimensional functions:

- $f_1(x, y) = \frac{sin(x).sin(y)}{x.y}$ (120 samples),
- $f_2(x) = 10(e^{-5|x|} + e^{\frac{-3|x-0.8|}{10}} + e^{10|x+0.6|})$ (201 samples),
- $f_3(x) = 8sin(10x^2 + 5x + 1)$ (201 samples).

The independent variables were normalised to the interval [0; 1]. Even though in all cases the data size was small, the obtained results demonstrate that it is possible to combine the **SVM** training methods with optimisation strategies for kernels via **EC**. The results are presented in table 2.

Table 2. Configuration parameters and results for f_1, f_2, f_3

GA parameters									SVM parameters		
Net.	*Gen.*	*Indi*	*Elite*	*Comp.*	SV_{min}	SV_{max}	w_{min}	w_{max}	C	ϵ_0	ϵ_{tube}
n_1	50	50	1	10	1	80	$1 \cdot 10^{-3}$	0.5	∞	$1 \cdot 10^{-6}$	0.05
n_2	50	50	1	10	1	80	$1 \cdot 10^{-3}$	0.5	∞	$1 \cdot 10^{-6}$	0.075
n_3	50	50	1	10	20	80	$1 \cdot 10^{-3}$	0.5	∞	$1 \cdot 10^{-6}$	0.1
n_4	100	50	1	10	201	201	$1 \cdot 10^{-3}$	0.5	∞	$1 \cdot 10^{-6}$	0.05
n_5	50	50	1	10	1	80	$1 \cdot 10^{-3}$	0.5	∞	$1 \cdot 10^{-6}$	0.05
n_6	50	50	1	10	1	60	$1 \cdot 10^{-3}$	0.5	∞	$1 \cdot 10^{-6}$	0.2

Results						
Net.	*func.*	SV	w_{min}	w_{max}	Out_{tube}	MSE
n_1	f_1	28	0.01	0.49	5	0.00058
n_2	f_1	17	0.006	0.47	0	0.00067
n_3	f_2	42	0.018	0.49	16	0.017
n_4	f_2	74	0.1	0.49	0	0.0003
n_5	f_3	39	0.007	0.49	1	0.0001
n_6	f_3	31	0.2	0.49	0	0.0035

4 Conclusions

In this paper we presented two different heuristics based on **SVM**, **EC** and **LP** for adjusting the kernels widths on classification and approximation networks. The aim has been to find networks with a given generalisation error and a minimum number of support vectors. In the case of classification networks, the heuristic is based on monitoring the **SV** coefficients while the kernels widths are increased in small steps. The method was tested on well known benchmark data and the results obtained are better than the ones reported in literature. Regarding functions approximation, we proposed a method which avoids the usage of the whole data at the same time for evaluating the activation matrix while it optimises the widths of each possible kernel. Even though this implementation takes longer to find the **SV**, it has the advantage of being independent of the data size and to represent the data with more accuracy and less number of support vectors. For most of the problems on which the method was tested, the support vectors were the same as those found using other training methods. On the other hand, it was noticed that the algorithm finds easily most of the **SV** to set the output within the tube whereas it takes longer for some particular data. Therefore, it could also be possible to speed up the training process by optimising the tube in order to obtain a network with different and minimised generalisation error depending on the regions of the input space.

References

1. Campbell C., Cristianini N., Shawe-Taylor J. Dynamically Adapting Kernels in Support Vector Machines. NeuroCOLT2 Technical Report Series, NC2-TR-1998-017. May, 1998.
2. Friess T.T, Cristianini N., Campbell C. The Kernel-Adatron Algorithm: a Fast and Simple Learning Procedure for Support Vector Machines. Dept. of Automatic Control and Systems Engineering. University of Sheffield, UK.
3. Gorman R.P., Sejnowski T.J. Analysis of Hidden Units in a Layered Network trained to classify sonar targets. Neural Networks 1:pp 75-89.
4. Grauel A., Renners I., Ludwig L.A. Optimizing Fuzzy Classifiers by Evolutionary Algorithms. Conference on knowledge-Based Intelligent Engineering System and Allied Technologies, Brighton, Proceeding IEEE, pp. 353-356, 2000.
5. Kecman V., Hadzic I. Supports Vector Selection by Linear Programming. Proc. of the international Conf. on Neural Networks (IJCNN). Vol. 5, pp. 193-198. Como, Italy, 2000.
6. Kecman V., Hadzic I. Supports Vector Machines Trained by Linear Programming: Theory and Application in Image Compression and data Classification. 5th. Seminar On Neural Network Applications in Electrical Engineering. NEUREL 2000. pp. 18-23, Belgrade, Yugoslavia. 2000.
7. Mangasarian O.L., R. Setiono, and W.H. Wolberg. Pattern recognition via linear programming: Theory and application to medical diagnosis. in: "Large-scale numerical optimization", Thomas F. Coleman and Yuying Li, editors, SIAM Publications, Philadelphia 1990, pp 22-30.
8. Renners I., Grauel A., Saavedra E.Methodology for Optimising Fuzzy Classifiers based On Computational Intelligence. Proceedings 7th Fuzzy Days. Dortmund, Germany. October, 2001.
9. Vapnik V. The Nature of Statistical Learning Theory. Springer Verlag. 1995.
10. Vapnik V., Golowich S., Smola A. Support Vector Method for Function Approximation, Regression Estimation and Signal processing. Advances in Neural Information Processing Systems, Vol 9. MIT Press, Cambridge, Mass, pp. 281-287.

A Study of the Effectiveness of Detailed Balance in Avoiding Convergence in PBIL

E.S. Correa and J.L. Shapiro

Department of Computer Science, University of Manchester, Oxford Road, Manchester, M13 9PL, UK, {esc,jls}@cs.man.ac.uk

1 Introduction

Estimation of distribution algorithms (EDAs) are a class of evolutionary algorithms that use statistical information to guide the exploration of the search space. A prominent problem in EDAs is the loss of variability as the search progresses. This occurs because at each iteration the probabilistic model reinforces the probability of generating the best solutions found in the previous populations. This process may accelerate convergence to local optima. This paper investigates a method to diminish this convergence pressure by applying "detailed balance" to the Population Based Incremental Learning (PBIL) algorithm [4]. Detailed balance is a well-known condition in Markov chains. Basically, it says that, on a flat fitness landscape, the probability of going from a state i to a state j must be the same as the probability of going backwards from state j to state i. This condition slows the rate of convergence of the probability parameters when the landscape is flat. As a result, the algorithm requires more evidence from the fitness function to drive the search to a single point in the search space and maintains variability for longer.

2 Population Based Incremental Learning (PBIL)

PBIL introduced by Baluja [1], combines mechanisms of generational genetic algorithm with competitive learning. It assumes independence between variables of the problem. PBIL does not maintain a population of solutions. Instead, it uses a probability vector, $p_t(x) = (p_t(x_1), ..., p_t(x_l))$, to represent characteristics of the good solutions found in previous populations. This vector is then used to sample new solutions. In the case of binary variables, $p_t(x_i)$, $i = 1, 2, ..., l$, refers to the probability of obtaining a value of 1 in the ith component of a solution at time t and l is the problem size. PBIL works as follows: a population of N solutions is randomly generated. Based on their fitness value, the k ($k < N$) best solutions from the current population are selected. They are denoted by $S_t = x^t_{1:N}, x^t_{2:N}, ..., x^t_{k:N}$. These solutions are then used to update the probability model by using the equation: $p_t(x) = p_{t-1}(x)(1-\alpha) + \alpha \frac{1}{N}\sum_{i=1}^{k} x^{t-1}_{i:N}$, where $\alpha \in (0, 1]$ is a parameter of the algorithm. A new population is then sampled from $p_t(x)$; S_t is selected; $p_t(x)$ updated; a new population sampled; etc.

2.1 Detailed Balance PBIL (DBPBIL)

Introduced by Shapiro [4], DBPBIL modifies the dynamics of the standard PBIL in order to make it obey the detailed balance condition. This algorithm imposes reversibility to the transition probabilities and by doing that it significantly diminishes the convergence pressure imposed by PBIL's probabilistic model during the search. As a result, the algorithm becomes less dependent on the starting conditions. Shapiro's work also shows that, regarding the value of the parameter α, DBPBIL is much more robust than PBIL. For a more detailed explanation and results on Detailed Balance PBIL the reader is referred to the paper by Shapiro [4].

3 Satisfiability problem (SAT)

The SAT problem is to determine, for a formula of propositional calculus, if there is an assignment of truth values to its variables for which that formula evaluates to true. A general SAT problem can be defined as follows: given a set of m clauses $\{C_1, C_2, ..., C_m\}$ on n Boolean variables $v = (v_1, v_2, ..., v_n)$ with $v_i \in \{0, 1\}$, and a Boolean formula in conjunctive normal form, $f(v) = C_1 \wedge C_2 \wedge ... \wedge C_m$, determine an assignment of truth values to v so that $f(v)$ evaluates to true. Each clause consists of the disjunction of variables in v or their negation. In the MAX-3-SAT problem, every clause C_i has exactly three distinct literals and the goal is to determine an assignment that maximises the number of satisfied clauses in $f(v)$. Random SAT problems are known to undergo a phase transition between satisfiable and unsatisfiable [3]. The vicinity of the phase transition contains the formulae which are hardest to determine whether they are satisfiable.

3.1 PBIL & DBPBIL on SAT

Here PBIL and DBPBIL are compared on random MAX-3-SAT problems. Three sets of random 3-SAT problems, of 15, 20 and 30 variables respectively were generated. Each set contains 10 random generated instances, all satisfiable and phase transition. In all experiments carried out in this work the same sets of problems were used. Every problem in a set was submitted to 100 trials. That means that for every problem size 1000 trials were performed for every single value of α. The fitness of a solution is given by the formula: $fitness = \frac{8 \cdot S - 7 \cdot C}{C}$, where S is the number of satisfied clauses and C is the total number of clauses of the problem. Figure 1 presents the mean of the best fitness obtained over all trials.

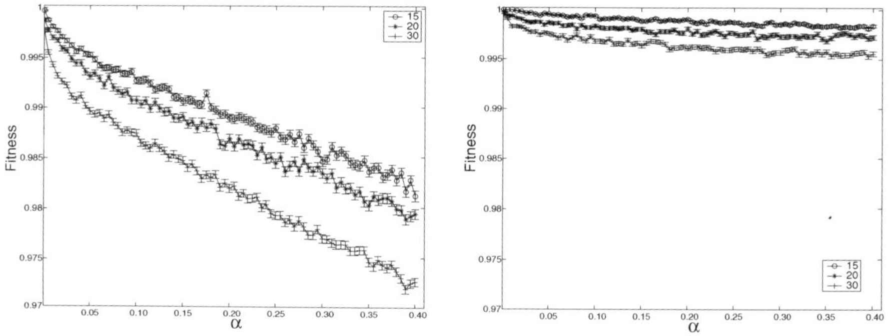

Fig. 1. PBIL and DBPBIL on SAT problem. Mean of the best fitness found on 1000 trials for 80 different values of α. **Left**: PBIL. **Right**: DBPBIL.

DBPBIL outperforms PBIL, in terms of the quality of the solution found, for practically all values of α. These results support the idea that by applying the detailed balance condition to PBIL, the performance of this algorithm is improved. They also indicate that DBPBIL is less sensitive to the value of α - since the fitness variation to DBPBIL is smaller than it is to PBIL. The stopping criterion used in these experiments was: the search stops when all probability parameters converge to either 0.05 or 0.95. For all experiments reported in this work the population size is always equal to 20. Figure 2 shows that, on average, DBPBIL takes more time to converge. This is not surprising - since the algorithm was developed to slow down convergence. More specifically, the more the parameters approach 0 or 1, the more DBPBIL rejects movements towards these values. As a consequence, to meet the stopping criterion adopted DBPBIL has to run for longer. Figure 3 shows that DBPBIL found the global optimum in more than 60% of the trials for literally all values of α and problem sizes. By contrast, for the problems with 30 variables, PBIL found the global optimum in 60% of the trials only for very small values of α, falling to less than 10% when α is equal to 0.4.

4 The P-median problem

We define a median as a facility selected to serve other facilities. The goal of the P-median problem is to select p medians in a set with n facilities ($n > p$). These medians must be selected so that the sum of the weighted distances between every facility not selected as median and its nearest median

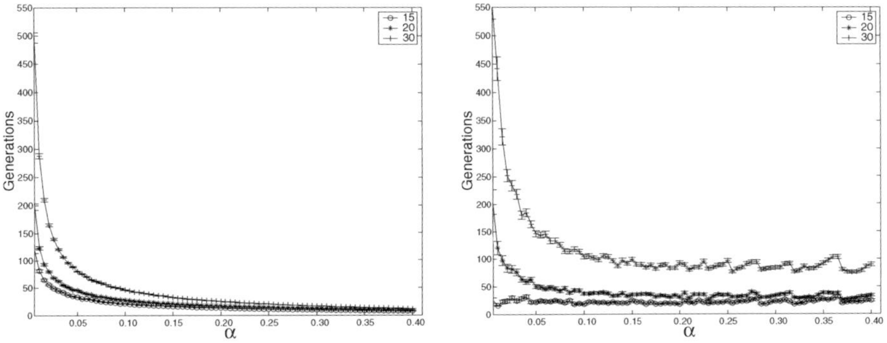

Fig. 2. PBIL and DBPBIL on SAT problem. Mean of the total number of generations until convergence on 1000 trials. **Left:** PBIL. **Right**: DBPBIL.

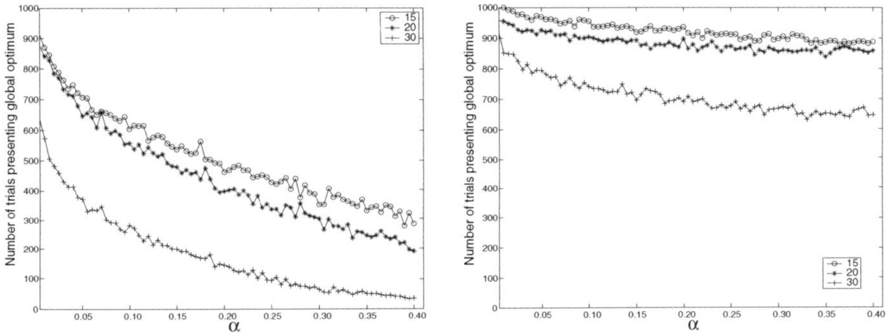

Fig. 3. PBIL and DBPBIL on SAT problem. Mean of the total number of trials in which the global optimum was found. **Left:** PBIL. **Right**: DBPBIL.

is minimised. Assuming all vertices of a graph are potential medians, this problem can be defined as follows: let $G = (V, A)$ be an undirected graph where V represents the set of the n vertices belonging to the graph and A represents the edges (each edge represents the distance between a pair of vertices). The goal is to find a subset of vertices $Vp \subseteq V$ (the median set) with cardinality p, such that the sum of the weighted distances between each remaining vertex in $\{V - Vp\}$ (the demand set) and its nearest vertex in Vp is minimised. For more details on the p-median problem the reader is referred to Correa et al. [2].

4.1 PBIL & DBPBIL on the P-median problem

We need to extend PBIL and DBPBIL to a non-binary encoding. To do this a solution is represented in the following way: each solution contains exactly p sites, where p is the desired number of medians, and each site can receive an index (a unique identity number) of a facility selected as median. A solution is represented as a set of facilities indices - there are no duplicated indices and there is no ordering among them. The fitness of a solution is given by the sum of the weighted distances between every facility not selected as median and its nearest median composing the solution. On this application PBIL and DBPBIL use a vector of probabilities with l parameters, $p_t(x) = (p_t(x_1), p_t(x_2), ..., p_t(x_l))$, where $p_t(x_i)$ represents the proportional probability of the ith facility be selected as median. At the start, the parameters $p_t(x_i)$ are initialised to 0.5. Every parameter $p_t(x_i)$ is then multiplied by a random number uniformly chosen in the interval (0, 1). After that, the set of the selected medians is defined by sorting the parameters $p_t(x_i)$ in a decreasing order. The first p indices (facilities) corresponding to the first p values $p_t(x_i)$ in the ranking are selected as medians. To illustrate, suppose that one wants to select 2 medians out of 4 and the multiplied initial probability vector is $p_0(x) = (0.13, 0.01, 0.27, 0.41)$. Sorting $p_0(x)$, (0.41, 0.27, 0.13, 0.01), one would obtain a solution in which the 2 selected medians are the facilities 4 and 3 respectively. For both algorithms only the fittest solution from the current population is used to update the probabilistic model.

We have not, as yet, found a set of equations that satisfies the detailed balance condition on non-binary encoding. Therefore, we used essentially the same equations as in [4]. Although this version of the equations still slows down the convergence, it does not truly obey detailed balance. Hence, we call it the pseudo DBPBIL algorithm (PDBPBIL).

For the P-median problem, 4 instances, with respectively 20, 30, 40 and 50 facilities, were randomly generated on a 100 by 100 square. The weight corresponding to each facility was also randomly determined by a real number between and including 0 and 100. The same data base was used for all experiments. Figure 4 shows that for the P-median problem the difference between the performance of the algorithms was not so apparent as it was for 3-SAT problems. Perhaps because we have not implemented the detailed balance condition to PDBPBIL exactly. But yet there is a difference and once again PDBPBIL outperformed PBIL. For problems with 20 and 30 facilities, PDBPBIL has always converged to a solution near to the best solution ever found by both algorithms. For problems with 40 and 50 facilities, it has not only found better solutions than PBIL for small values of α, but has also been more robust in all simulations.

5 Conclusions

One of the claimed advantages of EDAs over other evolutionary algorithms is that they involve less parameters to be chosen by the user. Although this generalisation may be true it is undeniable that there still exist decision parameters in EDAs. In particular, PBIL was shown to be strongly sensitive to

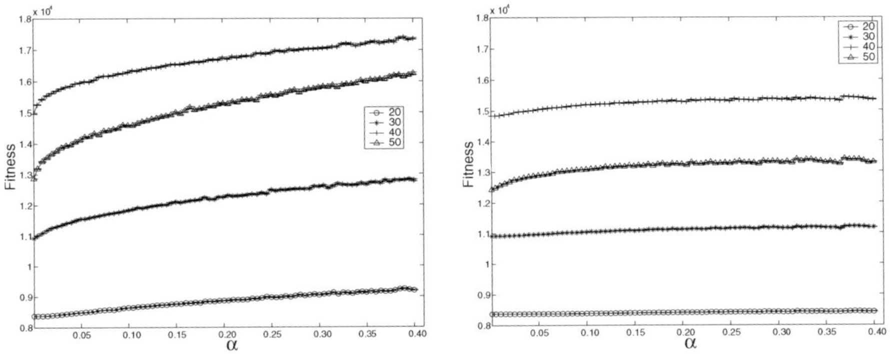

Fig. 4. PBIL and PDBPBIL on the P-median problem. Mean of the best fitness found on 1000 trials for 80 different values of α. **Left:** PBIL. **Right**: PDBPBIL.

such a parameters. As important as diminishing the number of parameters in an algorithm is making the algorithm robust regarding changes in its parameters. This work investigated the performance of the Detailed Balance PBIL on two different test problems and also extended its application to a non-binary encoding. On both test problems DBPBIL was shown to be more robust than the standard PBIL. The results obtained suggest that the addition of detailed balance to PBIL is indeed an efficient method for diminishing the convergence pressure imposed by the probabilistic model in PBIL. DBPBIL has not only found the global optimum a greater number of times, but has also found better solutions than PBIL for practically all values of the parameter α. Turning to other EDAs, an interesting question would be: does the convergence pressure imposed by the probabilistic model affect the performance of the more complex EDAs models?

References

1. Baluja S. Population based incremental learning: A method for integrating genetic search based function optimisation and competitive learning. Carnegie Mellon University, 1994. Pittsburg, PA.
2. Correa E.S., Steiner M.T.A., Freitas A.A., Carnieri C. A Genetic Algorithm for the p-median Problem In: Proceeding of the Genetic and Evolutionary Computation Conference, 2001. San Francisco, CA. Morgan Kaufmann Publishers.
3. Mitchell D.G., Selman B., Levesque H.J. Hard and easy distributions of SAT problems. In W.R. Swartout, editor, Proceeding of the 10th National Conference on Artificial Intelligence, pages 459-465, San Jose, CA, Jan. 1992. AAAI Press/MIT Press.
4. Shapiro J.L. The sensitivity of PBIL to its learning rate, and how detailed balance can remove it. Foundations of Genetic Algorithms 7, pages 115-133. De Jong K.A, Poli R., Rowe J.E. (editors), 2003. Morgan Kaufmann Publishers.

Part V

Soft Computing Applications

Application of Computational Intelligence Methods for Intelligent Modelling of Buildings

Alexander Gegov

Department of Computer Science and Software Engineering,
University of Portsmouth, United Kingdom, e-mail: alexander.gegov@port.ac.uk

Abstract. This paper considers the application of soft computing techniques for predictive modelling in the built sector and presents the extention of the results from previous works of the author. While the latter considers only short-term modelling which is used mainly for control purposes, the present paper discusses also long-term modelling results that may be used for efficiency evaluation in buildings. Three different types of buildings are considered, an air-conditioned zone, a naturally ventilated room, and an endothermic building. The are subjected to their normal occupancy effects and the natural external climatic disturbances which are difficult to incorporate in accurate modelling using conventional quantitative methods. The approach adopted here uses fuzzy logic for modelling, as well as neural networks and genetic algorithms for adaptation and optimisation of the fuzzy model. Takagi-Sugeno fuzzy models are built by subtractive clustering to provide initial values of the antecedent non-linear membership functions parameters and the consequent linear algebraic equations coefficients. A method of extensive searching the possible solution space is presented which explores all the possible permutations for a specified range of orders to derive the initial fuzzy model. This model is an extension of the traditional ARMAX (Auto Regressive Moving Average Exogenous) model where the effect of the moving average term has been accounted for by the fuzziness and its ability to represent uncertainty. The fuzzy model parameters are further adjusted by a back-propagation neural network and a real-valued genetic algorithm in order to obtain a better fit to the measured data. Model validation results using data from the three buildings are presented where the initial (fuzzy) and the improved (fuzzy-neuro and fuzzy-genetic) models are compared and analysed with each other and with conventional (non-fuzzy) models.

Keywords. Fuzzy logic, fuzzy modelling, neural networks, model adaptation, genetic algorithms, genetic optimisation.

1 Introduction

Soft Computing (SC) is a heuristic methodology which has attracted significant interest in recent years and has shown to be successful in many areas such as modelling, control, fault diagnosis and pattern recognition. It is based on the co-operative implementation of different approaches such as Fuzzy Logic (FL), Neural Networks (NN), and Genetic Algorithms (GA) [5, 6]. Each of these techniques is suited for solving specific types of problems. In this respect, FL is powerful for approximate reasoning and logical inference-based modelling, NN are well suited for learning-based adaptation while GA are efficient for evolution-based optimisation. The underlying idea of SC is to use these heuristic approaches in synergetic way with each other as well as combining them with conventional approaches, rather than using each of them separately.

The research presented here is aimed at developing efficient control methods to regulate the internal climates in buildings. The authors develop good quality models which can enable proactive control strategies to be developed in preference to the currently used reactive ones. In other words, instead of applying a control action only on the basis of the currently measured sensor values, the concept is to take into account future model based predictions of thee values as well. The advantage of such a proactive control philosophy lies in the possibility to apply heating and cooling control efforts more efficiently as a result of which the variations of the controlled parameters become smoother as well as better regulation performance is possible. This will lead to lower energy consumptions and reduced environmental pollution. However, obtaining predictive models for these buildings is not a trivial task because the thermal behaviour is affected by climatic and occupancy effects which are characterised by significant complexity and uncertainty as already mentioned.

The notion of a proactive control philosophy is the following: the control action to be applied at the current time instant k is computed not only on the basis of measured variables at the time k and the past $(k\text{-}1)$, $(k\text{-}2)$, etc, but also by taking into account of predictions at the future time instants $(k+1)$, $(k+2)$, etc.

It would be interesting to see, if the SC methodology can provide models of sufficient quality for the built environment so that such proactive model based strategies can be developed. Some investigations have recently been carried out in this domain but they have been narrow and have lead to limited conclusions. In most cases, they are focused on one modelled parameter and make use of individual FL, NN and GA techniques, rather than adapting a coordinated SC methodology. In other words, the potential of the SC methodology for buildings has still not been explored.

This paper presents results from a research project aimed at investigating the capabilities of the SC methodology for predictive modelling the temperature and relative humidity in buildings. The results are based on preliminary works of the research team [1, 2, 3, 4, 7]. Three types of buildings are considered in the study, namely: air conditioned, naturally ventilated and endothermic buildings. The three specific buildings under investigation are: the Anglesea Building (University of

Portsmouth), the Portland Building (University of Porsmouth) and St Catherine's Lighthouse (Trinity House, Isle of Wight). These buildings differ substantially and the intention is to identify how each type is suited to benefit from the SC methodology.

2 Methodology

The approach adopted uses Takagi-Sugeno (TS) fuzzy models which have received considerable attention over the last 10 years because of their suitability for processing information from input-output measurements [6]. In BMS applications, the main information source is the measured data from sensor readings making TS FL models easily realisable rather than having to rely on expert knowledge which is difficult to obtain because of the complex multivariable nature of the application [6]. Another advantage of the TS fuzzy model is its capability to approximate non-linear input-output mappings by a number of linearised models at a number of suitable operating points.

The TS fuzzy model consists of linguistic if-then rules in the antecedent part and linear algebraic equations in the consequent part. There are two types of parameters in this model: non-linear ones in the membership functions in the antecedent part and linear ones in the algebraic equations in the consequent part. The initial fuzzy model is obtained by determining the values of both parameters on the basis of the input-output data. The method used in the paper is based on subtractive clustering, i.e. by assuming that each data point is a potential cluster centre and using all the data to gradually find the final clustering.

The Takagi-Sugeno fuzzy model for a system with two rules, two inputs (u_1, u_2) and one output (y) is presented by Equation *(1)*. The linguistic labels (membership functions) of the inputs are denoted by A_i, B_i, $i=1, 2$ and their parameters are the non-linear antecedent parameters. The coefficients a_i, b_i, $i=1, 2, 3$ are the linear consequent parameters used for the computation of the output.

$$\begin{aligned}&\textbf{If } u_1 \text{ is } A_1 \textbf{ and } u_2 \text{ is } A_2 \textbf{ then } y = a_1.u_1 + a_2.u_2 + a_3\\&\textbf{If } u_1 \text{ is } B_1 \textbf{ and } u_2 \text{ is } B_2 \textbf{ then } y = b_1.u_1 + b_2.u_2 + b_3\end{aligned} \tag{1}$$

Equation *(1)* represents a static Takagi-Sugeno fuzzy model which does not contain the time argument. However, in order to dynamically predict the temperature and the relative humidity, it should be included in the equation, i.e. the model must be a dynamic one. In this respect, two types of models are introduced and investigated here, namely Regression Delay (RD) and Proportional Difference (PD). Examples of such models are represented by Equations *(2)* and *(3)*, respectively.

$$\begin{aligned}&\textbf{If } y_k \text{ is } A_1 \textbf{ and } y_{k-1} \text{ is } A_2 \textbf{ and } u_{1,k} \text{ is } A_3 \textbf{ and } u_{1,k-1} \text{ is } A_4 \textbf{ and } u_{2,k-1} \text{ is } A_5\\&\textbf{then } y_{k+1} = a_1.y_k + a_2.y_{k-1} + a_3.u_{1,k} + a_4.u_{1,k-1} + a_5.u_{2,k-1} + a_6\end{aligned} \tag{2}$$

$$\begin{aligned}&\textbf{If } y_k \text{ is } A_1 \textbf{ and } Dy_k \text{ is } A_2 \textbf{ and } u_{1,k} \text{ is } A_3 \textbf{ and } Du_{2,k} \text{ is } A_4\\&\textbf{then } y_{k+1} = a_1.y_k + a_2.Dy_k + a_3.u_{1,k} + a_4.Du_{2,k} + a_5\\&\text{where } Dy_k = y_k - y_{k-1},\ Du_{2,k} = u_{2,k} - u_{2,k-1}\end{aligned} \tag{3}$$

It can be seen that Equation *(2)* contains two auto-regression terms of the output *y,* two regression terms of the input u_1 and one delay term for the input u_2. In contrast, Equation *(3)* contains one proportional and one difference term of the output *y,* one proportional term of the input u_1 and one difference term of the input u_2. For simplicity purposes, each of the equations includes only one rule, but in general the number of rules is higher; it is equal to the number of the linearised sub-models applicable to the respective local regions of the whole operating range.

Equations *(2)-(3)* represent examples of fuzzy model structures. Usually, these structures are obtained on the basis of evaluating a number of model structures in accordance with a performance criterion (usually the Root Mean Squared Error - RMSE). The latter is a measure of the closeness of the model to the plant and is expressed by Equation *(4)*.

Equations *(2)-(3)* represent examples of fuzzy model structures. Usually, these structures are obtained on the basis of evaluating a number of model structures in accordance with a performance criterion (usually the Root Mean Squared Error - RMSE). The latter is a measure of the closeness of the model to the plant and is expressed by Equation *(4)*.

$$\text{RMSE} = [(y^P_1 - y^M_1)^2 + \ldots + (y^P_N - y^M_N)^2] / N \tag{4}$$

where N is the number of data points, and the superscripts P and M stand for the plant and the model, respectively.

The majority of identification techniques apply partial rather than extensive model structural searching to obtain the best structures. Usually, the approach is to represent the plant dynamics by either regression or delay terms. As far as proportional or difference terms are concerned, they are usually considered up to the most recent data point in the past. This might be a serious disadvantage as the dynamics of the plant are not fully explored and some important terms in the model structure may remain unaccounted for giving non-optimal results.

This paper presents a method of searching the range of model structures so that all the possible permutations of a specified order are explored. This method is demonstrated here for Takagi-Sugeno fuzzy models but can be also applied to any other types of models, including traditional ones. For a plant with m input variables and backward horizon equal to b, the whole number of investigated model structures S is given by Equation *(5)*.

$$S = (2.b)^m - 1 \tag{5}$$

It is evident from Equation *(5)* that the number of investigated models is an exponential function of the number of inputs and the backward horizon. This can lead to long processing because of the extensive searching involved. However, this is important to do as getting the model structure right is crucial to the success of the whole modeling procedure.

The task of neural adaptation is to adjust the fuzzy model parameters in order to obtain a better fit to the measured data. The method used in the paper is based on the idea of back-propagation, i.e. by iteratively propagating of the model error (the difference between the real and the modelled plant output) from the consequent to the antecedent part of the fuzzy rules until a specified number of iterations is achieved [6].

The main options for the neural network adaptation are the number of iterations, the error goal, the initial step size, the decreasing and the increasing learning rates. The number of iterations gives the number of new sets of parameter values evaluated, the error goal is a measure for the desired model accuracy, the initial step size shows the magnitude of change in the initial parameter values, while the decreasing/increasing learning rates take into account the negative/positive changes in the model's accuracy in the last iteration in comparison to the previous one.

The purpose of genetic optimisation is to adjust the model parameters in an alternative manner; the method used here is based on the idea of real-valued coding, i.e., by representing the individuals with real-valued genes and sequentially evaluating the modelled error until a pre-specified number of generations is reached [5].

3 Results and discussion

This section presents modelling results for the three buildings under consideration over the heating season. The results show that the outputs (temperature and relative humidity) can be accurately predicted on the basis of measurements from the buildings using SC modelling techniques. The variables are chosen to be the temperatures and relative humidities in the same and neighbouring zones. For the purpose of systematic modelling, both RD and PD types of fuzzy models have been considered. The best model structure corresponding to the smallest prediction error was chosen by evaluating all possible structures of models for order 2, i.e. involving all combinations of up to two input terms. Afterwards, the quality of the models was evaluated by using a new data set and the respective validation error was obtained. The errors calculated were both one-step ahead predictions and longer-term forecasts. The membership functions of the inputs in the fuzzy model were chosen to be of Gaussian types while the selected options for the neural network and the genetic algorithm were 500 iterations and 50 generations with 10 individuals each, respectively. These options seem to give a fair comparison between the two considered fuzzy model improvement techniques as the number of investigated fuzzy membership function parameters is equal to 500 in both cases.

Each one of the above model structures was chosen from a set of 1023 possible models, representing all combinations of (auto-) regressional and (auto-)delayed terms. The backward (dynamical memory) horizon was chosen to be 2, i.e., the prediction of the internal temperature at time $k+1$ is obtained on the basis of measurements of two previous values, namely, at times k and $k-1$. The plant and the model outputs for each of these one-step ahead prediction models and for each building are shown in figs 1-3. In the longer-term horizon case, the respective model root mean square errors are not shown in the paper due the limited space available.

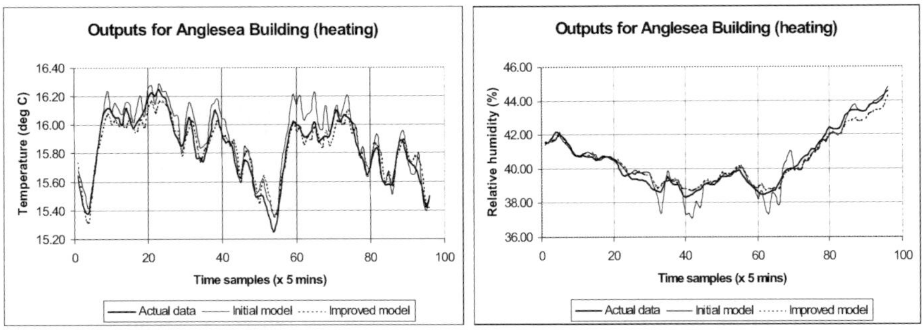

Fig. 1. Temperature and relative humidity outputs for Anglesea Bulding.

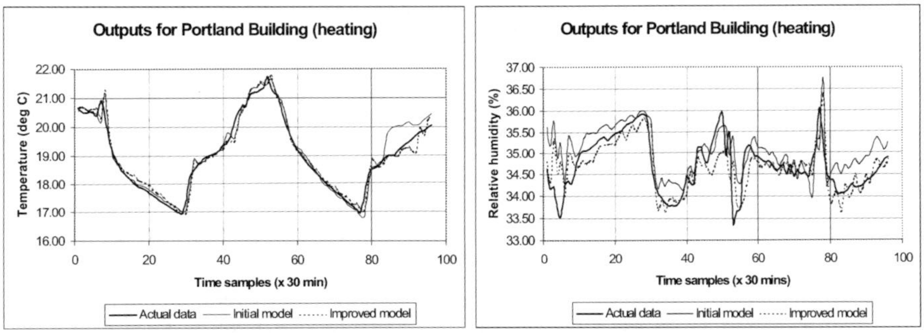

Fig. 2. Temperature and relative humidity outputs for Portland Bulding.

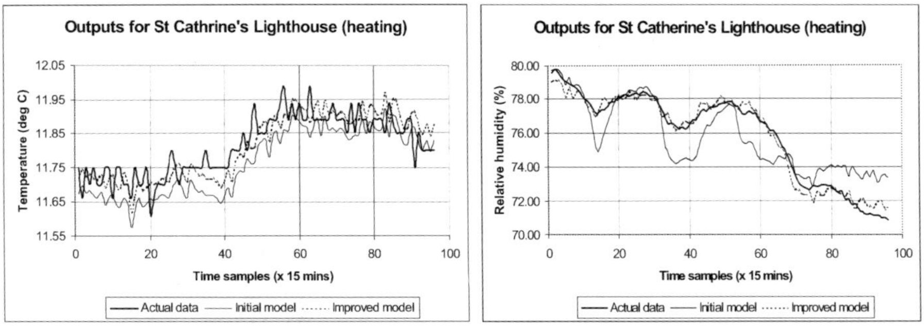

Fig. 3. Temperature and relative humidity outputs for St Catherine's Lighthouse.

It is evident for Anglesea Building and St Catherine's Lighthouse that the best model structures are of a regression-delay type, while for Portland Building they are of a proportional-difference type.

4 Conclusion

The best prediction results have been obtained for the naturally ventilated area, while the results for the air-conditioned zone and the endothermic building are not as good but comparable to each other. The temperature prediction is better than the humidity prediction for the three buildings because of the smaller variation range and smoother profile. The proportional-difference fuzzy model performs better than the regression-delay model in most cases for all the buildings.

In most cases, when the one-step ahead prediction of the fuzzy model is significantly improved by neural networks and/or genetic algorithms, the long-term prediction results are also improved. The long-term prediction error obtained is usually comparable to or is even bigger than that obtained before adaptation/optimisation of the fuzzy model. It is interesting to point out that in some cases the long term prediction error is smaller than the one-step ahead error either before or after model improvement. This happens in the case of Potland Building and is possibly due to the cyclic behaviour of the building which makes it possible for the model to predict more distant future values of temperatures and relative humidities very well because these values are close to some of the previous values of these two internal variables.

In general, there is a substantial improvement of the prediction capabilities of the initial fuzzy model for both short and long term prediction intervals, especially in the cases when this model is not very good. In this respect, genetic algorithms are in most cases superior to neural networks for both temperature and humidity predictions, for both types of regression-delay and proportional-difference fuzzy models (for all three building types). Genetic algorithms seem to show considerable better convergence properties than neural networks as the best model parameters are usually obtained from the last genetic generations while these parameters are in most cases found from the first neural iterations.

References

1. Azzi, D., Gegov, A., Virk, G., Haynes, B., and Alkadhimi, K. (2000) Application of Soft-computing Techniques in Modeling of Buildings, International Workshop on Recent Advances in Soft Computing, Leicester, UK, vol.1, pp. 143-150.
2. Gegov, A., Virk, G., Azzi, D., Haynes, B., and Alkadhimi, K. (2001) Soft-Computing Based Predictive Modelling of Building Management Systems, International Journal of Knowledge Based Intelligent Engineering Systems, vol.5, no.1, pp.41-51.
3. Gegov, A., Virk, G., Azzi, D., Haynes, B., and Alkadhimi, K. (1999) Soft-Computing Based Predictive Modelling of Building Management Systems, International Workshop on Recent Advances in Soft Computing, Leicester, UK, vol.1, pp.69-77.
4. Gegov, A., Virk, G., Azzi, D., Haynes, B., and Alkadhimi, K. (1999) Soft-Computing Based Modelling of the Internal Climate in Office Buildings, UK Workshop on Fuzzy Systems, Uxbridge, UK, vol.1, pp.145-152.

5. Haupt, R., and Haupt, S. (1998) Practical Genetic Algorithms, John Wiley & Sons, New York.
6. Jang, J., Sun, C., and Mizutani, E. (1997) Neuro-Fuzzy and Soft Computing, Prentice Hall, Upper Saddle River.
7. Virk, G., Azzi, D., Gegov, A., Haynes, B., and Alkadhimi, K. (2003) Intelligent Soft-Computing Based Modelling of Naturally Ventilated Buildings, International Journal of Solar Energy, to appear.

An Adaptive Multi Embedded-Agent Architecture for Intelligent Inhabited Environments

Elias Tawil and Hani Hagras

Department of Computer Science, University of Essex

1 Introduction

Intelligent Inhabited Environments (IIE) are personal spaces such as cars, shopping malls, homes, and even our bodies that work intelligently to satisfy the users' needs by maintaining tasks such as maximizing their comfort and safety, and minimizing energy costs. IIE embody embedded-agents that could be situated inside or outside them.

Embedded-agents are networked physical products that contain intelligent processes that enable them to co-operate to achieve common goals. In IIE, embedded-agents are heterogeneous. They can be static agents which can have a specialized task such as maintaining a desirable temperature within a room. They can be mobile in the form of a mobile robot that can deliver meals. They can also be portable and can reside within a mobile phone or a PDA. In IIE, embedded agents are expected to come in and leave or breakdown anytime. Therefore, there is a need for online life-long learning and adaptation of the co-ordination of the embedded agents in IIE that aim to satisfy many users with conflicting needs, and with the ability to operate with minimum user-interaction. Embedded multi-agent systems in IIE should be scalable to the whole world, or even beyond- just like the Internet. Hence, the need to address issues such as user safety, security and privacy arises.

There are different kinds of constraints and objectives that the agents have to deal with [9]. They have soft constraints that would be preferable but not necessary to satisfy, as they may conflict with others. An example of a soft constraint is maintaining the users' comfort, for different people with different needs. The agents should have hard constraints as well, such as ones related to the users' safety. People have to be kept warm, yet the temperature inside a room is not to exceed beyond a level that is harmful to its inhabitants. The agents have local goals and constraints as well as global ones.

In order to realize the IIE, technologies must be developed that will support ad hoc and highly dynamic (re)structuring and co-ordination of such artefacts and embedded agents to satisfy the different goals and constraints whilst shielding non-technical users from the need to understand or work directly with the technology "hidden" inside such artefacts or systems of artefacts. This situation is likely to get much worse as the number, varieties and uses of computer based artefacts increase. One major problem with co-ordinating a system of embedded agents in IIE is that too little information going to a single agent would reduce its potential and might even render it useless. On the other hand, too much information would clog up the network by increasing the probability of collisions.

In this paper, we propose a distributed architecture that learns to co-ordinate online and in real-time, a system of heterogeneous multi embedded-agents that operate in IIE. They are to retain their local and global objectives without the use of a central control or repository mechanism.

The Neural Network house [8], The Aware Home [5], and HIVE [7] are some of the numerous examples of related work. According to the authors' knowledge, there is no distributed system that has been developed that can learn online multi agent co-ordination in IIE satisfying local, global, soft, and hard goals and constraints.

2 A Novel Architecture

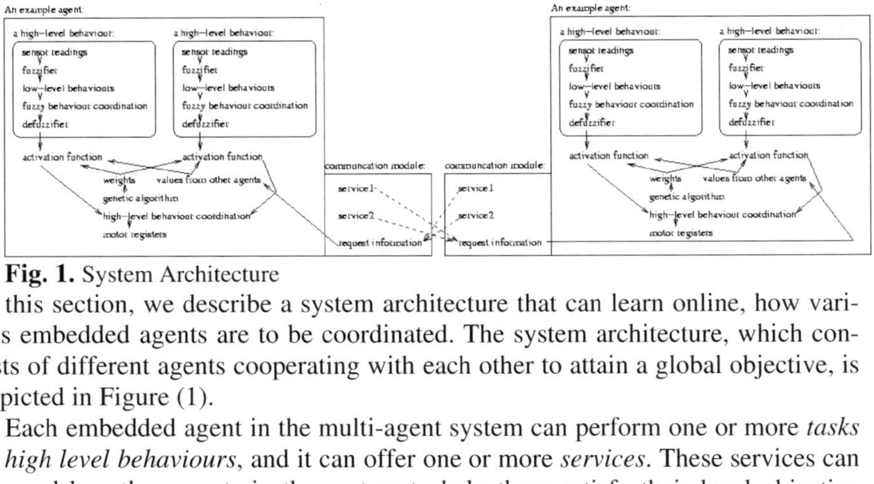

Fig. 1. System Architecture

In this section, we describe a system architecture that can learn online, how various embedded agents are to be coordinated. The system architecture, which consists of different agents cooperating with each other to attain a global objective, is depicted in Figure (1).

Each embedded agent in the multi-agent system can perform one or more *tasks* or *high level behaviours*, and it can offer one or more *services*. These services can be used by other agents in the system to help them satisfy their local objective functions and the system's global objective functions. For example, in ideal IIE, a user can buy an embedded robotic agent which has two high level tasks: surveillance of an area and cleaning it. It can offer services like locating different objects in a room, and reporting its proximity to human beings and obstacles. In this ideal IIE, the user should only specify his or her high level objectives and then plug the robotic agent into its environment and let it determine which services it requires from the other embedded agents to do its job (e.g. it might need the time from an embedded agent's clock). Simultaneously, the other embedded agents will deduce which services they need from the newly purchased mobile robotic agent.

This automatic learning and life-long adaptation of the system is important in a dynamic IIE. The system should adapt to users' needs instead of operating on a hardwired set of rules. As the number of embedded agents increase, human beings will not be able to figure out the optimal configuration of the multi-agent system co-ordination. Excess information being communicated would waste network and processing resources, and too little information coming to an agent might not be enough for it to play a role in the multi-agent system [1].

In our architecture the high level behaviours of an embedded agent are a coordinated set of *low level behaviours.* Obstacle avoidance for mobile robots is an example of a low level behaviour. The low level behaviours are Fuzzy Logic Controllers (FLCs) connected to the internal sensors and actuators of an agent. Fuzzy logic offers a framework for representing imprecise, uncertain knowledge, and it can deal with vague and incomplete information. In addition fuzzy controllers exhibit robustness with regard to noise and variations of system parameters. The problem of how to coordinate the simultaneous activity of several independent behaviour-producing units to obtain an overall coherent high level behaviour within an embedded agent have been discussed by many authors [4]. The work described in this paper suggests a solution based on using fuzzy logic to both implement individual behaviour elements and the necessary arbitration, allowing both fixed and dynamic arbitration policies to be implemented. We achieve this by implementing each behaviour as a fuzzy process and then using fuzzy logic to co-ordinate them. In the resultant architecture, a Hierarchical Fuzzy Logic Controller (HFLC) takes a hierarchical tree structure form. Using fuzzy logic for behaviour coordination allows several behaviours to be active at differing degrees, thereby avoiding the drawbacks of an on-off switching schema (i.e. deals with situations where several criteria need to be taken into account). In addition, using fuzzy coordination provides a smooth transition between behaviours with a consequent smooth output response. More information about HFLC can be found in [4].

The *activity* of a high level behaviour depends on the services it acquires from other embedded agents. For every high level task, each service will be scaled by a *fuzzy communication link weight* that reflects how important the service is for the high level task. The activity level of a high level behaviour can be *directly proportional* to a service, i.e. an increase in the value from a service causes an increase in the activity of the high level behaviour. The activity of a high level behaviour can be *inversely proportional* to a service's value, i.e. an increase in the service's value causes a decrease in the activity of the high level task. A high level task can also be *independent* of a particular service. In this case, an embedded agent does not establish a link to the one supplying this service.

There is a need for a distributed online learning mechanism that can express which services each embedded agent needs, and the fuzzy communication link weights associated with them. This defines how important the services are to the embedded agent as well as whether each of the agent's high level tasks' activity levels are directly proportional, inversely proportional or independent of the values from the provided services.

The communication links that carry the services' values between the agents are TCP/IP socket connections between two agents. If the corresponding weight for a service suggests that it will be taken into consideration, then the agent will establish a communication link to the one that provides this service and request it periodically. Otherwise, no connection would be made and precious time and resources would be saved. A communication link acts as a passageway for DIBALS, which is the protocol used to convey these requests and their replies.

Requests, commands, and information are exchanged using the Distributed Intelligent Building Agent Language- Simplified (DIBALS) protocol. It is a simplified version of DIBAL [2]. A DIBALS message has three parts which are separated by a delimiter, which is also used to mark the end of a transmission. The first part of a message is the name of the sender. This could also store a group name that the sender might belong to, should there ever be a need for that. The middle part is the purpose of the message, and indicates whether the message is a command, request, or for transmitting sensor data. The last part of the message is for details, such as sensor readings or the name of the receiver. A DIBALS message would look like this: SENDER_PURPOSE_DETAIL_. Example usage: Marvin_RequestBeacon_5, which means that Marvin wants the receiver to send him its alignment with beacon number five.

For the online learning system we have chosen Genetic Algorithms (GA) as their parallel nature is well-suited for online and distributed learning, and they do not need any mathematical representation of a problem to attain its optimal solution [3]. Our online system uses a GA for each high level behaviour in each embedded agent to optimise the other agents' services' fuzzy communication link weights, which also determine whether each high level task is directly proportional, inversely proportional or independent of each service. According to our GA representation, the weight for each link and its degree of proportionality is portrayed inside a GA chromosome. The GA, through the processes of selection, recombination and mutation, will be able to achieve the optimal values for the weights of the links and their degree of proportionality for each high level behaviour in each embedded agent.

The crossover and mutation probabilities play a major role in the fast convergence of the GA. The crossover probability, P_c, controls the rate at which the solutions are subjected to crossover. As P_c increases, more novel solutions will be introduced into the population at each stage. If P_c becomes too large, solutions would be disrupted before the selection process can exploit them. A reduced P_c, on the other hand, degrades the productivity of the GA since there is less recombination between individuals; hence it takes a longer time to obtain a good solution. As detailed by Linkens [6], we need a high P_c to guarantee fast convergence for online learning.

The choice of mutation probability P_m is critical to the GA performance. While the use of large values of P_m transforms the GA into a purely random search algorithm, some mutation is required to prevent the premature convergence of the GA to sub-optimal solutions. The traditional role of mutation has been that of restoring lost or unexplored genetic material into the population to prevent the premature convergence of the GA to sub-optimal solutions. However, recent investigations have demonstrated that high levels of mutation could form an effective search strategy when combined with conservative selection methods [6]. As we are using small chromosome and population sizes, it is necessary to use a sufficient mutation rate to allow for a wider variation in the search, and hence the ability to jump out of any local minima, without reducing the search to a random process [6].

3 Experiments and Results

In many real-world applications such as IIE, there are different kinds of agents that have different capabilities and different tasks, i.e. *heterogeneous agents*. This is why the experiments were conducted on a wide variety of embedded agents.

One of the embedded agents used in the experiments is a *μDorm* shown in Figure (2-a). μDorms are small intelligent inhabited environments that are managed by TINI boards. TINI (for Tiny InterNet Interface) is a complete TCP/IP node with a Java JVM. Each μDorm has at least two devices: one for lighting and another for heating. A basic lighting device consists of light sensors and light bulbs, while a heating device is made of a thermometer, a fan, and a heating bulb.

The other types of embedded agents used in the experiments are in the form of mobile robots. We have used different kinds of them. They are all controlled by VME buses with Motorola processors running the VxWorks real-time operating system. One of the mobile robots used is a small *ROVA* shown in Figure (2-b). ROVAs have a 68000 processor, eight bump sensors, two motors, and a keypad. Another type of mobile robots used are bigger ones called Marvins shown in Figure (2-c) that run on a 68040, have eight bump sensors, eight ultra-sound range finders and an infra red sensor. There are four Marvins that are slightly different from each other: Marvin, Arthur, Ford, and Trillian.

We have implemented a system which is based on our architecture to learn the degrees of association and proportionality for the fuzzy communication link weights using real world embedded agents operating in IIE.

The Marvins have a high-level surveillance behaviour that is composed of four low-level fuzzy controllers, namely obstacle avoidance, left wall-following, right wall-following, and beacon seeking. They are coordinated using a mechanism that employs fuzzy logic weights, i.e. a HFLC [4]. This enables them to patrol between two beacons while relaying *processed* sonar and beacon sensor readings to other agents as services. By processed sensor readings we mean averaged and normalised or fuzzified. This would reduce network traffic, and render the data more useful to the receiving agent.

Different activation levels of the surveillance behaviour correspond to the robots moving at different speeds. This controls the probability of locating a user and the rate of power consumed. For example, if the activation level of the surveillance behaviour increases, the robot will patrol more vigorously consuming more energy with a better chance of locating an inhabitant.

Individually, the low-level behaviours aim to deal with the Marvin's local objectives and constraints, while all of them work together along with some information from other agents to achieve the whole multi-agent system's global objectives and constraints, which are to maximise user-comfort and minimise energy consumption in our case.

The ROVA have the high level behaviour of patrolling a designated area using a reactive controller. It provides its bump sensor readings as a service.

The μDorm operates on one of two reactive high level behaviours at a time. This deals with the μDorm's local objectives and constraints. The *Life Support* behaviour is for providing light and heat for inhabitants. *Power Save* is for maintaining a system standby while no inhabitants are being detected by the multi-robot system. The μDorm provides processed temperature and light sensor readings, as services, for the other agents. For the system to attain its global objectives and constraints, the μDorm has to deal with commands from the Marvins that state which of the two μDorm low-level reactive behaviours should be active.

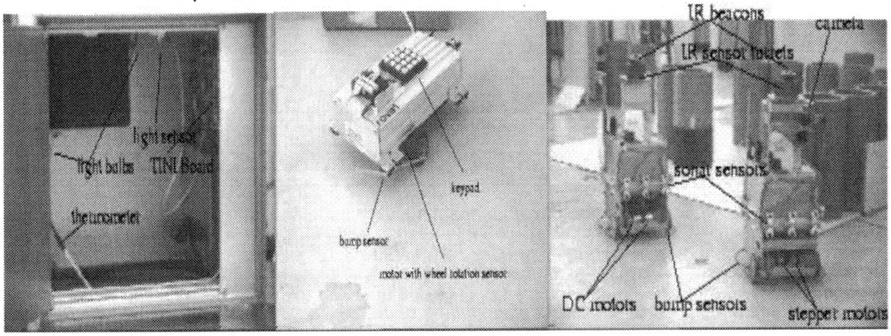

Fig. 2. a μDorm, **b** ROVA, **c** Marvins

The Marvins' high-level behaviours are directed by a fuzzy activation level that is determined by fuzzy values from services provided by other agents. These values are weighted based on whether or not they are needed by the robot. The weighting system determines what communication links are to be established. For example, if Arthur's fuzzy communication link weight for Ford's ultra-sound range-finders suggests that the activation of the surveillance behaviour is inversely proportional to Ford's sonar readings, Arthur will more readily perform this high-level task when there are obstacles in front of Ford (time-of-flight sensors such as ultra-sound range-finders return a lower value the closer an object is). If the weighting system suggests that Arthur's task is independent of the temperature, no communication link will be established between Arthur and the μDorm; therefore, the request for the temperature reading will not be made. These weights are optimised using a GA. Only the Marvins learn their fuzzy communication link weights, while the ROVA and μDorm have fixed ones.

When a user *kicks* a Marvin, i.e. pushes the rear-right bump sensor, it will automatically send messages to other Marvins in its vicinity, commanding them to immediately penalize the chromosome that they would be running on at the time the Marvin was kicked.

Chromosomes represent the fuzzy communication link weights for the services provided by other agents. Our fitness function is related to the amount of time a chromosome is active. The worst punishment is when a Marvin is kicked as soon as a chromosome starts running. The longer a chromosome is left running, the fitter it is assumed to be.

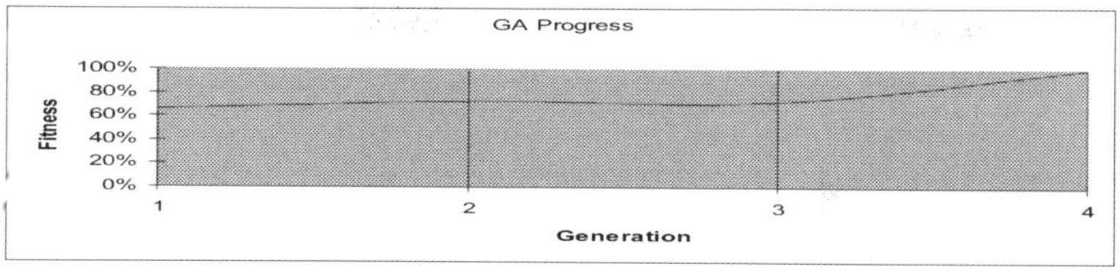

Fig. 3. GA Progress

We have performed the experiments using different combinations of Marvins, ROVAs and μDorms. Different random seed numbers were used for each experiment. Figure (3) shows how the fitness of the optimal individual, during one of the experiments, rises as one of the Marvin's GA progresses. The GA converged during the fourth iteration, which ended in about 10 minutes from the time the experiment had begun. For our experiments, we have pretended to be users that want at least one of the robots to move more quickly and command the μDorm to switch to Life Support mode when it is dark, a Marvin is aligned to a beacon where a user might be, and its ultra-sound range finders detect proximate objects.

The optimal chromosomes for our experiments suggest that the activation level for a Marvin's surveillance behaviour should be independent of the ROVA's bump sensors and room temperature, and it is directly proportional to the light levels in the room (the μDorm's light sensors return a higher value when it is dark), and indirectly proportional to the sonar readings and beacon alignment of the other Marvin. On average, at least one of the Marvins had an optimal chromosome in about ten minutes. Our experiments had demonstrated a system that is totally distributed and adaptive. This is why it can be easily modified to be scalable to the whole world. The system meets hard constraints while optimising for soft ones. Each agent has its own local objectives, while the whole system tries to minimise energy consumption and maximise the users' comfort and safety. All this is done with minimal user-interaction, i.e. prodding one of the Marvins every now and then.

4 Conclusions and Future Work

In this paper, we have expressed the need to look into how a system of heterogeneous embedded agents that would automatically coordinate themselves. We have presented an architecture that can learn the configuration for a set of embedded IIE and robotic agents. We demonstrated this using a construction that utilised a distributed GA structure to successfully learn, online, how to coordinate a system of heterogeneous embedded agents. For future work, we intend to experiment with more complicated scenarios with different kinds of embedded agents, and to exploit technologies such as Bluetooth or LonWorks to detect available agents and services. This would help achieve the unlimited scalability that we desire. We are also planning to experiment with Genetic Programming, which might be a better solution for optimising the weighting system as more complex rules may be devised.

References

[1] J Carmena, J Hallam. Improving Performance in a Multi-Robot Task Through Minimal Communication. In *Proceedings of the 7th Symposium on Intelligent Robotic Systems (SIRS),* pages 329-337, Coimbra, Portugal, July 1999.

[2] F Cayci, V Callaghan, G. Clarke. DIBAL- A Distributed Intelligent Building Agent Language. In *The 6th International Conference on Information Systems Analysis and Synthesis (ISAS 2000),* Orlando, Florida, July 2000.

[3] D Goldberg, "Genetic Algorithms in Search, Optimisation and Machine learning", Addison-Wesley, Reading , MA , 1989.

[4] H Hagras, V Callaghan, M Colley, Prototyping Design and Learning in Outdoor Mobile Robots operating in unstructured outdoor environments, *IEEE International Robotics and Automation Magazine*, Vol. 8, No.3, pp.53-69, September 2001.

[5] C Kidd, R Gregory, A Christopher, T Starner. The Aware Home: A Living Laboratory for Ubiquitous Computing Research. In *Proceedings of the Second International Workshop on Cooperative Buildings (CoBuild'99),* October 1999.

[6] G Linkens, O Nyongeso, "Genetic Algorithms for Fuzzy Control, Part II: Online System Development and Application", *IEE proceedings Control theory applications*, Vol.142 (1995) 177-185.

[7] N Minar, M Gray, P Maes. Hive: Distributed Agents for Networking Things. In *Proceedings of ASA/MA'99, the First International Symposium on Agents Systems and Applications and Third International Symposium on Mobile Agents,* 1999.

[8] M Mozer. The Neural Network House: An Environment That Adapts to Its Inhabitants. In M. Coen, editor, *Proceedings of the American Association for Artificial Intelligence Spring Symposium on Intelligent Environments,* pages 110-114, Menlo Park, CA, 1998. AAAI Press.

[9] Y Qu, S Beale. A Constraint-Based Model for Cooperative Response Generation in Information Dialogues. In *AAAI/IAAI,* pages 148-155, 1999.

Constraint Neuro-Model based Predictive Control

Majeed.Soufian[*], Adel A. Rahman[**], Mustapha Soufian[**], and A. A. Mokbel[**]

[*] Claydon Bio-Informatic, Manches ter Business Park,Manchester, UK.
email: m.soufian@cbi.uk.com
[**] Department of Engineering, Manchester Metropolitan University, Manchester, UK.

1 Introduction

Constraints are always present in the real life process. Control of processes without considering the constraints can lead to sub-optimal behaviour and may result in instability and violation of safety. Lack of a solid and applicable closed loop control theory in dealing with constraints cause that in the usual process-control-practice either to ignore the constraints or the constraints are dealt with ad-hoc fix ups using split range controllers, overrides, min-max selectors with some logic and etc.

The conventional Model Predictive Control (MPC) has enjoyed widespread acceptance and success as an effective technique for dealing with difficult control problems of industrial importance (see e.g. Henson, 1998; Morari and Lee, 1997). The success of MPC comes from its ability to deal with time delay, multivariable interactions and constraints, and it also allows a set point change to be announced in advance and facilitate feed-forward control. Model-based control strategies for non-linear processes have traditionally been based on local linearisation and linear controller design based on the linearised model (Richalet 1976; Mehra et al., 1979; Cutler and Ramaker, 1978; and Clarke 1987a,b). The standard approach for handling strong non-linearities in the Linear MPC (LMPC) framework is to sacrifice performance by detuning the controller (Sharidhar and Cooper, 1997). Non-Linear MPC (NLMPC) requires the availability of a suitable non-linear dynamic model of the process. Consequently, the development of non-linear process models is of paramount importance. MPC with non-linear dynamic models is used for processes where process conditions can vary significantly, or for processes that are expected to experience wide-ranging operating conditions. Various NLMPC have been developed, and their application has emerged (Garcia 1984; Gattu and Zafiriou 1992; Ali and Zafiriou 1993). Brengel and Seider (1989) developed an algorithm using a multi-step predictor instead of the algorithm limited to a single predictive step. In this algorithm, linearisation of the differential equations at several instants within a sampling interval was used to lead to recursive algebraic equations that relate the predicted output to future and past values of the manipulated variables. This multi-step predictor has shown superior behaviour over the single step predictor. This approach depends on linearising the non-linear model around a nominal trajectory, and solving the control problem using a Quadratic Programming (QP) optimisation once over the prediction horizon.

It is extremely difficult and expensive to obtain an accurate model from first principles. Therefor several investigators have studied the use of empirical models for NLMPC such as polynomial ARMA model (Hernandez and Arkun 1993; Mackay et al. 1996; Sriniwas and Arkun 1997), Hamerstain models (Fruzetti et al. 1997) and second order Voltera series (Maner et al. 1996; Maner and Doyle 1997). Artificial Neural network (ANN) have the potential of describing the behaviour of extremely nonlinear/complex systems (Sain-Donat et al., 1991; Chen and Weigand, 1994; Sorensen et al., 1999). They can produce a model with an acceptable accuracy and simplicity for use in MPC schemes (Sain-Donat et al. 1991; Hunt and Sbarbaro 1991; Chen and Weigand 1994; Turner et al. 1996; De Souza et al. 1996; Galvan and Zaldifan 1998: Sorensen et al. 1999; Hussain, 1999). Apart from dynamic programming approach (Soufian 2002) so far in all applications of the ANN based predictive control the ANNs were used as identification and prediction tool for finding numerical solution of ***unconstraint*** optimisation problem. Here a Non-Linear Programming (NLP) technique is developed for **constraint** predictive control problem using ANNs.

2 Constraint Neuro-MPC (CNMPC)

The MPC scheme is based on the explicit use of a process model and process measurements to generate values for process input as a solution of an on-line (real-time) optimisation problem subjected to input and output constraints to predict the future process behaviour. Usually, the objective function is the sum of square errors of the residuals between predicted outputs and the set-point values over the prediction horizon, a term penalising the rate of change of the manipulated variable is often included as well. Mathematically, the NNMPC problem can be stated in vector form as follows:

$$J(t, U(t)) = \Gamma^{y} E^{T}(t+1)E(t+1) + \Gamma^{u} \Delta U^{T}(t)\Delta U(t) \quad (1)$$

where Γ^u and Γ^y are weighting coefficients matrices, $\Delta U(t)$ is the rate of change of the manipulated variable ($\Delta u(t) = u(t)-u(t-1)$),

$E(t+1) = [e(t+1 \mid t) \ldots e(t+p \mid t)]^T$,

$e(t+k \mid t) = r(t+k) - \square(t+k \mid t)$ for $t = 1, \ldots, p$,

$\Delta U(t) = [\Delta u(t), \ldots, \Delta u(t+m-1)]^T$

m is the control horizon and p is prediction horizon. The first step in receding horizon control is to predict the system over the range of future times. This could be done by using a one-step ahead predictor. The one-step ahead prediction is given by:

$$\hat{y}(t) = \hat{y}(t \mid t-1) = f_{nn}[y(t-1),\ldots,y(t-n_y),u(t-t_d),\ldots,u(t-t_d-n_u)] \quad (2)$$

where f_{nn} is some function realised by a neural network, and t_d is the time delay, which is assumed to be at least one. The k-step ahead prediction of the system's output can be calculated by shifting the expression forward in time while substituting predictions for actual measurements where these do not exist. As it is written in the following Equation:

$$\begin{aligned}\hat{y}(t+k) = f_{nn}[&\hat{y}(t+k-1),\ldots\ldots\ldots,\hat{y}(t+k-\min[k,n_y]),\\ &y(t-1),\ldots\ldots\ldots\ldots\ldots,y(t-\max[n_y-k,0]),\\ &u(t+k-t_d),\ldots\ldots\ldots\ldots,u(t+k-t_d-n_u)].\end{aligned} \quad (3)$$

It is assumed that the observation of the output is available up to time t-1 only; for this reason □(t) enters the expression instead of the real output y(t). The NLP technique chosen here to solve this optimisation problem is a Sequential Quadratic Programming (SQP) for the event if the problem is subjected to constraints on the input and output as follows:

$$y_{min} \le \hat{y} \le y_{max}, \qquad u_{min} \le u \le u_{max}$$

Introducing constraints to the problem improves the computational aspects of non-linear optimisation by eliminating significant regions of the area in which to search for a solution, and improving the ability to generate only feasible iterates, where inequality constraints are satisfied. The aim is to transform the problem into an easier sub-problem (Lagrangian function) that can be solved and used as the basis of an iterative process. For doing so, define Lagrangian function as:

$$\begin{aligned}L(t,U(t),\lambda) = J(t,U(t)) + &\sum_{i=1}^{p} \lambda_{1,i}(y_{min}-\hat{y}) + \lambda_{2,i}(\hat{y}-y_{max}) + \\ &\sum_{j}^{m} \lambda_{3,j}(u_{min}-u) + \lambda_{4,j}(u-u_{max})\end{aligned} \quad (4)$$

The above Lagrangian function L consists of the objective function and all terms of input and output constraints. For abbreviation we can rewrite equation (4) as:

$$L(t,U(t),\lambda) = J(t,U(t)) + \lambda C(U(t)) \quad (5)$$

$$C = \begin{bmatrix} y_{min}-\hat{y} \\ \hat{y}-y_{max} \\ u_{min}-u \\ u-u_{max} \end{bmatrix} \le 0 \quad (6)$$

For this problem to be solved, the Karush-Kuhn-Tucker (KKT) condition should be hold for U to be a local optimum. The KKT condition is stated as:

$$\nabla L = \frac{\partial L(U)}{\partial U} = 0$$
$$\lambda \nabla C(U) = 0$$
$$\lambda \geq 0$$

where the first equation describes a cancelling of the gradients between the objective function and active constraints at the solution point, i.e. the Lagrange multipliers are necessary to balance the deviations in magnitude of the objective function and constraints gradients. The last two equations describe the inclusion of only active constraints; the constraints that are not active are not included and so are given multipliers equal to zero. In implementation of SQP, if the constraints are non-linear then a sequence of linear constraints of the form:

$$Ad + b \leq 0 \tag{7}$$

where $A = \nabla C(U)^T$ and $b = C(U)$, are considered to approximate the non-linearity in the constraints. The optimisation problem is then transferred to a sequence of QP sub-problem:

$$\text{Minimise } q(d) = \frac{1}{2} d^T H d + (\nabla J)^T d \tag{8}$$

subject to constraint 7. The solution to the QP sub-problem produces a vector d, which is used to form a new iteration towards the optimum:

$$U_{i+1} = U_i + \alpha_i d_i \tag{9}$$

where α_i is a step length parameter. If a step of unity can be taken without constraints violation, this leads to the QP sub-problem's solution. Otherwise the step length parameter is determined by an appropriate line search procedure. The H in QP sub-problem is an approximation made for the Hessian of the Lagrangian by using a quasi-Newton updating procedure:

$$H_{i+1} = H_i + \frac{q_i q_i^T}{q_i^T s_i} - \frac{H_i^T s_i^T H_i}{s_i^T H_i s_i} \tag{10}$$

where
$$q_i = [\nabla J(U_{i+1}) + \lambda \nabla C(U_{i+1})] - [\nabla J(U_i) + \lambda \nabla C(U_i)]$$
$$s_i = U_{i+1} - U_i$$

3 Application to a continuous polymerisation reactor

Figure 1, illustrates the control configuration of the case study. Details of this case study and the simplifying assumptions were obtained from Maner et al. (1996). The control problem focuses on manipulating the volumetric flow-rate of the initiator ($u = F_I$) in order to regulate the Number-Average Molecular Weight (NAMW). The open loop simulation entailed the use of uniform random signal as a forcing input for generating data. The training data were generated by forcing the volumetric flow-rate of the initiator with a uniform random signal of 0.0001 minimum value and maximum value 0.1. The ANN architecture was chosen to be a feed-forward, multi-layer perceptron, with one hidden layer having five neurons. The activation functions in the hidden layer are hyperbolic tangent sigmoid transfer function. Once the ANN model was obtained, step changes of ±0.008392 m^3/h in F_I from its nominal value of 0.0016783 m^3/h, as validation tests were carried out. The open loop response of the system, ANN model, and a developed ARX model with two past inputs and outputs to these step changes are presented in Figure 2, which demonstrates the long-term prediction accuracy of the identified models. It can be noticed that how the ANN model captures the asymmetric behaviour of the system output subjected to input steps of the same magnitude and opposite sign. The linear model fails to capture this asymmetric behaviour of the process.

The tuning parameters were chosen so that the Integrated Square Error (ISE) between the simulated output and set point is minimised, as follows: p = 25, m = 2, $\Gamma^u = 0.95$ and $\Gamma^y = 1$. The constraints were chosen so that a feasible solution can be found that gives minimum ISE too. The actuation value was bounded at the maximum to 0.0001, and free at the minimum. The output constraints were free at the minimum, and set to the specified set point at the maximum. A quantitative summary of the system response to a different number of set point changes is given in Table 1. The NNMPC algorithm produced stable response for all set point changes compared to the algorithm developed by Maner (1996), which showed unstable response and poor performance at set points below 18000 kg/kmol.

Figure 6 presents the closed loop response of the system for LMPC and constraint neuro-MPC algorithms, and a quantitative summary of the system response to these algorithms is given in Table 2.

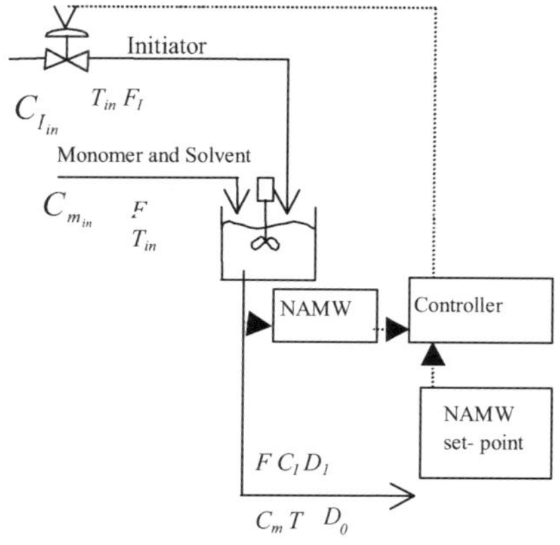

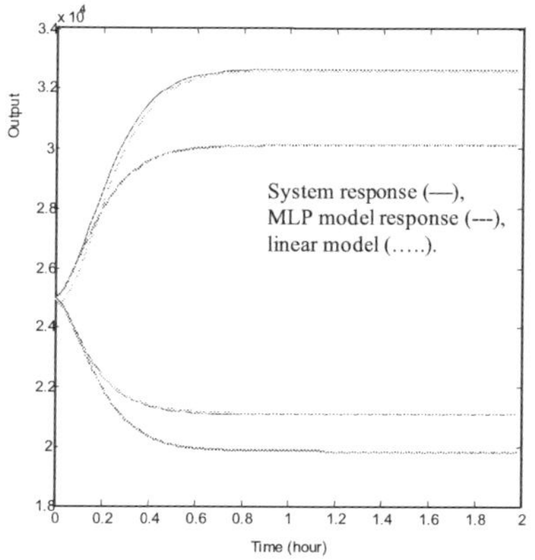

Fig. 1. Control Configuration for the Process. **Fig. 2.** Results of the validation tests.

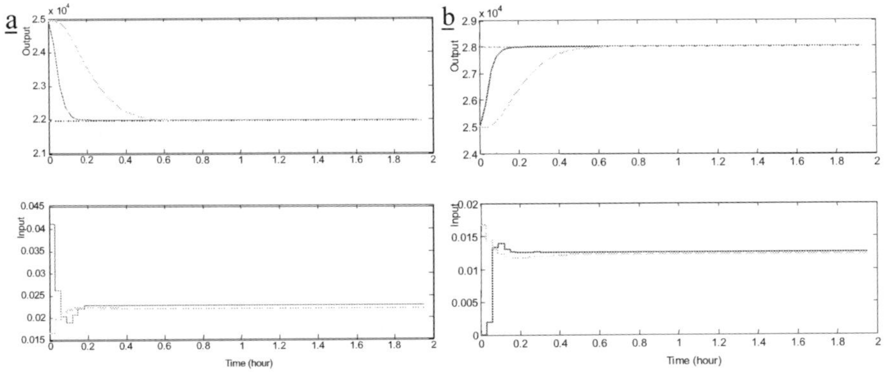

Fig. 3. The closed loop response for set point change from a) 25000 to 22000, and from b) 25000 to 28000. Key: Constraint Neuro-MPC (—), LMPC (----).

4 Conclusion

A Non-Linear Programming (NLP) technique is developed for constraint predictive control problem using an ANN model. The proposed non-linear controller methodology is applied to a highly non-linear polymerisation reactor. It was shown that the hybrid method of the ANN and the SQP algorithm can be successfully incorporated into the MPC framework to cope well with the constraint control problems.

As shown in Figure 6, both controllers bring the reactor to the new set point, and the two controllers perform the task in a short time. However, the constraint neuro-MPC is faster than the LMPC. In all situations, the constraint neuro-MPC controller outperforms the linear controller by considerably reducing the ISE between the system output and set point tracking. However compare to a non-linear unconstrained optimisation, it scarifies the performance for respecting constraints and consequently safety issues in the region of high changes in setpoints (Table 1).

Table 1. Closed-loop performance for different set point changes

Set point changes kg/kmol	Integrated Square Error (ISE) Unconstrained	Constrained
38000	0.0011	0.0033
32000	3.69421×10^{-4}	8.3825×10^{-4}
18000	0.0012	0.0013
12000	0.0086	0.0064

Table 2. Closed-loop performance for LMPC and NNMPC algorithms.

Set point changes kg / kmol	Integrated Square Error (ISE) LMPC	CNMPC
28000	0.0022	0.000124
22000	0.0022	0.00016

References

Ali, E. and Zafiriou, E., (1993). 'Optimization-based tuning of non-linear model predictive control with state estimation', Journal of Process Control, vol. 3, no. 2, pp. 97-107.

Brengel, D. D. and Seider, W. D., (1989). 'Multistep predictive controller', Industrial Engineering Chemical Research, vol. 28, pp. 1812-1822.

Chen, C-L, Lin, S-H and Chen, C-K, (1996). 'Application of Taylor transformation to non-linear predictive control problem', Applied Mathematical Modelling, vol. 20, September, pp. 699-710.

Chen, Q. and Weigand, W. A., (1994). 'Dynamic optimization of non-linear processes by combining neural net model with UDMC', AIChE Journal, vol. 40, no.9, September, pp. 1488 - 1497.

Clarke, D. W., Mohtadi, C. and Tuffs, P. S., (1987a).'Generalised predictive control - Part I. The basic algorithm', Automatica, vol. 23, no.2, pp. 137-148.

Clarke, D. W., Mohtadi, C. and Tuffs, P. S., (1987b).'Generalised predictive control - Part II. Extensions and interpretations', Automatica, vol. 23, no.2, pp. 149-160.

Cutler, C. R. and Ramaker, B. L., (1978). ' Dynamic matrix control algorithm', AIChE 86th National Meeting. Houston, TX.

De Souza, M. B., Pinto J. C. and Lima, E. L., (1996). 'Control of a chaotic polymerization reactor: A neural network based model predictive approach', Polymer Engineering and Science, vol. 36, no. 4, February, pp. 448-457.

Fruzzetti, K. P., Palazoglu, A. and McDonald, K. A., (1997). 'Non-linear model predictive control using Hammerstain models', Journal of Procees Control, vol. 7, no. 1, pp. 31-41.

Galvan, I. M. and Zaldivan, J. M., (1998). 'Application of recurrent neural networks in batch reactors Part II: Non-linear inverse and predictive control of the heat transfer fluid temperature', Control Engineering and Processing, vol. 37, pp. 149-161.

Garcia, C. E., (1984). 'Quadratic/dynamic matrix control of non-linear processes. An application to a batch reaction process', AIChE Annual meeting, San Francisco, CA.

Henson, M. A., (1998). 'Non-linear model predictive control: current status and future directions', Computers and Chemical Engineering, vol. 23, no. 2, pp. 187-202.

Hernandez, E. and Arkun, Y., (1993). 'Control of non-linear systems using polynomial ARMA models', AIChE Journal, vol. 39, no. 3, pp. 446-460.

Hunt, K. J. and Sbarbaro, D., (1991). 'Neural networks for non-linear internal model control', IEE Proceedings-D, vol. 138, no. 5, pp. 431-438.

Mackay, M., Thomson, M. and Soufian, M., (1996). 'A bilinear non-parametric model based predictive controllers,' Proceedings IFAC 13th World Congress, June, California, pp. 459-364.

Maner, B. R. and Doyle III, F. J., (1997). 'Polymerization reactor control using autoregressive plus Volterra-based MPC', AIChE Journal, vol. 43, no. 7, pp. 1763-1784.

Maner, B. R., Doyle III, F. J., Oguunaike, B. A. and Pearson, R. K., (1996). 'Non-linear model predictive control of a simulated multivariable polymerization reactor using second-order Volterra models', Automatica, vol. 32, no.9, pp. 1258-1301.

Mehra, R. K., Rouhani, R., Rault A. and Reid, J. G., (1979). 'Model algorithmic control: Theoretical results on robustness', Proceedings JACC, pp. 387-401.

Hussain M. A., (1999). 'Review of the applications of neural networks in chemical process control--- Simulation and online implementation', Artificial Intelligence in Engineering, vol. 13, no. 1, pp. 55-68.

Morari, M. and Lee, J. H., (1997). 'Model predictive control: Past, present and future', European Symposium on Computer Aided Process Engineering (ESCAPE-7), 25-29 May, Trondheim, Norway.

Richalet, J., Rault, A., Testud, J. L. and Papon, J., (1976). 'Model predictive heuristic control: application to industrial processes', Automatica, vol. 14, pp. 413-428.

Saint-Donat, J., Bhat, N. and McAvoy, T. J., (1991). 'Neural net based model predictive control', International Journal of Control, vol. 54, no. 6, pp. 1453-1468.

Shridhar R. and Cooper, D. J., (1997). 'A tuning strategy for unconstrained SISO model predictive control', Industrial Engineering Chemical Research, vol. 36, pp. 729-746.

Sorensen, P. H., Norgaard, M., Ravn, O. and Poulsen, N. K., (1999). 'Implementation of neural network based non-linear predictive control', Neurocomputing, vol. 28, no. 1/3, pp. 37-51.

Soufian, M (2002). 'Hard and Soft Computing Techniques for Non-linear Modeling and Control with Industrial Application', PhD Thesis, MMU, Manchester, England.

An Efficient Hybrid Method of Regression Tree and Fuzzy Inference for Short-term Load Forecasting in Electric Power Systems

Hiroyuki Mori[*], Yoshinori Sakatani[*], Tatsurou Fujino[**] and Kazuyuki Numa[**]

[*] Department of Electrical and Electronics Engineering, Meiji University, Tama-ku, Kawasaki 214-8571, Japan, email: {hmori, ce22026}@isc.meiji.ac.jp
[**] Central Load Dispatching Center, Chubu Electric Power Co., Inc. Higashi-ku, Nagoya 461-8680, Japan

Abstract. This paper proposes an efficient hybrid method of the regression tree and simplified fuzzy inference for short-term load forecasting in electric power systems. Short-term load forecasting plays a key role in economic load dispatching, unit commitment, etc. However, it is not easy to handle due to the nonlinear relationship between input and output variables. This paper makes use of the regression tree to clarify the nonlinear relationship and extract important rules. In this paper, tabu search is used to optimize the attribute at the splitting node of the tree. Also, this paper applies the simplified fuzzy inference of the nonlinear approximation to one-step ahead load forecasting for data in each terminal node. The fuzzy model is optimized by tabu search to tune up the fuzzy membership functions. The proposed method is tested with real data.

1 Introduction

This paper proposes a hybrid method to combine the regression tree of data mining with simplified fuzzy inference (SFI) [9, 10] for short-term load forecasting in electric power systems. Short-term load forecasting plays an important role to smooth electric power systems [1]. More exact load forecasting brings about more secure and economic power system conditions. The conventional methods on short-term load forecasting may be classified as follows:

(a) regression model [2]
(b) Kalman Filtering [3]
(c) ARIMA model [4]
(d) expert system [5]
(e) artificial neural net (ANN) [6]
(f) fuzzy inference [7]
(g) neuro-fuzzy model [8]
(h) simplified fuzzy inference (SFI) [9, 10]

However, load forecasting is not easy to handle due to the nonlinearity related to the weather conditions such as temperature, humidity, etc. The sudden load change often results in a large error between actual and predicted values. As a result, ANN, neuro-fuzzy and SFI are more effective than others. The classical methods like Methods (a)-(c) do not handle the complicated behavior of loads. Namely, they give us erroneous results in short-term load forecasting. To overcome the problem, the intelligent systems have been developed to deal with short-term load forecasting. To improve the model accuracy, the supervised learning has been used in the learning process of Methods (e), (g) and (h) above. In other words, they correspond to the nonlinear approximator for one-step ahead load forecasting.

This paper combines a clustering method with SFI that determines the number and location of the fuzzy membership function with tabu search (TS). As a meta-heuristic method, TS is effective for solving a combinatorial problem [14-18]. In this paper, a clustering technique is introduced as a data prefiltering technique of SFI to classify input data into clusters in which the nonlinear approximator is constructed to improve the prediction accuracy. Recent studies has revealed that SFI is better than the multiplayer perceptron (MLP) of ANN [10] and that the regression tree of data mining works so well as a prefiltering technique for MLP [11-13]. This paper focuses on clustering by the regression tree for SFI. The conventional intelligent systems such as ANN, etc. are based on the black-box model and fail to explain the relationship between input and output variables appropriately. On the other hand, the regression tree classifies input data into the terminal node corresponding to the cluster of clustering. It allows to extract rules from data easily. Also, the tree structure is visually understandable in a sense of the variable importance. Thus, this paper makes use of the regression tree as a clustering technique. The effectiveness of the proposed method is tested with real data.

2 Data Mining

Data mining is useful for extracting rules from database. It requires the easiness to interpret the rules. Several techniques such as ANN, fuzzy inference and decision tree have been developed to handle data mining. Among them, the decision tree has advantage in a sense that the structure of rules is more understable through the tree

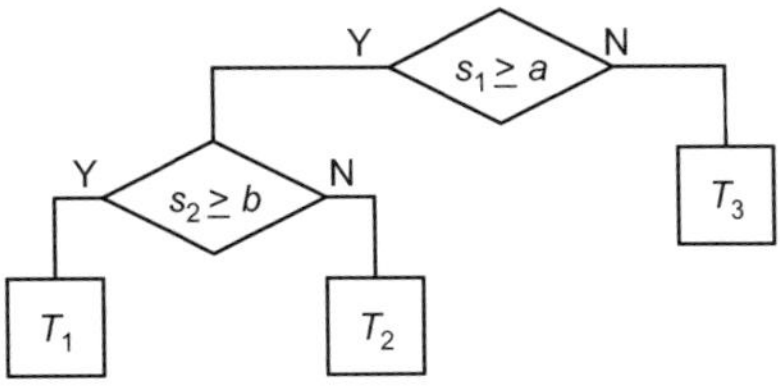

Fig. 2.1. Structure of Regression Tree

structure and the attribute at the splitting node have an if-then rule to extract features of data. The decision tree may be classified into the classification and regression ones. The former deals with data mining of discrete number while the latter handles with continuous one. Figure 2.1 shows the structure of decision tree, where the lozenge and square denote the splitting and terminal nodes, respectively. The splitting node of the starting point is referred to as the root node. The conditions are assigned to the splitting node. For example, if given data meets conditions $s_1 \geq a$ and $s_2 \geq b$, then it belongs to terminal node T_1 in Figure 2.1. As a data mining technique, this paper makes use of CART to classify data and to extract rules from continuous data of short-term load forecasting. CART has two phases to construct the decision tree.

Phase I

This phase grows the tree to determine an if-then rule at the splitting node and classify input data into the terminal node. CART separates a set of data into two groups with the attribute that provides the smallest error. Specifically, the criterion of the attribute selection may be written as

$$\Delta R(s,t) = R(t) - R(t_L) - R(t_R) \tag{2.1}$$

where, ΔR: degree of error reduction in case where attribute s is employed as the splitting conditions at Node t, s: attribute, t: parent node, $t_L(t_R)$: left-side (right-side) child node, $R(t)$: sum of squared errors between data i and the average of data i (i=1, 2, …, N) such as

$$R(t) = \sum_{i=1}^{N} (y_i - \bar{y})^2 \tag{2.2}$$

where, N: total number of data, y: data i, $\bar{y}$: average of y_i

The attribute at the root node is determined by selecting the optimal attribute that maximizes Eqn. (2.1). The same process is repeated at the child node until the convergence criterion is satisfied.

Phase II

This phase carries out pruning of the tree obtained at Phase I. It is useful for avoiding the overfitting of the model. In other words, that contributes to the easiness to create simple rules. After constructing the regression tree with CART, unnecessary branches are cut down to simplify the model. The pruning process is to find out the splitting node that gives the smallest error reduction and replace it with the terminal one. The criterion may be written as

$$r(t_p) = r^{CV}(t_p) + \sigma(t_p) \tag{2.3}$$

where, r: error, r^{CV}: cross-validation error, σ: standard deviation of cross-validation error, t_p: pruned tree number

3 Proposed Method

This section proposes a hybrid method of the optimal structured regression tree and SFI for short-term load forecasting. The proposed method makes use of a prefiltering technique to improve the prediction accuracy. Figure 3.1(a) shows the conventional model while Figure 3.1(b) denotes the proposed method. The proposed method constructs a predictor at each cluster as shown in Figure 3.1(b). As the prefiltering technique, this paper uses the regression tree. The conventional CART allows to construct the regression tree. However it has a drawback that it has locally optimal structure of the regression tree because of successfully growing the tree through the relationship between the parent and child nodes. Thus, it is necessary to consider the relationship between the nodes as a while in growing the decision tree systematically. In this paper, a TS based method is proposed to determine the optimal attribute at the splitting node. That results in a combinatorial optimization problem to minimize the sum of the variances of variables in the terminal nodes. In other words, the combination of the attributes at the splitting node influences the performance of the decision tree. After constructing the decision tree, this paper makes use of SFI to predict one-step ahead short-term load forecasting. SFI is applied to each terminal node to construct the predictor model. The proposed method is referred to as TSCART+SFI in this paper. The algorithm of the proposed method may be summarized as follows:

Algorithm

Step 1: Set the initial conditions.

Step 2: Carry out data clustering for given data and calculate the splitting value at the splitting node.

Step 3: Prune the tree to evaluate the optimal tree and the cross-variation error.

Step 4: Repeat Steps 4A-4E until the algorithm criterion of TS is satisfied.

Step 4A: Create the neighborhood solutions around the initial solution.

Step 4B: Carry out data clustering for each solution candidate and calculate the splitting value at the splitting node.

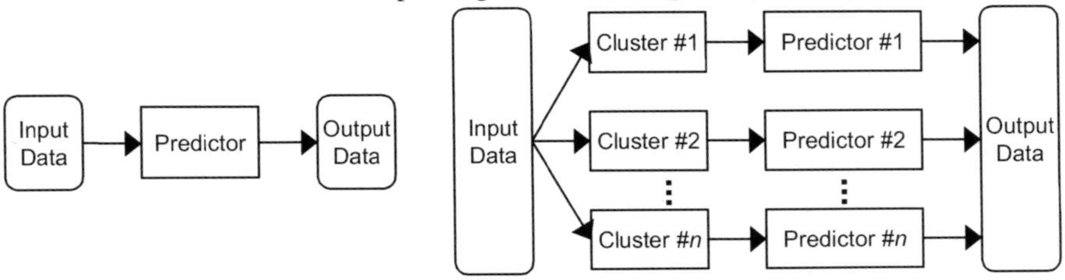

(a) Conventional Method (b) Proposed Method

Fig. 3.1. Structure of Regression Tree

Step 4C: Prune the tree to calculate the optimal tree and the cross-validation error.

Step 4D: Evaluate the solution with the smallest cross-validation error.

Step 4E: Go to Step 5 if the criterion is satisfied. Otherwise, set the solution as the initial one at the next Step and return to Step 4A.

Step 5: Optimize the fuzzy membership function of SFI with TS for the accumulated data assigned to each terminal node and construct the SFI model.

Step 6: Give test data to the proposed method and identify the terminal node corresponding to test data.

Step 7: Predict the one-step ahead load with SFI at the terminal node.

4 Simulation

4. 1 Simulation Conditions

The proposed method is applied to one-step ahead daily maximum load prediction

Table 4.1. Attributes

Attributes	Contents
s_1	Day Type on day *d+1*
s_2	Predicted Maximum Temperature on Day *d+1*
s_3	Predicted Minimum Temperature on Day *d+1*
s_4	Predicted Average Temperature on Day *d+1*
s_5	Predicted Minimum Humidity on Day *d+1*
s_6	Predicted Discomfort Index on Day *d+1*
s_7	Maximum Load on Day *d*
s_8	Difference between Maximum Load on Days *d-1* and *d*
s_9	Difference between Average Temperature on Days *d-1* and *d*
s_{10}	Average of Maximum Load on Days *d-2*, *d-1* and *d*
s_{11}	Average of Average Temperature on Days *d-2*, *d-1* and *d*

in the Chubu Electric Power Company. In particular, the proposed method was tailored for the summer weekdays from June to September. That is because the summer season has the largest power demand in a year and the weekday power demand is larger than weekend one. The learning data was accumulated from 1991 to 1998 while the test data was collected from 1999.

The effectiveness of the proposed method is demonstrated and compared with the conventional methods. For convenience, the following methods are defined:

Method A: MLP
Method B: SFI
Method C: CART+MLP
Method D: CART+SFI
Method E: TSCART+MLP
Method F: TSCART+SFI (Proposed Method)

Table 4.1 shows the attributes at the splitting node of the decision tree in the proposed method that are related to the maximum load of the next day. Also the prediction model of MLP and SFI employs the predicted maximum temperature (s_2), the predicted average temperature (s_4), the predicted minimum humidity (s_5) and the predicted discomfort index of the next day (s_6). They are determined by the variable importance.

4. 2 Simulation Results:

Figure 4.1 shows the regression tree obtained by the proposed method, where the lozenge shows the splitting node and has an if-then rule. Also, the square means the terminal node to which input data is finally assigned. The regression tree has 18 terminal nodes and the left side of the tree is more complicated than the right side. The left side has 10 terminal nodes (T_1- T_{10}) while the right possesses 8 (T_{11}-T_{18}).

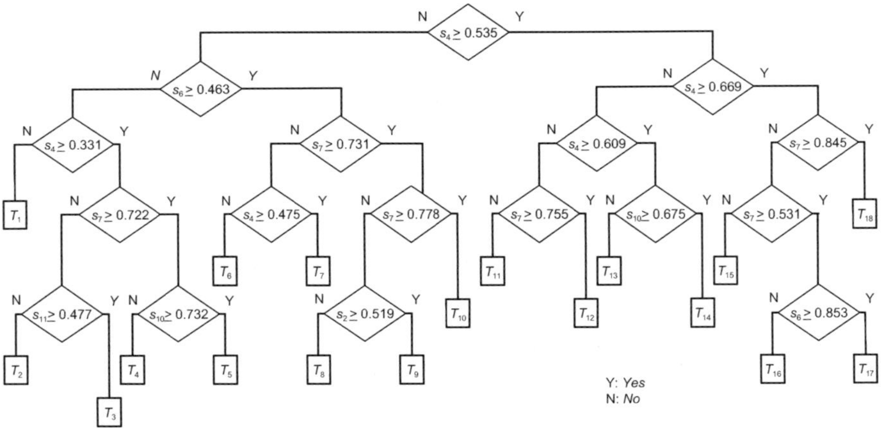

Fig. 4.1. Regression Tree

Actually, the regression tree has six attributes such as s_2, s_4, s_6, s_7, s_{10} and s_{11}. Attribute s_4 is the most important due to the attribute at the root node of the starting point at the regression tree. Apart from the root node, attribute s_4 is used at the five splitting nodes. Table 4.2 shows the number of nodes and cross-validation error of CART and TSCART. It can be seen that the proposed method improves the cross-validation error of 2.38 %.Thus, the proposed method gives better results in

Table 4.2. Number of Nodes and Cross-validation Error

Method	Nomber of Nodes	Cross-varidation Error (%)
CART	11	8.33
TSCART	18	5.95

Table 4.3. Average and Maximum Errors of Each Method

Method	Average (%)	Maximum (%)
A	1.92	7.17
B	1.88	6.50
C	1.68	5.84
D	1.68	5.32
E	1.63	5.21
F	1.62	5.17

spite of the increase of nodes. Table 4.3 gives the comparison of the methods in terms of the average and maximum errors. The proposed method is better than

others in terms of the average and maximum errors. That shows that the scheme of Figure 3.1(b) works so well. Compared with the Method A, the proposed method improves the average error of 0.30% and the maximum error of 2.38%. That implies that the proposed method appropriately extracts the similarity of data.

5 Conclusion

This paper has proposed an efficient hybrid method of the optimal regression tree of data mining and SFI for short-term electric load forecasting. The optimal regression tree serves as a data prefiltering technique and classifies input variables into a terminal node. The proposed method globally optimizes the structure of the regression tree with TS to classify input variables into the terminal nodes in a globally optimal way. Also, SFI was constructed to predict one-step ahead electric load forecasting at each terminal node.

The proposed method was applied to real data in the utilities. The simulation results have shown that simple rules are extracted through the regression tree. Compared with the conventional MLP, the proposed method improved the average error of 0.30% and maximum error of 2.00%. Therefore, the proposed method is more acceptable to the system operators in power systems.

References

1. G Gross and FD Galiana (1987). Short-term Load Forecasting. In *Proc. of the IEEE*, vol. 75, no. 12, pp. 1558-1573, Dec. 1987.
2. CE Asbury (1975). Weather Load Model for Electric Demand Energy Forecasting. In *IEEE Trans. on Power Systems*, vol. PAS-94, no. 4, pp. 1111-1116, July 1975.

3. J Toyoda, *et al.* (1970). An Application of State Estimation to Short-Term Load Forecasting, I and II. In *IEEE Trans. on Power App. and Syst.*, vol. PAS-89, no 7, pp. 1678-1688, Sep. 1970.
4. M Hagan and R Klein (1977). Identification Techniques of Box and Jenkins Applied to the Problem of Short Term Load Forecasting. In *IEEE PES SM*, Paper A 77 168-2, Mexico City, Mexico, July 1977.
5. S Rahman and R Bhatnagar (1988). An Expert System Based Algorithm for Short Term Load Forecast. In *IEEE Trans. on Power Systems*, vol. 3, no. 2, pp. 392-399, May 1988.
6. DC Park, *et al.* (1991). Electric Load Forecasting Using An Artificial Neural Network. In *IEEE Trans. on Power Systems*, vol. 6, no. 2, pp. 442-448, May 1991.
7. YT Park and JK Park (1989). An Expert System for Short Term Load Forecasting by Fuzzy Decision. In *Proc. of IFAC, International Symposium on Power Systems and Power Plant Control,* pp. 831-836, Soul, Korea, Aug. 1989.
8. H Mori and H Kobayashi (1994). A Fuzzy Neural Net for Short-term Load Forecasting. In *Proc. of ISAP'94*, pp. 775-782, Monpellier, France, Aug. 1994.
9. H Mori and H Kobayashi (1996). Optimal Fuzzy Inference for Short-term Load Forecasting. In *IEEE Trans. on Power Systems*, vol. 11, no. 1, pp. 390-396, Feb. 1996.
10. H Mori and Y Sone (1998). Simplified Fuzzy Inference Method with Tabu Search for Short-term Load Forecasting in Power System. In *Proc. of IEEE Conference on Decision and Control*, vol. 4, pp. 4240-4245, Tampa, FL, USA, Dec. 1998.
11. H Mori and N Kosemura (2001). Optimal Regression Tree Based Rule Discovery for Short-term Load Forecasting. In *Proc. of 2001 IEEE PES Winter Meeting*, vol. 2, pp. 421-426, Columbus, OH, USA, Jan. 2001.
12. H Mori, *et al.* (2001). Short-term Load Forecasting with Fuzzy Regression Tree in Power Systems. In *Proc. of 2001 IEEE SMC Meeting*, pp. 1948-1953, Tucson, AZ, U.S.A., Oct. 2001.
13. H Mori, *et al.* (2002). Data Mining for Short-term Load Forecasting. In *Proc. of 2002 IEEE PES Winter Meeting*, vol. 1, pp. 623-624, New York, NY, USA, Jan. 2002.
14. F Glover (1989). Tabu Search, Part I. In *ORSA Journal on Computing*, vol. 1, no. 3, pp. 190-206, Summer 1989.
15. H Mori and T Usami (1996). Unit Commitment Using Tabu Search with Restricted Neighborhood. In *Proc. of ISAP'96*, pp. 422-427, Orlando, FL, U.S.A., Feb. 1996.
16. H Mori and T Hayashi (1998). New Parallel Tabu Search for Voltage and Reactive Power Control in Power Systems. In *Proc. of IEEE International Symposium on Circuit and Systems (ISCAS'98)*, vol. 3, pp. 431-434, Monterey, CA, U.S.A., May 1998.
17. H Mori and O Matsuzaki (1999). Tabu Search Based Approach to Meter Placement in Static State Estimation. In *Proc. of ISAP'99*, pp. 365-369, Rio de Janeiro, Brazil, Apr. 1999.
18. H Mori and Y Sone (1999). Tabu Search Based Meter Placement for Topological Observability in Power System State Estimation. In *Proc. of 1999 IEEE Transmission and Distribution Conference*, vol. 1, pp. 172-177, New Orleans, LA, USA, Apr. 1999.

An Iterative Learning Control Method for a Level Control Problem

Michael H. Schwarz and Chris S. Cox.

Control Systems Centre, University of Sunderland, United Kingdom

Abstract. This chapter is concerned with a filter-based iterative learning control method (ILC) for a repetitive coupled-tank experiment. Within a few iteration-cycles of the experiment, the ILC method improves the input tracking performance significantly. The ILC method is able to learn new reference-signals within few iterations without resetting the feed-forward ILC vector and can cope with model changes.

1 Introduction

The roots of iterative learning control (ILC) go back to the year 1984, where the method was originally proposed in the robotics community [1] as an intelligent teaching method for robot manipulators. Initially used in robotics, the technique can nowadays also be applied to manufacturing processes such as chemical batch reactors and batch distillation, or any other repetitive industrial process [4,7].

The basic idea of ILC is to improve the current control process by using the knowledge of the previous result (iteration-cycle) as an additional feed-forward signal. Every time the same batch process is repeated, the tracking performance should improve until it converges to a final result. If an entirely new batch process is prepared, then the ILC should simply learn the new trajectory tracking [4].

ILC differs from many other feed-forward compensation methods in that, it efficiently exploits input-output relationships (system-input, system-output/error) of previous experiments. The method learns and improves through the repetition of the experiment and is not based on any process model [3,4,7].

Traditionally, the PI(D) controller is one of the most commonly used instruments in the process control industries [5]. Its structure is simple and many sophisticated tuning methods incorporate its main features. Nevertheless, its design is still a quite challenging task if good or even perfect trajectory tracking is required. However, it is straightforward to design a PI controller with an ILC method as an additional feed-forward controller. In this chapter, a filter-based ILC method is examined to improve the tracking performance of a two tank system. A simple controller was designed using the auto-tuner technique described by Åstrom and Hägglund [2]. When the ILC approach is introduced the tracking performance improves, while the controller burden is relaxed. The filter itself provides an extended moving average [7] which incorporates system input information as well as the tracking errors.

2 Controller Design

The auto-tuning method, proposed by Åstrom and Hägglund [2], was carried out for the first tank of the coupled tank system, to obtain the ultimate crossover-frequency and peak amplitude of the resulting oscillation. The designed PI-Controller transfer function is presented in the Laplace domain in Eq.1.

$$G_c = \frac{2VP \cdot \sin(\theta)}{\pi^2 A}\left(\frac{(P \cdot \tan(\theta))\, s + 2\pi}{(P \cdot \tan(\theta))\, s}\right) \tag{1}$$

Where 'V' is the relay amplitude, 'P' is the ultimate period, 'A' is the peak amplitude of the oscillation at the input to the non-linearity and 'θ' a design parameter within the range of [0°-90°]. The same controller structure was used for the second tank.

3 Filter Based ILC

Since the first introduction of ILC, many different algorithms and methods have been proposed. The generalised update law may be given by [3]:

$$U_{k+1} = H_1 \cdot E_k + \mathrm{H}_2 \cdot \mathrm{U}_\mathrm{K} \tag{2}$$

Where U_{k+1} represents a vector, that contains every single ILC value for every sample time.U_k , E_k are the recorded system input signal and error, respectively. H_1 and H_2 are linear filters. As ILC update is normally an off-line procedure, H_1 and H_2 can be non-causal. The subscript k represents the entire k^{th} iteration cycle. Two simple moving average filters have been selected. Therefore, Eq. 2 can be rewritten as:

$$U_{k+1} = G_c \cdot \lambda \cdot h * E_k + \gamma \cdot \mathrm{h} * \mathrm{U}_\mathrm{K} \tag{3}$$

and finally yields:

$$U_{k+1}(a) = G_C \cdot \left(\frac{\lambda}{2N+1}\sum_{j=-N}^{N} E_k(a+j)\right) + \frac{\gamma}{2M+1}\sum_{i=-M}^{M} U_k(a+i) \tag{4}$$

To indicate a specific sampled value within the different vectors the variable *a* is used. γ and λ are filter gains and h* represents the convolution operator. M and N specify the filter-horizon. G_c characterises the controller transfer-function.

Finally, to improve U_{k+1} of the ILC-method a segmentation technique is implemented, introduced by Xu [7]. The data-set is divided into two sections, the first contains only information about the input/output relationships during the ramp signal, the second section includes only input/output relationships in steady state. Fig. 1 shows the process (P) controlled by a feedback controller augmented with the additional ILC-circuitry. In the experiments γ, λ, M and N were selected as 1, 0.4, 100, 50, respectively, which had been proven as a good choice during initial experiments.

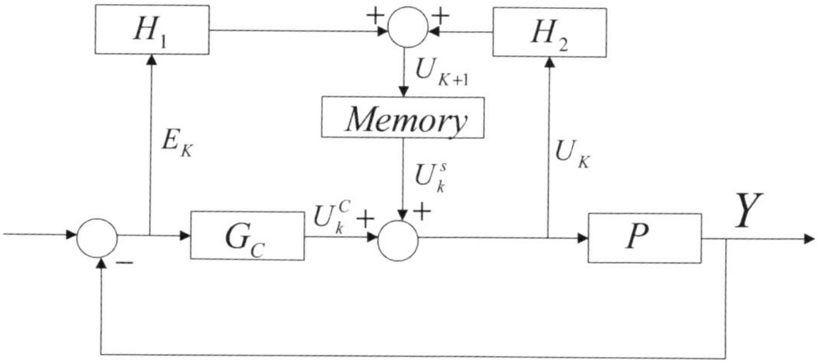

Fig. 1. Process System with Feedback-Controller and Iterative Learning Control

4 Stability

The input/output relationship can be written as:

$$Y = P \cdot U = P \cdot U_k^C + P \cdot U_k^S \Rightarrow Y = P \cdot G_C \cdot E_k + P \cdot U_k^S \tag{5}$$

The current error can be expressed as:

$$E_{k+1} = R - Y_{k+1} = R - P \cdot U_{k+1}^S - P \cdot G_c \cdot E_{k+1} \tag{6}$$

Inserting Eq. 2 in Eq.6 adding and subtracting $H_2 \cdot E_k$ yields:

$$E_{k+1}(1 + P \cdot G_c) = R - P \cdot H_1 \cdot E_K - P \cdot H_2 \cdot U_k^S + H_2 \cdot E_k - H_2 \cdot E_k \tag{7}$$

The error of the previous ILC cycle can be stated as:

$$E_k = R - P \cdot U_k^c - P \cdot U_k^S \tag{8}$$

Inserting Eq. 8 in Eq. 7 yields

$$\begin{aligned} E_{k+1}(1 + P \cdot G_c) = {} & R - P \cdot H_1 \cdot E_K - P \cdot H_2 \cdot U_k^S + H_2 \cdot E_k - H_2 \cdot R \\ & + H_2 \cdot P \cdot U_k^c + H_2 \cdot P \cdot U_k^S \end{aligned} \tag{9}$$

After cancellation and redefining Eq. 9 becomes:

$$E_{k+1} = \frac{(1 - H_2)}{(1 + P \cdot G_c)} \cdot R + \left(H_2 - \frac{P \cdot H_1}{(1 + P \cdot G_c)} \right) \cdot E_K \tag{10}$$

The stability criterion becomes [3,6]:

$$\left\| H_2(jw) - \frac{P(jw)\cdot H_1(jw)}{(1+P(jw)\cdot G_c(jw))} \right\| < 1 \forall w \tag{11}$$

5 Illustrative Examples

A laboratory coupled tank system, developed by TecQuipment Ltd, is used as a process to demonstrate the benefits of an ILC method. It consists of two tower-type rectangular tanks with an internal valve in between. The valve-position can be varied to regulate the tank leakage from one tank to the other; this changes the dynamics of the system. Each tank is equipped with a separate outlet and a separate pump, therefore, liquid (water) can be independently pumped into each tank. A piezo-resistive silicon pressure type depth transducer provides the level value in the range of 0-10volts. The system is configured as a SISO (Single Input Single Output) control problem. The outlet-valves are half open as is the internal valve which couples both tanks. Fig. 2 shows a schematic of the coupled-tank experiment.

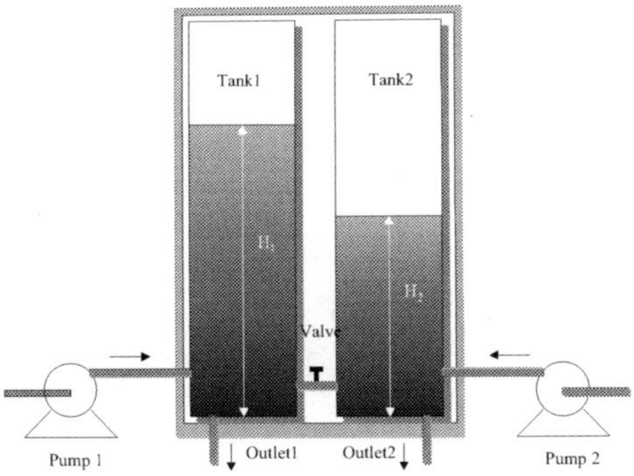

Fig. 2. Schematic of the coupled tank system

To demonstrate the usability and benefits of the proposed ILC method the following control targets have been considered. As reference signals only ramp and steady-state signals are accounted. The first task will show the learning capability of ILC techniques. The second task deals with a change in the dynamics of the system. This is accomplished by fully opening the outlet valve of the second tank. The final task discusses the learning-ability when the slope of the ramp-signal is changed. Task 1: The first reference signal (tank 1) is ramped up for 80seconds with a constant slope of 0.1V/second, where upon the system is kept at steady state for 70 seconds. The second reference signal (tank 2) is ramped up for only 60 seconds with a constant slope of 0.1V/second, where upon the system is kept at steady state for 90 seconds. For task 2 the first task is repeated until the ILC-method converged, then the second outlet-valve is regulated to fully open. The new task had to be learnt without resetting the ILC-vector. Task 3: After task 1 was again repeated until the ILC-method converged, the first reference signal was changed to a slope of 0.05V/second without resetting the ILC-Vector.

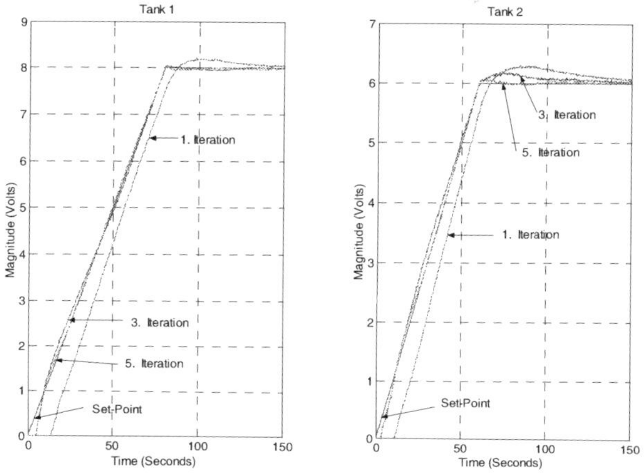

Fig. 3. Results for Task 1

The ILC-method enhanced significantly the tracking performance for the initial task within 5 repetition-cycles. Fig. 3 shows the initial tracking result and after the third and fifth iteration.

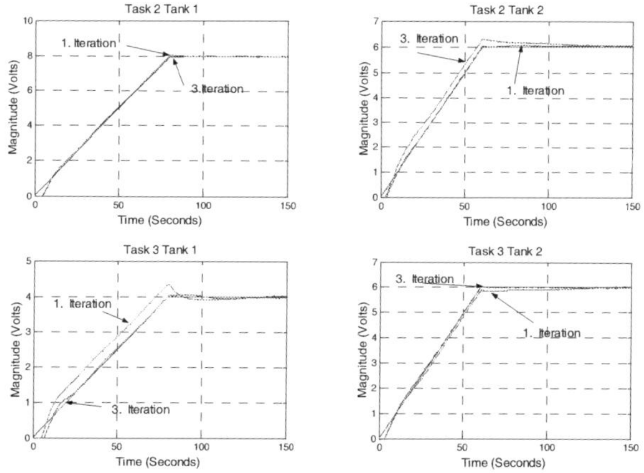

Fig. 4. Results for task 2 (upper Figures) and task 3 (lower Figures)

In task 2, the new valve position of the second outlet changed the dynamics of the system and the ILC-method had to learn a new task. Because the two tanks are coupled the different dynamics of tank 2 had an impact on tank 1. Within 3 iterations the learning method provided nearly perfect tracking, again. The ILC-vector did not have to be reset. Fig 4 (upper) demonstrates the tracking performance before the new iteration and after the final third repetition.

The last task - changing the slope of the second reference signal - did not represent a difficult task. The ILC-algorithm learnt the new reference signal within 3 iterations. Again, the change of the reference signal of tank 1 had also an impact on the dynamics of tank 2. Fig. 4 (lower) presents the initial tracking performance after changing the slope of the reference signal and the improved performance after the third iteration.

6 Conclusions

Three different examples illustrate how iterative learning control can improve tracking performance and how iterative learning control can learn different tasks without reset or special data-treatment. In particular the first task demonstrated how the tracking performance of a standard PI controller could be significantly improved. The ILC-algorithm learned up to two iterations faster, if the ILC-vectors were not reset. While the ILC-algorithm ensured good reference tracking the controller action is relaxed and is mainly responsible for disturbance rejection. The advantage of the method introduced, lies in how the algorithm exploits input/output relationships without the requirements for a model.

References

1. Arimoto S., Kawamura S., Miyazaki F. (1984) Bettering operation of robots by learning. Journal of Robotic Systems 1(2):23-140
2. Åstrom K., Hägglund T. (1984) Automatic tuning of simple regulators with specifications on phase and amplitude margins. Automatica 20(5):635-645
3. Gunnarsson G., Norrloef M. (1997) A Short Introduction to Iterative Learning Control. Technical Report Report No.: LiTH-ISY-R-1926
4. Moore K.L (1998) Iterative Learning Control: An Expository Overview. Applied and Computational Controls, Signal Processing, and Circuits
5. O'Dwyer A. (2000) A summary of PI and PID controller tuning rules for processes with time delay Part 1: PI controller tuning rules. In: Preprints IFAC Workshop on digital control: Past, present and future of PID control. Teressa, Spain, pp 175-180
6. Schwarz M.H. Cox C.S. (2002) An Iterative Learning Control Method for a Level Control Problem. Proceedings of the 4th Intern. Conf. on Recent Advances in Soft Computing. Nottingham pp 25-31
7. Xu, J.-X., Lee T.-H., Tan Y., (2002) Enhancing trajectory for a class of process control problems using iterative learning. Engineering Applications of Artificial Intelligence, 15:1-12

Face Verification Using SVM: Influence of Illumination

Cristina Conde [*], Antonio Ruiz [**] and Enrique Cabello [*]

[*] Computer Science, Statistics and Telematics Department, Universidad Rey Juan Carlos. C/ Tulipán s/n 28933 Móstoles (Spain) {cristina.conde, ecabello}@escet.urjc.es
[**] Departamento de Tecnología Fotónica Department, Universidad Politécnica de Madrid, Campus de Montegancedo, 28660 Madrid (Spain) aruiz@fi.upm.es

Abstract. Influence of illumination conditions in face verification using support vector machines (SVM) and k-nearest neighbours is analysed using an experimental set up in which images are acquired in controlled or uncontrolled illumination conditions. Principal components analysis (PCA) has been considered to perform dimensional reduction. SVM techniques offers better results even if linear kernels are considered.
Keywords. Human face verification, Principal Components Analysis, K-Nearest Neighbours, Support Vector Machine, Image Processing.

1 Introduction

The experiment presented in this paper was focused to test performance of a face verification system under real control access conditions. In control access environments it is possible to obtain an initial set of images of the subjects in good illumination conditions, the face is located in this images and the system is trained. But in normal operation mode, the face verification system has to work with images in which illumination conditions may have changed. This cause errors in final verification process because of the differences with the initial face estimation. In this paper influence of illumination conditions in face verification using SVM and k-nearest neighbours is analysed using an experimental set up in which images are acquired in controlled (Experiment one) or uncontrolled (Experiment two) illumination conditions.

2 Experimental set-up

A set-up was built to measure only illumination errors, so subject pose, gesture and distance camera–subject were maintained unchanged. Two diffuse lights offered controlled illumination conditions in Experiment one, two diffuse lights plus some ceiling fluorescent lamps were illumination conditions in Experiment two. CCD was placed firmly in front of the subject. Subjects were forced to change its pose between acquisition of two consecutive images. Figure 1 shows the acquisition set up, with diffuse lights and the CCD camera in front of the subject. Figure 2 represents one typical acquisition session in which the subject was placed in front of the camera.

The database is formed by 30 subjects (15 male and 15 female) with 12 images per subject. 8 images were used for train and 4 for test. The train images are captured with controlled illumination in both experiments. Test images are captured in controlled illumination conditions in case of Experiment one and in uncontrolled illumination conditions in case of Experiment two. Image size is 320 x 240 pixels with face covering great part of the image. Figure 3 shows one image of the subject in experiment one and Figure 4 shows the same subject in Experiment two, both images have been reduced for illustration purposes.

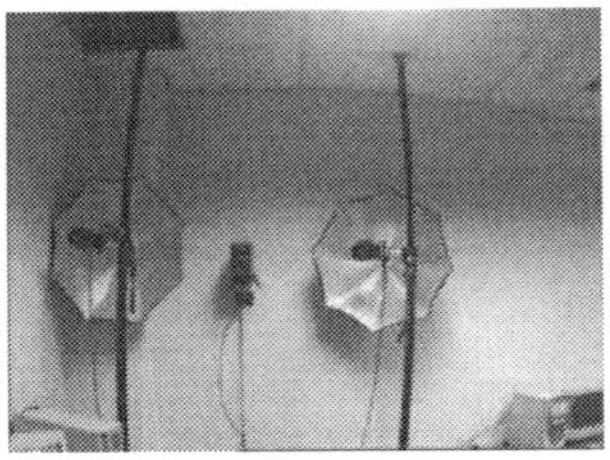

Fig. 1. Diffuse lights and CCD camera

Fig. 2. One subject in front of the camera

Fig. 3. Controlled illumination

Fig. 4. Uncontrolled illumination

3 State of the art

Human face verification or recognition is a wide field of research. Automatically recognizing faces may help in a wide variety of forensic and security applications. Several approaches have been taken: geometrical features with manual input, 3D images with laser measurements or 2D images with global or local extraction of features. Extensive reviews of approaches to face recognition were published in 1995 [2], 1999 [9], and in 2000 [10]. A workshop on face processing in 1985 [3] presented studies of face recognition. In 1998, lectures on face recognition using 2D face patterns were presented from theory to applications [8].

Global characteristics using principal components analysis has become an standard in face processing and is widely used in research and commercial systems. Turk and Pentland [6] developed a face recognition system that recognize the person by projecting face images into a feature space where only significant characteristics are represented. The significant features are known eigenfaces because they are eigenvectors of a set of images. This allows a great reduction of the information to process. Results of the experiments show a 96% of correct classification. P. Jonathon Phillips [5] compared rate of correct verification using SVM (77-78%) with a principal component analysis (PCA) (54%).

Our group has [1] presented several results of face recognition with neural networks. Two algorithms were tested: learning vector quantization (LVQ) and multilayer perceptron (MLP). Experimental results show higher recognition rates for LVQ tan for MLP (96.7% versus 83.3%)

4 Algorithm Description

In the design of a face verification system two different problems has to be solved: one is to built an on-line algorithm that has to work with high accuracy but in real time conditions (or at least with small response times) to perform verification tasks, second one is to find in a off-line procedure the values of all the parameters (i.e., to train the system). Off-line training algorithms has no time restrictions in industrial environments, computation time could be high without any critical impact on the system. But on-line verification algorithms has to be designed to be fast (response times bigger than few seconds are not allowed in any practical application) but with high accuracy (system is supposed to work correctly in the first attempt, a system in which the subject has to typed his code two times is allowed only if this is a very abnormal situation).

Our system has been split in three parts:

- Face extraction:

Our face location system cropped the face to a window of 130x140 pixels (as shown in figures 5 and 6). This process is made by selecting a region of the image in which the face is present. Using background subtraction and measuring the maximum correlation with a face template, faces are located and dimension is reduced to 130x140 pixels. As well image is converted to a grey level scale and equalized. Figures 5 and 6 show results of face location algorithm applied to Figure 3 and 4. To speed up this process convolution is done with a template representing only half face (so template size is 65x140 pixels). Using this template and considering face symmetry, convolution time is earned without any loss of information.

Fig. 5. Face location for figure 3

Fig. 6. Face location for figure 4

- Dimensional Reduction using PCA.

Principal Components Analysis has been used in face verification literature as a standard technique to reduce image dimensionality loosing as little information as possible. This approach has been considered in our work. The reduction depends on the train set variability, and also on the variance that we want to conserve in the reduced space. In our experiments, PCA projection is done using first 150 eigenvalues, which are enough to represent almost all database variance (as shown in figure 7). PCA matrix computation can be done off-line and PCA projection is a fast method performed in on-line operation.

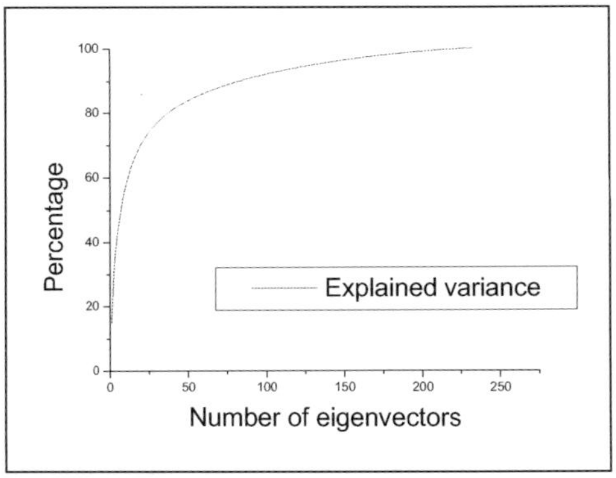

Fig. 7. Percentage of explained variance versus number of eigenvectors, 96.44% of variance is explained with 150 eigenvectors.

- Verification using SVM and KNN.

As classificators, an implementation of SVM by T. Joachims [10] with linear kernels were used. To test our results, a classical KNN was also considered. SVM offers excellent results in 2-class problems. This classifier could be easily used in verification problems (recognizing one subject against rest). In our experiment, linear kernel has been considered. Both algorithms has to be trained in off-line operation and applied to PCA coefficients of the image in normal verification operation.

5 Experimental Results

SVM results are reported as a receiver operator characteristic (ROC). The ROC curve was computed by averaging the ROC curve for each of the individuals in the Database. For person p_i the probe set consisted of one test image of person p_i and 29 faces of different people.

To calculate the ROC curve a threshold magnitude is needed. The function decision value is selected as threshold magnitude for SVM. If the output value of this function is big this means that confidence is high. So positive verification has been considered when output value is bigger than acceptance threshold. This acceptance threshold has to be set to obtain the optimum value that minimizes positive false and negative false, and maximizes the correct rate.

KNN results are reported as a correct rate versus threshold parameter curve. The magnitude used as threshold is the Euclidean distance. KNN classifier output is more reliable if distance is low, so a positive verification has been considered when output value is smaller than acceptance threshold. We have not analysed Positive False and Negative False for KNN classifier because this classifier is used just as a really simple comparative method, but the purpose of this paper is not to make a study of this classifier. KNN is a simple but powerful method to obtain an initial estimation of the faces distribution in PCA (reduced) space.

Figures 8 and 9 show experimental results of KNN classifier (with K=1), Figure 10 shows experimental results of SVM classifier.

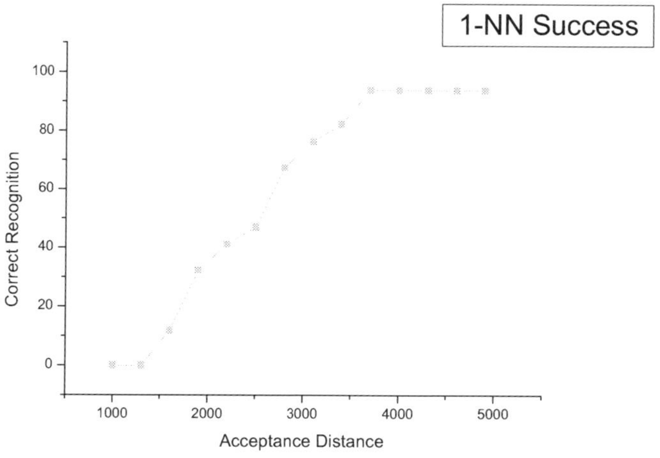

Fig. 8. Results for 1-NN, train and test with controlled illumination, test images acquired one day later.

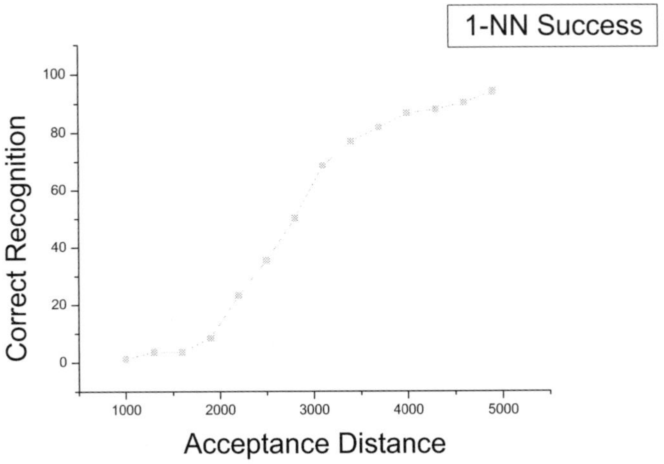

Fig. 9. Results for 1-NN, train and test sets with uncontrolled illumination conditions.

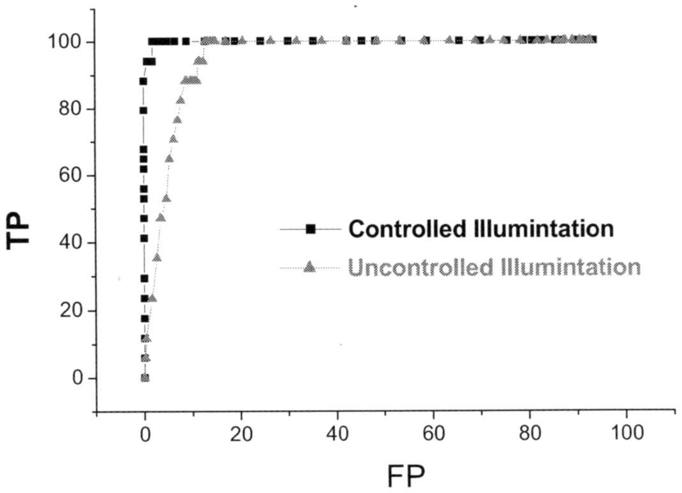

Fig. 10. Results for SVM for experiments one (controlled illumination) and two (uncontrolled illumination).

6 Conclusions

In our experiments, SVM achieves better recognition rates than kNN. Results show that SVM offers better results in both conditions (controlled-uncontrolled lights), kNN is more sensitive to this parameter.

The parameter we considered in SVM was the result of the decision function. Figures 7 and 8 show that acceptable working conditions are tuning the parameter with a value higher than zero. Knn parameter is Euclidean distance. SVM parameter is independent on light conditions but distance parameter is more sensitive to light conditions . In figure 7 and 8 for a correct recognition of 90% parameter is about 3500 with controlled light and 5000 with uncontrolled lights.

An important conclusion is also that achieving appropriate success rates needs the careful design of the image acquisition and preprocessing stages. Being faces 3D entities, differences in pose or position may affect projected shadows on the face to be analysed. Since it is unfeasible to limit pose and positions variations, it is advisable to use diffuse lighting to minimise this problem. In general, false positives are strongly increased when the amount of projected shadows is also increased.

Acknowledgements

Work presented in this paper was partially supported by the V EU Framework Program as part of "Video Sensor Object Request Broker open Architecture for Distribute Services" project (IST-1999-10808). Also, URJC financial support is gratefully recognized.

References

[1] E. Cabello, A. Sánchez and L. Pastor. *Some experiments on Face Recognition with Neural Networks.* Face Recognition: From theory to applications. Springer-Verlag. Eds: H. Weeechsler et al. Pp: 589-599. 1998.
[2] R. Chellappa, C. L. Wilson and S. Sirohey. Human and Machine Recognition of Faces: A Survey. Proceedings of the IEEE. Vol. 83. Nº 5. pp 705-740 .May 1995.
[3] H. D. Ellis, M. A. Jeeves, F. Newcombe ,and A. Young Eds., Aspects of Face Processing, Martinus Nijhoff Publishers, Dordrecht, Netherlands, 1985.
[4] T. Joachims, *Making large-Scale SVM Learning Practical.* Advances in Kernel Methods Support Vector Learning, B. Schölkopf and C. Burges and A. Smola (ed.), MIT-Press, 1999.
[5] P. Jonathon Phillips. *Support Vector Machines Applied to Face Recognition.* Advances in Neural Information Processing Systems 11, eds. M.J. Kearns, S. A. Solla, and D.A. Cohn, MIT Press, 1999.
[6] M. Turk and A. Pentland. Face Recognition Using Eigenfaces. Proceedings Computer Vision and Pattern Recognition. pp 586-591. 1991.
[7] V.Vapnik. *The nature of statistical learning theory.* Springer, New york, 1995.
[8] H. Wechsler, P. Phillips, V. Bruce, F. Soulie and T. Huang, Eds. Face Recognition: From Theory to Applications, Springer-Verlag, 1998.
[9] J. Weng and D. L. Swets, "Face recognition," in Biometrics: Personal Identification in Networked Society, A. K. Jain, R. Bolle, and S. Pankanti, Eds., pp. 67–86, Kluwer Academic, Boston, MA, 1999.
[10] W. Zhao, R. Chellappa, A. Rosenfeld and P. J. Phillips, "Face recognition: A literature survey," CVL Technical Report, Center for Automation Research, University of Maryland at College Park, Oct. 2000, ftp://ftp.cfar.umd.edu/TRs/CVL-Reports-2000/TR4167-zhao.ps.gz.

A Neuro-Fuzzy Technique for Scene Cut Detection from MPEG-1 Video

Ali M. Dawood and Arif S. Al-Hammadi

Etisalat College of Engineering, Sharjah, UAE.

Abstract. In this paper we propose a Neuro-Fuzzy (NF) algorithm that can automatically detect the type of MPEG frame at which clear scene cuts occurred. The algorithm is based on detection of the majority of Intra-macroblocks and the directionality of the two neighboring B-frames motion vector types from the bit stream. It is shown that the method is able to identify the location of scene cuts in I- P- and each one of B-frames with a very high success rate.

1 Introduction

An international effort on multimedia database under MPEG-7 has launched with the objective to specify a standard set of descriptors that can be used to describe various types of multimedia information, that include high-level feature extraction [8]. Scene cut detection is a key issue in high-level video processing that is required for applications such as video browsing, video retrieval, video indexing etc. An example of scene cut occurring between two video shots is illustrated in figure 1 between picture number 3 and 4.

Many scene cut detection methods proposed in the past were based on measuring the pixel intensity histogram of frame difference signals [1,7,9]. Although, these methods work well, they require the MPEG coded bit streams to be decoded first, which limits the usability of the proposed methods. It would make more sense if the scene cut could be detected from the information provided in the compressed bit stream.

Fig. 1. Example of a scene cut between two video shots.

On the compressed domain, Liu et al [5] have used a median filter-based algorithm for calculating threshold values, to detect peaks in the bit rate differences.

Although the results are good, some false detection have been reported. Scene cuts were also detected by extracting DC images from the bit stream, which offers significant computational savings [10]. Nevertheless, the detection was not 100%. Feng et al [3] proposed a detection algorithm for scene cuts from the MPEG compressed bit stream. Three different schemes were used to detect the cut in the I-, P- and B-frames. The decision of the scene cut location was made based on the result of the three detection schemes used for each frame type. I-frames were used as scene cut indication in the Group of Pictures (GOP), if the cut was detected in an I-frame, then the P and B-frames algorithms would be applied to the entire GOP, otherwise the GOP will be skipped and assumed to be free from any cuts.

In this paper we propose a neuro-fuzzy technique for detecting scene cuts from the MPEG coded bit stream. The technique exploits the various statistics of macroblock coding types in order to define the set of rules which represent scene cut conditions within the MPEG coded video.

2 The MPEG Video Standard

The MPEG video codec compresses video streams by exploiting spatial and temporal redundancies. Individual frames are coded as I (intra), P (predicted) and B (bidirectionally predicted) frames, where each consists of a number of macroblocks (16x16 pixels) which are the basic coding units in all types of frames. The existence of different frame types resulted into different macroblock (MB) coding modes. Since I frames are only intra-frame coded, their bit rates only depend on the texture of the image (spatial complexity). For P and B frames, the generated bit rate is primarily motion dependent, however, texture also affects the bit rate due to the uncovered background of the moving objects. P frame may contain MB type of Intra or Forward predicted (FWD), while B frame has two types extra: Backward predicted BWD and Bidirectionally predicted (BIP). However, the actual MB type among these types there are some specific modes to be selected such as intra/inter, motion compensated (MC)/no-MC, coded/not-coded and quantiser change or not. A detailed description of the MPEG coding standard can be found, for example in [4].

The N and M parameters defines the Group of Picture (GOP) structure in MPEG stream, where N is the distance between the successive I frames and M represents the frame distance between the I or P (*anchor*) frames. For example if the GOP defined as N=12 and M=3 then the frame type pattern of this MPEG stream is as follows: ...BBIBBPBBPBBPBBIBBPB... etc. Next section explains possible locations for scene cuts to occur and their impact on the behavior of MPEG encoder.

3 Scene Cut Conditions

There are three possible locations for scene cuts to occur, when the MPEG video is coded with GOP structure of N=12 and M=3, namely anchor, first B and second B frame types. Lets start by explaining the default condition where there are no scene cuts. In normal operation, I-frames contain only Intra MB type. P-frame is normally forward (FWD) predicted from the previous I- or P- (anchor) frames, but when the prediction fails, due to a violent motion or a scene cut, most or all of the macroblocks (MBs) will be of Intra type. In B-frames, MBs are usually predicted from both previous and future anchor frames, but at scene cuts they will mainly be predicted either FWD only or backward (BWD) only depending on the location of the cut. Figure 1-a shows the direction of predictions in a normal video shot within a GOP of 12 frames and 3 frame distance between the anchor frames.

Now consider that there is a scene cut at the location of a P-frame, say P_4. First of all since P_4 is the first anchor frame of the new shot, the majority of its MBs will be Intra-frame coded. Second, since the two preceding B-frames, B_2 and B_3 belong to the previous shot, they find a better prediction from the previous anchor frame (i.e I_1) than the scene cut frame. Hence, the type of prediction for these two B-frames will be FWD only, as shown in figure 1-b. Also since, in MPEG, the anchor frames are coded prior to B-frames, then a scene cut in a P-frame implies a high level of Intra-coded MBs in a frame, followed by the B-frames, which are mainly FWD predicted.

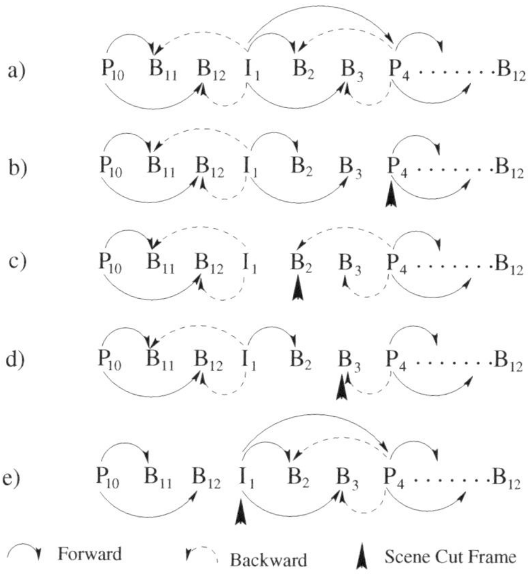

Fig. 2. Prediction directions when scene cut occurs

When the position of a scene cut at the first B-frame after an anchor frame, B_a, such as B_2 in figure 1-c. Again the first P-frame in the next shot (P_4) will be mainly Intra-coded. Since now both B_2 and B_3 belong to a new shot, they will use P_4 as a prediction reference and therefore the prediction for these B-frames will be mainly BWD. Note that a scene cut at a B-frame does not cause entire Intra coding for this frame.

If scene cut occurs in B-frame preceding an anchor frame, B_b, such as B_3 in figure 1-d, then as before, P_4 will be mainly Intra-coded, but since now each one of B_a and B_b belong to different scenes, B_a will be using FWD prediction, but B_b will be BWD.

It should be noted that if the scene cut occurs at an I-frame then the condition of P-frame explained above applies since I and P are considered as anchor frames. Figure 1-e shows a condition where scene cut occurs at $I_{1.}$, here, since the two preceding B-frames B_{11} and B_{12} belong to the previous scene, they use FWD prediction from P_{10}. This is not the same as the normal scene of figure 1-a, hence a frame with mainly Intra-coded MBs followed by two B-frames of FWD predicted (due to coding order B_{11} and B_{12} will follow I_1) is an indication of scene cut, which is the same as scene cut in a P-frame of figure 1-b.

To summaries the conditions for the scene cut locations, considering that, in the coding (or transmission) order the two B-frames B_n and B_{n+1} will follow the anchor frame I_{n+2} or P_{n+2} [9], then: a high percentage of Intra-MB in the bit stream in a given frame, followed by a high percentages of the FWD/BWD macroblocks in the next two frames indicates existence of scene cuts where the scene cut frame type is decided according to table 1. In table 1, a value of "1" indicates that the percentage of that particular item (i.e. Intra, FWD or BWD macroblock) is higher than threshold and "0" is below threshold.

Table 1. Decision Table – For MPEG video Coded with GOP structure of N=12 M=3

Frames under consideration									Scene Cut
Frame (n+2)			Frame (n)			Frame (n+1)			Type?
Intra	Fwd	Bwd	Intra	Fwd	Bwd	Intra	Fwd	Bwd	
1	0	0	0	1	0	0	1	0	Anchor
1	0	0	0	0	1	0	0	1	B_a
1	0	0	0	1	0	0	0	1	B_b

4 Why Neuro-fuzzy systems?

In order to overcome some of the limitations associated with neural network and fuzzy systems, combinations of both systems have been proposed in the literature. NEFPROX as a function approximation system was proposed in [6]. In NEFPROX, if certain prior knowledge is available in the form of fuzzy rules, the Neuro-Fuzzy (NF) system can be initialized with these rules. The remaining rules can be found by learning. If there is no prior knowledge, the NEFPROX can be started without hidden units, and then incrementally learns all the rules (self-structure). The triangular membership functions were used for simplicity [6].

The structure learning properties of the NEFPROX makes it attractive in the problem of scene cut detection. This is because of the benefits to knowing the rules and membership functions generated from scratch using NEFPROX. These rules can show the conditions that cause scene cuts and hence verifies the assumptions made in table 1.

5 Creating the Training and Testing Files

Selecting the training and testing data sets for the NF is very important. The selection must cover the problem domain, however, if the training file is large, it might cause unsuccessful training.

Three files where created in the format shown in table 2 while each file contains 2500 entries. The training file contains 23 scene cuts, the generalisation file contains 21 scene cut and the test file contains 23 scene cuts. The training file was used to train the NF system, the generalisation file was used to test the NF against generalization where the weights are set accordingly and the third one was used to test the final system.

Table 2. Input and outputs to the NF system

Input									Output:
Frame (n+2)			Frame (n)			Frame (n+1)			Scene Cut?
Intra	Fwd	Bwd	Intra	Fwd	Bwd	Intra	Fwd	Bwd	Yes / No

6 Neuro-Fuzzy scene cut detector

The system settings employed in the approximation of the Mackey-Glass equation [10] were used for the scene cut detection problem. The number of fuzzy sets per input and output variable was chosen to be 2.

In the first scenario, NEFPROX uses 9 inputs representing the frames under consideration. Since the number of inputs is 9, with 2 fuzzy sets defined for each input variable, the number of possible rules is 512 (2^9). However, the numbers of rules generated by the NEFPROX were 35. The NEFPROX starts with all rules that can be created and then deletes insufficient rules from the rule base. Samples from the rules produced by the NF system were selected[1] and listed in table 3.

It is clear from table 3 that rules number 4, 7 and 10 are the conditions for the scene cut as expected in section 3. For example, an interpretation of rule 4 in table 3 would look like this: "Rule 4: if any MPEG frame contains majority of its Mac-

[1] Not all rules are listed because of the limited space. Selections were done to include the three important rules indicating scene cuts plus few other non-scene cut rules.

roblocks coded with intra type and it is proceeded by two frames both contain majority of their Macroblocks coded with BWD type, then there will be a scene cut".

The Mean Square Error (MSE) curve produced during the generalisation process is shown in figure 2. An example of the membership functions produced by the NEFPROX with 9 inputs and one output are shown in figures 3 and 4. Figure 3 shows the membership functions of two sets, weak and strong, for input 1 (Intra Frame (n+2)). Due to the nature of our problem a fuzzy region cab noticed in figure 3 between the two sets. The membership function for the output is shown in figure 4 and it has two sets representing the scene cut and non-scene cut conditions. From figure 4 it can be seen that there are two clearly distinguished regions, i.e. scene cut and non-scene cut locations are successfully detected. The Neuro-Fuzzy detector was developed based on the 35 rules produced earlier to detect scene cut locations from a given MPEG coded video stream. The system was able to detect all of the scene cuts in the training and generalisation files and 86% of the testing files.

Table 3. Rules Produced by the NF system

	Frames under consideration									
Rule	Frame (n+2)			Frame (n)			Frame (n+1)			Scene
No.	Intra	Fwd	Bwd	Intra	Fwd	Bwd	Intra	Fwd	Bwd	Cut?
1	1	0	0	0	0	0	0	0	0	No
2	0	0	0	0	0	0	0	0	0	No
3	0	0	0	1	0	0	0	0	1	No
4	1	0	0	0	0	1	0	0	1	Yes
5	0	0	0	0	0	1	0	0	0	No
6	0	0	1	0	0	0	0	0	0	No
7	1	0	0	0	1	0	0	1	0	Yes
8	0	1	0	0	1	0	0	1	0	No
9	0	0	1	1	0	0	0	0	0	No
10	1	0	0	0	1	0	0	0	1	Yes
11	0	1	0	0	0	1	0	0	0	No

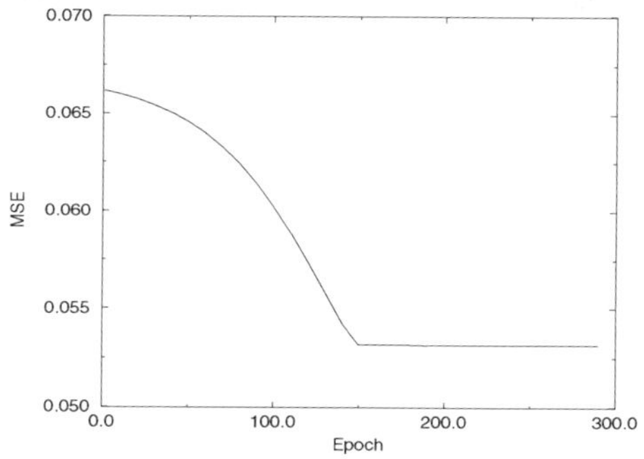

Fig. 3. Mean Square Error (MSE) curve.

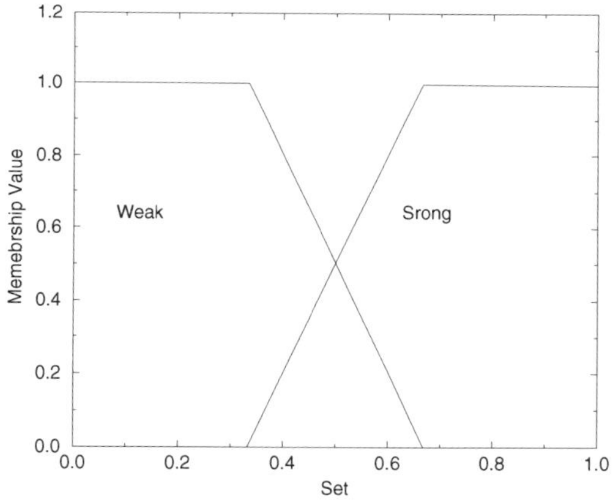

Fig. 4. Input 1 Membership Functions

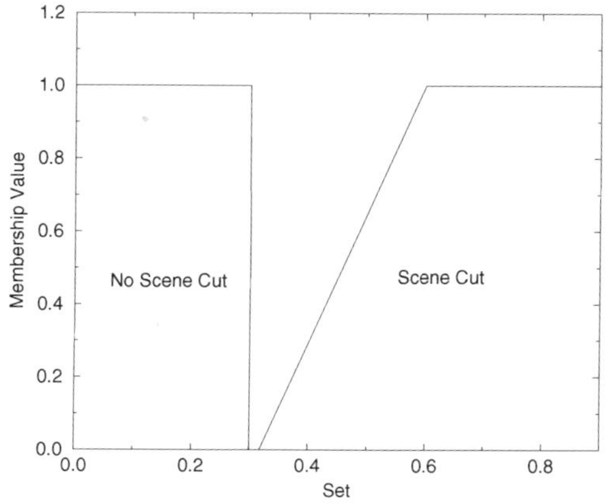

Fig. 5. Output Membership Functions

7 Conclusion

We have proposed a Neuro-Fuzzy scene cut detection algorithm that can be applied directly to the MPEG coded bit stream without the need to decode the frames. The detection accuracy of the NF technique on the test sequences was 100% in the training and generalisation files and 86% of the testing files. In addition, the rules produced by the NF technique complies fully with the conclusions of [2] and it will be used in the future to detect gradual scene changes as an extension to the work presented in this paper.

The advantage of using NEFPROX for the scene cut detection is that rules and membership functions can be generated from scratch. This helps in understanding the scene cut condition and hence better understanding of the final model is achieved. It also helps in making the system adaptive and raises the possibility of real time implementation.

References

1. Ahn HS, Oh IS (1995) Fast shot detection from video images using large and adaptable skip factors. In: Proceedings of the ACCV'95 conference. Singapore, 2:489-493
2. Dawood A, Ghanbari M (1999) Clear scene cut detection directly from MPEG bit streams. In: the proceedings of the IEE seventh international conference on image processing and its applications IPA'99, Manchester, UK, 1:285-289
3. Feng J, LO KT, Mehrpour H (1996) Scene change detection algorithm for MPEG video sequence. In: Proceedings of the international conference in image processing ICIP'96, 2:821-824
4. ISO/IEC (1994) Generic coding and moving pictures and associated audio. Recommendation H.262
5. Liu HC, Zick G (1995) Automatic determination of scene change in MPEG compressed video. In: Proceedings of the International Symposium on circuits and Systems, Seattle, USA, 1:764-767
6. Nauck D, Klawonn F, Kruse R (1997) Foundation of Neuro-Fuzzy systems. John Wiley & sons Ltd.
7. Rachidi T, Maelainin SA, Bensaid A (1997) Scene change detection using adaptive skip factor. In: Proceedings of the international conference in image processing ICIP'97, Santa Barbara, CA, USA, vol.1
8. Sezan MI, Qian RJ (1998) MPEG-7 standardization activities. In: Proceedings of the international conference in image processing ICIP'98, Chicago, USA
9. Smoliar S, Zhang HJ (1994) Content-Based Video Indexing and Retrieval. IEEE Multimedia Magazine. 1:62-72
10. Yeo BL, Liu B (1995) A unified approach to temporal segmentation of motion JPEG and MPEG compressed video, In: Proceedings of the ICMCS'95 conference, pp 81-88

Fuzzy Audio-Visual Feature Maps for Speaker Identification

Claude C. Chibelushi

School of Computing, Staffordshire University, Beaconside, Stafford ST18 0DG, UK. e-mail: C.C.Chibelushi@staffs.ac.uk

Abstract. Speech-based person recognition by machine has not reached the level of technological maturity required by some of its potential applications. The deficiencies revolve around sub-optimal pre-processing, feature extraction or selection, and classification, particularly under conditions of input data variability. The joint use of audible and visible manifestations of speech aims to alleviate these shortcomings, but the development of effective combination techniques is challenging. This paper proposes and evaluates a combination approach for speaker identification based on fuzzy modelling of acoustic and visual speaker characteristics. The proposed audio-visual model has been evaluated experimentally on a speaker identification task. The results show that the joint model outperforms its isolated components in terms of identification accuracy. In particular, the cross-modal coupling of audio-visual streams is shown to improve identification accuracy.

1 Introduction

Speaker recognition based on acoustic speech is a fairly well established research area. However, this approach to speaker recognition suffers from sensitivity to speech variability, inadequate recognition accuracy for some prospective applications, and susceptibility to impersonation (see [7] [11] for reviews of speaker recognition). The configuration of visible articulators during speech is used less often than voice, for speaker recognition. However, fuelled by the falling costs of high-performance computing, real-time recognition based on audio-visual processing or analysis is gradually becoming a feasible proposition. The joint use of audible and visible manifestations of speech is a promising avenue for alleviating the deficiencies of unimodal recognition, but the development of effective combination techniques is difficult [3].

Relatively few studies have investigated fuzzy data fusion for bimodal speaker recognition [3]; among the few examples, fuzzy clustering has been used for decision-level data fusion in [2]. This paper proposes and evaluates a combination approach, for speaker identification, based on fuzzy modelling of acoustic and visual speech streams. Both intra-stream characteristics and inter-stream coupling are modelled as speaker-specific fuzzy sets. The rest of the paper is organized as follows. First, the feature extraction, feature mapping, and classification techniques are presented. Thereafter, an experimental study of the proposed combination technique is given.

2 Speaker Identification Approach

In pattern recognition, the fusion of multiple information sources can be performed at the data level, feature level, decision level, or at several levels [3]. An investigation into speaker recognition based on post-categorical Bayesian or linear combination of acoustic and visual speech was reported in [4]. This paper presents further work, which explores an alternative approach cast as a multi-level amalgamation based on fuzzy sets (see Fig. 1). In the approach proposed herein, speaker modelling based on vector-quantisation [12] [13] is extended by including fuzzification in the feature-space quantisation process, and by adding a fuzzy audio-visual coupling between the fuzzy feature-space partitions. Some of the recent advances in audio-visual fusion are anchored on models of interacting processes, using variants of factorial hidden Markov models (HMMs), which have been shown to be superior to conventional HMMs [6] [9]. The coupling scheme proposed herein also follows the joint modelling approach.

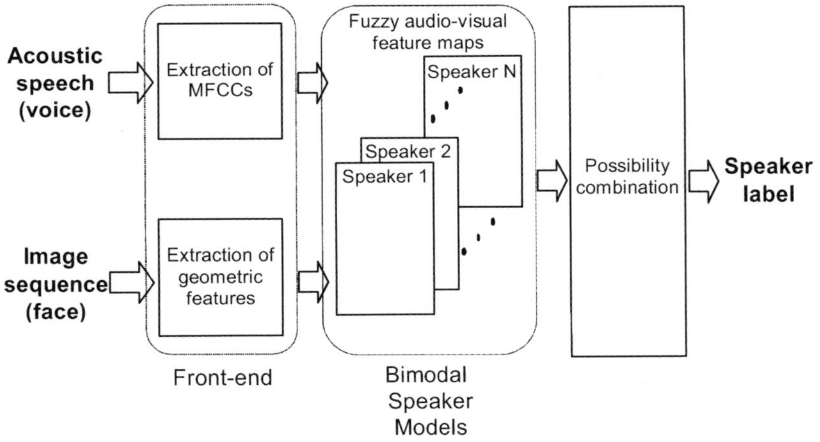

Fig. 1. Architecture for speaker identification based on multi-level fuzzy fusion of acoustic and visual speaker characteristics

2.1 Features

After temporal and spatial segmentation of the speech signals, 14 mel-frequency cepstral coefficients (MFCCs) [5], and 12 geometric measurements taken on the outer lip margins, are extracted from segments along the temporal paths of the acoustic and visual speech. The acoustic and visual feature vector rates are 100 vectors/s and 25 vectors/s, respectively. The mismatch is caused by the lower video digitisation frame rate (see Section 3.1). The geometric features are normalised against translation, rotation, and scaling, through a regularisation of lip margin position, orientation, and size with respect to a line joining the eye centres.

MFCC features have been chosen in this work because they often yield very good speech or speaker recognition accuracy. The decision to use geometric features is based on the discriminability, of lip-margin shape and size across speakers, which is noticeable in [8]. The geometric features, used in the experiments given in this paper, are: the area between parabolic curves approximating the outer lip margins, the perimeter of the parabolas, the mouth width, the mouth heights taken at three equidistant positions along the width of the mouth, and two sets of three coefficients from the equation of each outer lip-margin parabola.

2.2 Fuzzy feature maps

Each speaker is allocated a coupled model of the distribution of his or her acoustic and visual speech. The model consists of two unimodal sets of fuzzy partitions (one set for the acoustic space and another for the visual space), together with a joint possibility map representing the co-occurrence of the acoustic and visual partitions that correspond to each temporal segment of the input speech. In the following, u_i is the hypothesis that a test pattern belongs to speaker *i*.

2.2.1 Unimodal fuzzy sets

The unimodal fuzzy partitions are estimated using the fuzzy C-means clustering algorithm [1]. The degree of membership of a test pattern in the unimodal cluster, which corresponds to the cluster prototype nearest to the pattern, is regarded as the possibility that the speaker whose model contains the given fuzzy partition uttered the pattern. The unimodal possibility distributions are therefore denoted as $\pi_j(u_i)=\mu_j(u_i)$ and $\pi_k(u_i)=\mu_k(u_i)$ respectively for acoustic cluster *j* and for visual cluster *k*, whose prototypes happen to be nearest to the test pattern. $\mu_*(u_i)$ is the membership function of fuzzy subset '$*$' in the model of speaker *i*. These membership functions are estimated by applying the fuzzy C-means clustering algorithm to the training data of speaker *i*.

2.2.2 Bimodal fuzzy sets

In addition to the unimodal possibility distributions, described above, cluster co-occurrence frequencies are used as joint possibility measures. If n_{jk} denotes the count of joint occurrences of acoustic cluster *j* and visual cluster *k*, in the coupled model for speaker *i*, then the corresponding joint possibility distribution is

$\pi_{jk}(u_i) = \mu_{jk}(u_i) = n_{jk} / \max_k n_{jk}$. Effectively, $\mu_{jk}(u_i)$ is the membership function for the audio-visual fuzzy set that corresponds to the activation of acoustic cluster *j* in model *i*. n_{jk} is calculated using the training data of speaker *i*.

2.3 Classification

During a recognition trial, the similarity $s(i)$ between an audio-visual test pattern and a speaker model *i* is computed using the following equation, proposed by Dubois and Prade [10], for the combination of possibility distributions.

$$\forall u_i \in U\,,\; s(i) = \frac{\min(\pi_j(u_i), \pi_k(u_i), \pi_{jk}(u_i))}{\sup\limits_{u_i \in U} \min(\pi_j(u_i), \pi_k(u_i), \pi_{jk}(u_i))} \qquad (1)$$

In Eq. 1, $\pi_j(u_i)$, $\pi_k(u_i)$, and $\pi_{jk}(u_i)$ are as defined in Section 2.2, and U is the universe of discourse, which is the set of all identity labels. The classification decision logic, for closed-set speaker identification, is implemented as a 'winner-take-all' decision procedure applied to the similarity calculated for each speaker model.

3 Performance Assessment

3.1 Data set

The data set used in the experiments contains 25 male and female Caucasian native English-language speakers (all are university students). The utterance set for each speaker comprises 4 versions of each English digit {*1, 2, 3, 4, 5, 6, 7, 8, 9, 'zero'*}, all recorded in one session. The recording environment was a studio with low levels of acoustic noise. The video data consists of head-and-shoulder images with a plain scene background, under constant illumination. Image data has been digitised as monochrome images of resolution 640 x 480 x 8 bits per frame, at 25 frames/s. Voice data has been digitised as mono sound, at 16 bits per sample and sampling frequency of 11025 Hz.

3.2 Experimental method

The experiments aim to assess the speaker identification performance of the proposed fuzzy audio-visual model, and the performance of its individual components, at various levels of acoustic speech degradation. All speaker models are trained with undegraded ("clean") speech, whereas the test set contains speech that is either undegraded or whose audio signal-to-noise ratio (SNR) is 0, 12, 24, or 36 dB. The acoustic speech is degraded with additive Gaussian noise. Three utterance versions are used for speaker model training, and 1 version for testing. Hence, two distinct data sets, comprising 750 and 250 utterances, are used for speaker model training and testing, respectively. Test and training version rotation is applied so that each version is used once for testing, thereby yielding a total of 1000 test utterances. Thirty-two clusters are used in each speech space; this is based on preliminary investigations, which found no substantial improvement beyond this number.

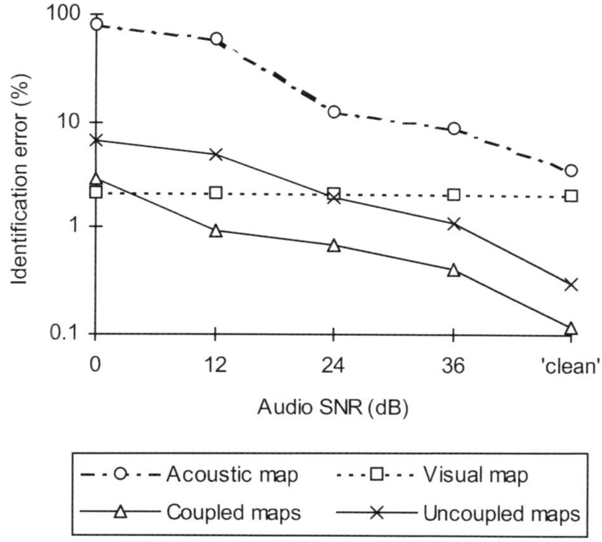

Fig. 2. Speaker identification error rates for various acoustic speech noise conditions and clean visual speech. All speaker models are trained with clean speech, whereas the test set contains speech at the stated SNR

3.3 Results

The speaker identification results are summarised in Fig. 2. Audio-only speaker identification has been adopted as the baseline for a comparative evaluation of the fusion strategy. Overall, the speaker identification results show that at most levels of mismatch between training and test data SNR, the combination of personal characteristics from the acoustic and visual domains outperforms isolated acoustic or visual modalities. In particular, the results indicate that some speaker identity information is embedded in the audio-visual coupling.

4 Conclusions

This paper has proposed and evaluated a multi-level combination approach, for speaker identification, based on fuzzy modelling of acoustic and visual speaker characteristics. The experimental results show that bimodal recognition generally outperforms unimodal recognition in terms of speaker identification accuracy. Under matched or mismatched training and test SNR conditions, joint recognition generally enhances recognition accuracy. Current work focuses on evaluating other combination approaches, such as the fuzzy integral, possibly incorporating some scheme for adapting to the recognition environment. This aims to achieve more optimal fusion, and hence obtain better bimodal speaker identification accuracy.

References

1. JC Bezdek (1981). Pattern Recognition with Fuzzy Objective Function Algorithms, New York: Plenum, 1981.
2. V Chatzis, AG Bors, I Pitas (1999). Multimodal Decision-Level Fusion for Person Authentication, IEEE Trans SMC-Part A:, Vol 29, No 6, pages 674-680, 1999.
3. CC Chibelushi, F Deravi, JSD Mason (2002). A Review of Speech-Based Bimodal Recognition, IEEE Trans Multimedia, Vol 4, No 1, pages 23-37, 2002.
4. CC Chibelushi, F Deravi, JSD Mason (1999). Adaptive Classifier Integration for Robust Pattern Recognition, IEEE Trans SMC—Part B:, Vol 29, No 6, 902-907, 1999.
5. SB Davis, P Mermelstein (1980). Comparison of Parametric Representations for Monosyllable Word Recognition in Continuously Spoken Sentences, IEEE Trans Acoustics, Speech, and Signal Processing, Vol ASSP-28, pages 357-366, 1980.
6. S Dupont, J Luettin (2000). Audio-Visual Speech Modeling for Continuous Speech Recognition, IEEE Trans Multimedia, Vol 2, No 3, pages 141-151, 2000.
7. Qi Li, Biing-Hwang Juang, Chin-Hui Lee, Qiru Zhou, FK Soong (1999). Recent Advancements in Automatic Speaker Authentication, IEEE Robotics & Automation Magazine, Vol 6, No 1, pages 24-34, 1999.
8. AA Montgomery, PL Jackson (1983). Physical Characteristics of the Lips Underlying Vowel Lipreading Performance, J Acoust Soc of Am, Vol 73, pages 2134-2144, 1983.
9. C Neti, et-al (2000). Audio-Visual Speech Recognition, Tech Rep, Center for Language and Speech Processing, Johns Hopkins University, 2000.
10. H Prade (1985). A Computational Approach to Approximate and Plausible Reasoning with Applications to Expert Systems, IEEE Trans Pattern Analysis and Machine Intelligence, Vol PAMI-7, pages 260-283, 1985.
11. DA Reynolds (2002). An Overview of Automatic Speaker Recognition Technology, IEEE Int'l Conf Acoustics, Speech, and Signal Processing, Vol 4, 4072-4075, 2002.
12. FK Soong, AE Rosenberg, LR Rabiner, BH Juang (1987). A Vector Quantization Approach to Speaker Recognition, AT&T Tech. J, Vol 66, No 2, pages 14-26, 1987.
13. K Yu, J Mason, J Oglesby (1995). Speaker Recognition Using Hidden Markov Models, Dynamic Time Warping, and Vector Quantisation, IEE Proc-Vision, Image, and Signal Processing, Vol 142, No 5, pages 313-318, 1995.

Adaptive User Agents for Intelligent Biometric Applications

N.J. Mavity, F. Deravi, and M.C. Fairhurst

Dept of Electronics, University of Kent at Canterbury, Canterbury, Kent, CT2 7NT, United Kingdom

1 Introduction

Interest in biometric systems for identity authentication has been growing in recent years. The strengths of biometric systems are well known [2], however, the success of any biometric system is measured ultimately by its usability and user acceptance. Serious obstacles to wider adoption of biometric systems include user interface issues that result in significant failure to enroll and failure to verify rates. Poor templates and template aging are important factors that need to be addressed by any biometric system developer.

This paper considers strategies for improving the performance of single-modality biometric systems through the adoption of intelligent agent architecture in the design of the system.

In the description given below the example of a specific commercial fingerprint verification system is used. The general principles of this approach, however, could be applied to other biometric products and modalities, and it is hoped that further work will demonstrate this in a multi-modal context. This particular system has two main parameters that affect enrolment and verification procedures.

- Security Level. This parameter (threshold) is related to the False Accept/False Reject rates of the system. The higher the security level, the lower the chance of false accept, although it does increase the chance of false reject. There are 9 levels of security settings.
- Quality Setting. This parameter is related to the image quality that is acquired by the device sensor. The value of this quality setting ranges from 0 - 100.

2 System Architecture

The system components are shown in Figure 1. These components are intended to act as an intermediate layer between the biometric device interface API, responsible for sample acquisition, enrollment and verification procedures, and the user interface itself.

Some terms need to be defined in order to further explain the mechanics of the system. We define a *transaction* as a single attempt by a user to verify his/her identity. A *session* consists of one or more transactions conducted within a certain timeframe.

2.1 Software Agents

An agent is described as anything that can be viewed as perceiving its environment through sensors and acting upon that environment through effectors [6]. Much work on software agents has been performed in various fields [3,4,8,9,10], however, it is the approach to combine biometrics and the use of software agents in this manner which is considered novel. Some of the properties of the software agents readily lend themselves to this type of application.

Our system exhibits *adaptivity* "the ability to modify an internal representation of the environment through sensing of the environment in order to change future sensing, acting and reacting for the purpose of improving assistance". *Autonomy* "the ability to sense, act and react over time within an environment without direct intervention". Each component as described in the following sections has its own goal which through its operation attempts to satisfy.

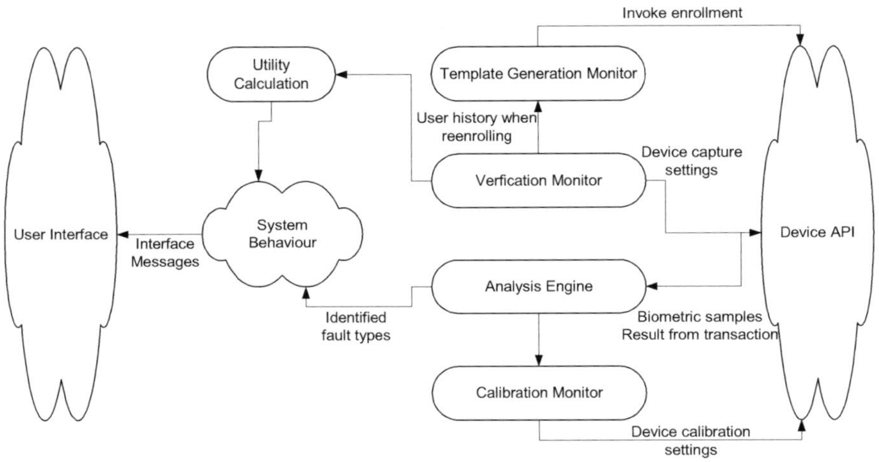

Fig. 1. System Architecture

2.2 The Role of Utility

Generally most biometric systems have some form of verification threshold. This threshold is usually related to the false accept rate of a system and ultimately the equal error rate (EER). Successful verification at a higher security level indicates that the system has increased confidence in the identity of the user.

The use of utility for user modeling in adaptive systems is not new [1]. We propose to employ the notion of *utility* in a different manner. In our system it refers to the agent's utility with respect to the levels of security that the user is achieving. In other words the agents 'happiness' is directly related to the confidence in which it has in the identity of the user.

The goal of the system is to try and maximise its utility over the period of time that the system is being used. The value of utility is directly used to change the behaviour of the agent with respect to the user. I.e. an agent with a low utility score will aggressively attempt to aid the user with the donation of higher quality samples through the use of extended user assistance. Whilst an agent with a relatively high utility score is less likely to offer the degree of assistance the low utility agent is exhibiting, however, will still be able to offer help if the agent determines the user is experiencing difficulty in donating samples.

2.3 Calculation of Utility

Since the modality we are dealing with gives us 2 attributes when dealing with samples donated from the user, Multi Attribute Utility Theory (MAUT) [7,11] can be used in this instance to generate an overall utility score from these 2 factors. According to MAUT, the overall evaluation $v(x)$ of an object x is defined as a weighted addition its evaluation with respect to its relevant *value dimensions*. In our system we are evaluating the user interaction based on value dimensions, Security level and Quality. The overall evaluation is defined by the following *overall value* function which gives us the utility value (Eq. 2.1):

$$v(x) = \sum_{i=1}^{n} w_i v_i(x) \tag{2.1}$$

Here, $v_i(x)$ is the evaluation of the object on the i-th value dimension and w_i the weight determining the impact of the i-th value dimension on the overall evaluation (also called the *relative importance* of a dimension), n is the number of different value dimensions, and

$$\sum_{i=1}^{n} w_i = 1 \tag{2.2}$$

We need to construct a scale for each of the attributes to represent the *value function* of an attribute. The value function for the security level attribute follows a simple linear scale where the maximum security level, in this case 9 equates to a value of 1. The quality attribute follows a similar linear scale. Further work may include the investigation of non linear value functions and what effect this has on system performance. Table 1 illustrates 2 different biometric transactions.

Table 1. Utility Calculation

Attribute	Transaction 1	Transaction 2	Relative Weight
Security	0.55	0.99	0.8
Quality	0.40	0.45	0.2

To calculate the utility of Transaction 1 for instance the following equation is used (Eq. 2.3)

$$(0.55 * 0.8) + (0.4 * 0.2) = \mathbf{0.52} \tag{2.3}$$

A similar analysis of Transaction 2 would yield a utility score of **0.88**.

2.4 System Behaviour

The behaviour of the overall agent entity is based upon the current utility value for the user. After each session the user logs will be mined to determine the overall utility score. The method with which this session utility score is calculated is described below. An overall mean utility score is calculated for the session, this value is then scaled using a utility transformation function which generates a percentage figure for the session under examination. This figure is scaled based on the maximum figure of utility that the user can produce based on the security level and quality setting at which the user templates were acquired. This normalized session utility score represents the performance of the user over the last session. This figure is used to determine one of the following 4 behaviour bands which determine the degree of user feedback the system exhibits in order to attempt to increase the performance of the user :-

Table 2. Behaviour Levels

Normalized Session Utility %	Behaviour Band
0 – 25	1
25.1 – 50	2
50.1 – 75	3
75.1 - 100	4

The behaviour levels shown in table 2 describe the level of feedback exhibited by the system in order to improve user performance. Level 1 is the lowest performance band and the system will examine any identified fault types and will enter into a dialog with the user in an attempt to determine the cause of this poor recognition rate. The highest behaviour band possible is Level 4, this represents good verification scores from the user. The system does not need to actively aid the user in the donation of samples, however, is still monitoring for specific fault types and will offer assistance on these if they occur. Sustained scores in this range indicate to the template generation agent that there may be a possibility of increasing the user's security level of enrollment templates (if possible) at the appropriate time.

2.5 Verification Monitor

This agent is responsible for the monitoring and setting of the 2 parameters that correspond to the utility rating. It is recommended that the level of security that is used for verification purposes is kept at the level at which enrollment occurred or below, attempting a match at higher levels may lead to incorrect results. The matching procedure automatically attempts to match up to the enrollment security level. The result from this operation is used as one parameter to calculate the utility of the transaction. The 'Quality' setting of any acquired sample is handled in a different manner. Since this parameter plays a role in the calculation of utility the agent here has the opportunity to adjust this factor in order to increase the utility rating. This is achieved in the following manner:-

The recommended value of the quality setting for verification transactions is 40. If the user was enrolled at a much higher level of quality (say 60), the system assumes that user should be able to produce a sample of at least this value at some stage. In order to proactively adjust this quality setting this agent examines the results obtained from the session logs.

If a user is achieving successful verification, at a rate above 75% (see Section 2.4) over the last session then the agent will attempt to increase the Quality setting for the next user session. There is a roll back mechanism in place here as mentioned below so if the analysis agent believes that the image capture settings are unrealistic for the user then these settings can be reduced, to a level which the user can achieve. The speed at which this agent increases the Quality setting is determined obviously by user performance itself. However, there are 2 more factors which govern the rate at which this process can develop, these are:

1. The number of concurrent sessions in which the user has achieved the specified percentage of successful transactions.
2. The actual percentage of successful transactions that the user must perform in order that the session is marked as a valid session to be counted in (1) above.

2.6 Analysis Engine

This component is responsible for the examination of any user-donated sample. This is independent of whether the user is enrolling or verifying. By examining the biometric sample and knowing the result of the transaction (i.e. pass or fail) this component can attempt to classify the nature of any failure using the error types described below.

TYPE A User Fault

These types of fault involve the poor usage of the device. The user is not donating a sample of sufficient quality (i.e. the image is too dark or too light). The system will have some initial quality limits for establishing this fault type, although through monitoring the quality values that are obtained, both for images which fail and the ones that pass, these limits are gradually adjusted over time.

TYPE B Device Error

To classify image acquisition problems we can take image samples from the sensor and calculate simple parameters to determine optical irregularities. An important test here is to attempt to determine whether the optical device is clean, this being an important factor that can influence the quality of the acquired image.

TYPE C Template Aging

The aspect of template aging here of course rests on what particular modality is being used. However, the majority of biometric modalities have to ensure that templates are kept current as the user characteristic upon which the modality is based gradually changes over time. These types of error will be difficult to classify. The only constraint we have used is based on a time limit for the validity of the template. If we are approaching the time when templates need reacquisition (set in software) then we must be aware that verification errors around this time may be due to template aging.

TYPE D Acquisition Parameter Error

This type of error may be flagged if the system believes that the current image acquisition parameters are unrealistic for the current user. This type of fault can be identified by keeping track of the current and previous acquisition parameters. After an unsuccessful transaction event the user is made aware of the possible nature of the fault as determined by the analysis engine.

The results from the analysis agent are logged in an XML (e**X**tensible **M**arkup **L**anguage) file. This file is mined periodically and is used to determine the overall system behaviour.

2.7 Template Generation Monitor

This component is responsible for the generation of user templates and also the re-enrollment of the user at the appropriate time. During enrollment the goal of the agent is to attempt to acquire the best quality templates a user can realistically donate. At this time the quality setting is set to the manufacturers recommended setting for enrollment, this setting is used as the default to ensure that the largest number of users can actually enroll in the system. The templates are enrolled at the highest security level that is possible. This is in order to minimize the false reject rate by generating high quality templates.

The reenrollment of a user at the necessary time is also dealt with by this component. Data is available for the time between physical changes in bodily characteristics [5], upon which many biometrics are based. These values are used teristics [5], upon which many biometrics are based. These values are used for the default time between template reacquisition. Another mechanism for invoking re-enrollment is triggered by the Analysis Engine described above. These methods of template management are intended to alleviate any possible template aging issues the user may develop with the system. Naturally the user can opt to reenroll at any time without having to be prompted by the application itself.

At the time of reenrollment, however, a different process is invoked. Since we know what levels the user has been achieving over the period of time since enrollment, we can analyse these results and determine whether there is a possibility that either the quality or security settings for this user can be modified. This process is augmented by another feature of this component that can accept samples from the analysis agent obtained from the user during the verification phase. In this manner the agent can increase its confidence in the ability of the user to produce a sample that is capable of reenrollment at higher security/quality level, before actually asking the user to reenroll at these levels.

2.8 Calibration Monitor

Many biometric devices have the capability to be calibrated. In the fingerprint modality examined here, this is an important feature. Every subject will use the device in a slightly different way. Since the device requires the finger to be placed on the sensing area, each subject may apply a different amount of pressure. If the same calibration data is used for all the users this may lead to unacceptable image quality for a portion of the subject set. In an attempt to rectify this problem, this system records calibration details for each user individually. This calibration data is initially saved before the user can initiate enrollment, and is valid only for a specified amount of time. If the analysis agent determines that the samples acquired are failing due to old calibration data the user will be prompted to recalibrate the device.

3 Conclusions

A novel method has been presented in this paper to address some of the issues surrounding current biometric technologies, involving the use of software agents and focusing on a single modality. It is hoped that further work can demonstrate this concept in a multi modal context.

References

1. Brown SM, Santos E, Banks SB (1998) Utility theory-based user models for intelligent interface agents. In Lecture Notes in Artificial Intelligence 1418: Advances in Artificial Intelligence AI '98, pp. 378-392
2. Hong L, Jain A, Pankanti S (1999) Can Multibiometrics Improve performance? In *Procs AutoID'99,* Summit, New Jersey, pp. 59-64
3. Kephart JO, Hanson JE, Greewald AR (2000) Dynamic Pricing by Software Agents, Computer Networks, vol 32, pp. 731-752
4. Mea V, Roberto V, Conti A, Gaspero L, Beltrami CA (1999) Internet Agents for Telemedicine Services, Medical Informatics, vol 24, pp. 179-186
5. NPL/Biometrics Working Group (2000) Best practices in testing and reporting performance of biometric devices. www.cesg.gov.uk/technology/biometrics/media/Best%20Practice.pdf
6. Russell S, Norvig P (1995) Artificial Intelligence: A Modern Approach. Prentice Hall ISBN 0-13-360124-2 pp 31
7. Schäfer R (2001) Rules for Using Multi-Attribute Utility Theory for Estimating a User's Interests. ABIS-Workshop 2001, pp 21- 28
8. Shen W, Norrie D (1998) An Agent-Based Approach for Dynamic Manufacturing Scheduling. In Working Notes of the Agent-Based Manufacturing Workshop, Minneapolis, pp. 117-128
9. Thomas C, Fischer G (1996) Using agents to improve the usability and usefulness of the world wide web. *In Procs of the Fifth International Conference on User Modelling*, pp 5-12
10. Wilson LF, Burroughs D, Sucharitaves J, Kumar A (2000) An agent-based framework for linking distributed simulations, *In Procs of the 2000 Winter Simulation Conference*, pp. 1713-1721
11. Winterfeld D, Edwards W (1986) Decision Analysis and Behavioral Research. Cambridge, England. Cambridge University Press.

Gait Recognition by Moment Based Descriptors

J. P. Foster, M. S. Nixon, A. Prugel-Bennet

Department of Electronics and Computer Science, University of Southampton, Southampton, SO17 1BJ, United Kingdom
email: jeff.foster@sira.co.uk / msn@ecs.soton.ac.uk / apb@ecs.soton.ac.uk

Abstract. Gait receives increasing attention as a biometric. Given the requirement to process sequences, computational demands are high, motivating the development of basic approaches. Amongst these, change in area can be used as a recognition metric. We describe a new approach using moments, and its motivation from an area based approach. The technique derives from a spatial approach and is here extended to sequences. We show how it can be used to discriminate subjects by their gait. By its formulation, the new technique derives invariance advantages, but these can reduce performance.

1 Introduction

Recently, there has been much attention devoted to using gait as a biometric [1-6]. Approaches to automatic gait recognition can broadly be divided into two. Model based approaches aim to explicitly model gait [3, 7] as a series of equations. One of the main advantages of the model-based approaches is their handling of occlusion, which is of especial importance in gait as the walking human body is self-occluding. However, the model used to describe gait is often complicated, resulting in a high computational cost. Holistic / statistical approaches [2, 4, 6, 8] aim to process a gait sequence to find a set of measurements to distinguish between subjects. The disadvantage of traditional statistical approaches is they are not intimately related to gait and just produce raw numbers to distinguish between subjects. In addition, most current statistical techniques fail to take into account the temporal component of gait and instead focus on distinguishing between collection of silhouettes, rather than a sequence as a whole.

The holistic approaches do enjoy computational advantage: gait requires analysis of sequences of images, so computational demand is of much concern. Some of the approaches have aimed at simplicity, to provide "baseline" analysis [2, 4]. The penalty is that these approaches lose specificity to gait. In a similar approach, we included specificity by analysing gait from the changes in areas selected by masking functions [9]. Again, it is a generic approach to analysing periodic motion in image sequences. Our new approach enjoys similar computational advantage of other baseline approaches, with flexibility in deployment.

In order to capitalise further on these baseline approaches, we sought to determine a link between the area-based approach and more conventional analysis. Unfortunately, there were no links other than basic, but it became apparent that we could deploy moments in a similar vein, to analyse gait with specificity and speed using moments to acquire invariant properties. We use a family of moment descriptors to describe a single shape, relating to work from Sluzek [10]. Then, Sluzek's approach has been extended to describe a sequence of images. We discuss the invariance properties of this new approach before comparing results with those of area masks, prior to further work and conclusions.

2 Recognition via Moment Based Descriptors

Previously, we have used area masks to describe the dynamics of area change through a gait sequence and this has formed the basis of description. By using moments we can provide a geometric description of a shape. Lower order moments provide information similar to area masks, such as total area of the shape. We use moments to describe each image in the gait sequence and this forms the basis of description. Using Hu moments, the technique is invariant to rotation, translation and scaling of the silhouettes.

A gait sequence is represented as a collection of binary silhouettes. These binary silhouettes provide us with invariance to most lighting conditions and clothing colour, but do lose information. These silhouettes can be analysed singly (by current techniques) or as sequences, by new techniques.

Moment Based Descriptors for Single Images

Moments of order p, q of a discrete shape R are defined as

$$m_{pq} = \sum_{i,j \in R} i^p j^q \qquad \textbf{(1)}$$

Centralised moments are defined with co-ordinates translated to the center of a discrete shape R as:

$$M_{pq} = \sum_{i,j \in R} (i-a)^p (j-b)^q \qquad \textbf{(2)}$$

a and b are the centres of mass in the 2D co-ordinate system. The lower order moments are simple properties to describe the shape. Hu [11] derived moment expressions that are invariant to translation, rotation and scaling of shapes. Some examples are given below. The moments omitted, 3 to 7, are usually assigned to moment invariants of order 3.

$$I_1 = \frac{M_{20} + M_{02}}{(m_{00})^2} \qquad I_2 = \frac{(M_{20} - M_{02})^2 + 4M_{11}{}^2}{(m_{00})^2} \qquad I_8 = \frac{M_{20}M_{02} - M_{11}^2}{(m_{00})^4} \qquad \textbf{(3)}$$

Sluzek proposed a method to improve the quality of description by using a family of shape descriptors. By using additional descriptors, it is possible to improve the quality of descriptions. Let C_α be the circle defined as

$$C_\alpha = (x-a)^2 + (y-b)^2 = \alpha m_{00} / \pi \qquad \textbf{(4)}$$

The area of the circle is equal to the area of the shape multiplied by □. The value of α_{MAX} is chosen so the circle does not occlude the entire area of *R* and so a part of the shape *R* is visible. Let us denote the region *R* which is not occluded by the circle as *R(*□ *)* which is defined as

$$R(\alpha) = \{(x,y\} \in R \wedge \left|(x,y)-(a,b)^2\right| > \alpha m_{00} / \pi\} \qquad \textbf{(5)}$$

For a given shape *R* we can create functions $I_k(\alpha)$ for each moment *k* which operates on the region *R(*□*)*. These functions depend on both the region *R* and the value of □ and RST invariant. The function $I_k(\alpha)$ is the *k*-th Hu invariant on the region *R(*□*)*. Note that when calculating $I_k(\alpha)$ that *a* and *b* must be recalculated on the new region *R(*□*)*.

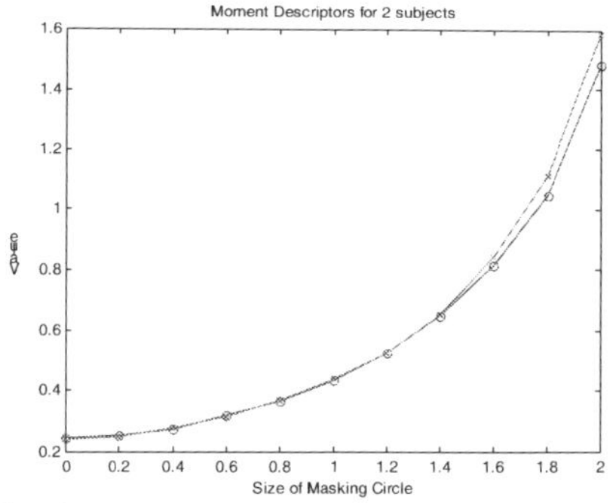

Fig. 1 Moment Descriptors for Two Subjects

A family of shape descriptors can now represent a given shape/image. Adequate sampling should be chosen to give a detailed representation of the shape. $I_k(\alpha)$ is then the signature of the image for a given range of circles α. These moment descriptors can be used to discriminate between two different images. Sluzek [10] showed that these shape descriptors are more reliable than traditional moment shape descriptors at identifying undamaged objects and/or to detect damaged ones. Sluzek analysed a database of 130 images and achieved a hundred percent accuracy both in the identification of objects and in the quality inspection. Fig. 1 shows the moment descriptors for two single images from the walking sequence of different subjects. As can be seen from the graph, there is minimal visible difference between the two graphs when the value of α is low, but as the value of αrises the differences between the two becomes more apparent. Gait consists of a temporal, as well as spatial component, so over a sequence the differences between subjects should become even more apparent as the small differences will accumulate over an entire sequence.

Recognising Gait using Moment Based Descriptors

Gait recognition differs from traditional biometrics because it is a function of both space and time. Let us consider a gait sequence as a collection of regions denoted by R_t where t labels the position of the region in the gait sequence. For each individual image R_t, we can once again define regions masked by circles as $R_t(\alpha)$. As the size of the masking circle increases, we describe areas on the periphery of the image, such as the subjects' feet. These are the areas of the image most likely to vary considerably over a gait sequence, but are not necessarily the areas that will allow distinction between subjects.

Original Image α=0.5 α=1.0 α=1.5

Fig. 2. Masking Circles for various values of Alpha

Fig. 2 illustrates how the masking circle affects a silhouette. We can now define a complete signature set for a gait sequence thus

$$S = \{I_k(t,\alpha) \mid k = 1,2,8; t = 0, N, \alpha = \alpha_{\min}, \alpha_{\min} + \alpha_{step}, \ldots, \alpha_{\max}\} \quad \textbf{(6)}$$

t is chosen such that a full gait cycle was sampled at intervals of $1/N$ starting from a known start point. In the case of the SOTON database, all subjects start from a heel-strike and exactly one gait cycle is present. In other databases, key frames could be located using techniques developed by Collins' et al [2]. A spline curve was used to interpolate between values to provide exactly thirty samples for a full gait cycle. Thirty samples were chosen, as this was approximately equal to the number of frames filmed. Each sequence can therefore be described as a set as above for values of k and α. Fig. 3 sample output for a single subject.

A suitable α range must be chosen to give adequate image representation. In practice if the value of α is too large then no information about the image is obtained and this can lead to poor recognition performance. To ensure that all sequences are the same length we use a cubic spline curve to represent the data and take the same amount of samples from each sequence $S_{\alpha,k}(t)$..

Invariant Properties

By using Hu invariant moments, the moment descriptors for each silhouette are invariant to rotation, translation and scaling. Invariance to rotation could be of benefit if the subject is walking uphill or downhill, but this implies that gait stays constant when walking at an incline, which is certainly not true [12]. Scaling and translation invariance is important because they enable silhouette sequences from different databases to be compared easily. Scale invariance is of particular importance as gait can be filmed from a great distance and therefore the resolution of such images may be lower than those used in the SOTON database. In addition, scaling invariance gives us immunity to camera zoom. Translation invariance removes sensitivity to the centring of the silhouette, which can be a problem when area masks are used [5].

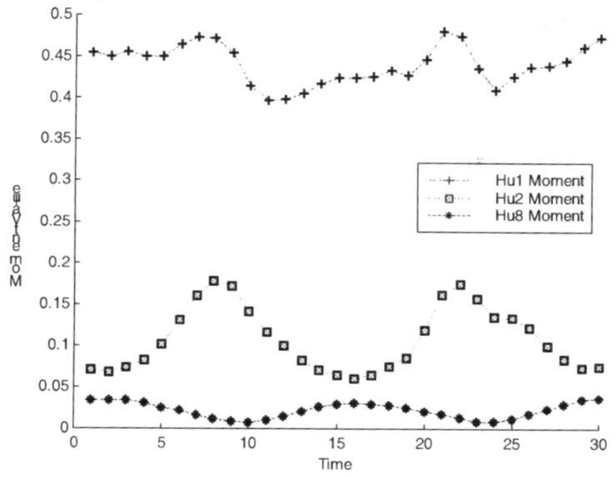

Fig. 3. Hu Moments for a Subject with α =0

3 Results

We analyse our results on part of the SOTON database consisting of 114 subjects with at least eight samples of each subject. Subjects walk normal to the camera's plane of view and are filmed under laboratory conditions. These conditions allow us to concentrate on the basic properties of gait as a biometric, rather than the problem of extracting clean data. The background of the image is removed using chroma-key techniques, and the image is cropped and resized to a 64x64 silhouette with the subject centred in the image. We have chosen to use a simple *k* nearest neighbour classifier, rather than a more sophisticated classifier such as canonical analysis, and Euclidean distance for basic comparison, together with leave one out cross validation.

Person Identification using Moments

Table 1 shows the performance using a selection of Hu moment invariants (without using a family of masking circles). The feature vector from each moment was combined by simply concatenating the feature vectors of each individual moment invariant to form one large feature descriptor. Performance levels using a single moment invariant may appear poor, but the chance recognition rate is 1/114 and the results are substantially better than that. By combining the feature vectors, the performance level was raised, indicating independent information from each of the moment descriptors. The moments were combined simply by concatenating the feature vectors. Note that no normalisation took place when feature vectors were combined.

Table 1. Performance of Moment Invariants for Person Discrimination

Moment Invariant	Recognition Rate
Hu1	44.7%
Hu2	35.9%
Hu8	22.7%
All 3 combined	51.4%

By using a family of shape descriptors, we can describe the shape with more accuracy and thus increase the performance of the description and therefore the recognition rate.

Table 2Table 2 demonstrates how we improve performance using a family of descriptors.

Table 2. Using a Family of Shape Descriptors for Person Discrimination

Values of α	Recognition Rate in %			
	Hu1	**Hu2**	**Hu8**	**Combined**
0	44.7	35.9	22.7	51.4
0 to 0.5 in steps of 0.05	63.3	47.6	29.1	63.3
0 to 1 in steps of 0.1	68.3	44.5	28.5	56.7
0 to 1.5 in steps of 0.25	65.9	34.3	28.8	42.1
0 to 2 in steps of 0.1	64.4	39.4	31.6	44.7

As can be seen from Table 2 the performance using a family of shape descriptors rises rapidly when low values of α are used but the performance falls away rapidly when large values of α are combined. This suggests that using large masking circles and thus describing information at the periphery of the silhouette can detract from recognition performance. The Hu1 moment performs consistently better than other moments at discriminating between subjects. Both the Hu2 and Hu8 moments perform poorly, achieving low recognition rates despite of the increased information present from using masking circles. The major difference between the Hu2 and Hu8 moments is the inclusion of the M_{11} term. We suggest from this that the M_{11} term is not very descriptive for discriminating between subjects because the moments that contain this term perform significantly worse than those that do not. The combined set of moments performs consistently poorly when a family of descriptors is used. This may be due to the poor performance on the Hu8 scheme that affects the overall recognition result heavily due to no normalisation of feature vectors.

Is Gait Symmetric?

Gait symmetry has been defined as the perfect agreement between actions of the lower limbs [13, 14]. Gabbard [15] has suggested using the term when no statistical differences are noted on parameters measured bilaterally. Historically, the psychologists' view is that gait is a symmetrical pattern of motion [16, 17]. We have assumed this true, and not taken into account the foot on which the subject starts, or their direction of travel. Does taking this information into account result in a significant difference in recognition rate? The SOTON database has labelled information including heel strikes and direction of walk. By using this information, we were able to undertake experiments where the direction of walk and the starting foot for the gait cycle were taken into account. As can be seen from Table 3 the performance gain when considering the starting leg is substantial. This suggests, once again, that there is an asymmetrical component of gait and by taking this into consideration performance can be improved.

In the following table the database is constrained so that only sequences where the direction of walk and starting foot are as indicated in the table are considered. As the table indicates, by constraining the database to specific starting foot and direction of walk we are able to dramatically increase the recognition rate. In application this indicates that if we could automatically determine the starting foot of the gait cycle and the direction of walk we could increase recognition rates.

Table 3. Recognition Rates Considering Starting Foot and Direction of Travel

Direction of Walk	Starting Foot	Recognition Rate using Hu1 Moments
Left	Left	83.2%
Left	Right	80.6%
Right	Left	81.9%
Right	Right	78.7%
Left and Right	Left	82.9%
Left and Right	Right	81.5%

Conclusions

We have presented a new technique for recognition of gait sequences using moment-based descriptors. A family of descriptors is formed for each silhouette in the sequence, by using a masking circle to define what areas of the image to describe. The technique uses Hu moments and is hence invariant to rotation, translation and scaling.

Results show that performance is dramatically improved if a family of masking circles is used, and we achieve a recognition rate of 68.3% on the SOTON database of 114 subjects, with eight samples each. We show that by considering the direction of travel and heel strike information performance can be substantially increased. By constraining the database to subjects using the same heel strike to start a gait cycle and walking in the same direction, performance is increased to over 80% on a databases of 114 subjects.

Future work will concentrate on evaluating the approach on a more varied database and examine the use of different ways of masking areas of the silhouettes. Further work will also analyse the affects of starting from a non-zero value of alpha (i.e. include only larger masking circles) and investigate the affects of using different moment descriptors.

Acknowledgement: support by the European Research Office of the US Army under Contract No. N68171-01-C-9002 is gratefully acknowledged.

References

[1] C. BenAbdelkader, R. Cutler, and L. Davis, "Stride and Cadence as a Biometric in Automatic Person Identification," *Proc 5th IEEE FGR 02*, pp. 372-378, 2002.

[2] R. Collins, R. Gross, and J. Shi, "Silhouette-based Human Identification from Body Shape and Gait," *Proc 5th IEEE FGR 02*, pp. 366-371, 2002.

[3] C.-Y. Yam, M. S. Nixon, and J. N. Carter, "Gait Recognition by Walking and Running: A Model-based Approach," *Proc. 5th Asian Conf .on Comp. Vision*, 1-6, 2002.

[4] P. J. Phillips, S. Sarkar, I. Robledo, P. Grother, and K. Bowyer, "Baseline Results for the Challenge Problem of Human ID Using Gait Analysis," *Proc 5th IEEE FGR 02*, pp. 137-143, 2002.

[5] J. P. Foster, M. S. Nixon, and A. Prugel-Bennett, "New Area Based Measures for Gait Recognition," *Proc. Audio- and Video-Based Biometric Person Authentication*, pp. 312-7, 2001.

[6] J. B. Hayfron-Acquah, M. S. Nixon, and J. N. Carter, "Recognising Human and Animal Movement by Symmetry," *Proc. IEEE International Conference on Image Processing*, pp. 290-293, 2001.

[7] D. Cunado, M. S. Nixon, and J. N. Carter, "Automatic Gait Recognition via Model-Based Evidence Gathering," *Proc. AutoID99: IEEE Workshop on Automated ID Technologies*, pp. 27-30, 1999.

[8] C. BenAbdelkader, J. Cutting, and L. Davis, "Motion-based Recognition of People in EigenGait Space," *Proc 5th IEEE FGR 02*, pp. 378-384, 2002.

[9] J. Foster, M. Nixon, and A. Prugel-Bennett, "New Area Based Metrics for Automatic Gait Recognition," *Proc. British Machine Vision Conference 2001*, pp. 233-242, 2001.

[10] A. Sluzek, "Identification and inspection of 2-D objects using new moment-based shape descriptors," *Pattern Recognition Letters*, vol. 16, pp. 687-697, 1995.

[11] M. Hu, "Visual pattern recognition by moment invariants," *IRE T. Inf. Th*, vol. 8, pp. 179-187, 1962.

[12] J. Sun, M. Walters, N. Svensson, and D. Lloyd, "The Influence of Surface Slope on Human Gait Characteristics" *Ergonomics*, vol. 39 (4), pp. 677-692, 1996.

[13] K. Soudan, "Standardization of Gait Kinematic data using a Gait Symmetry index and Fourier analysis," *In: Biomechanics: Principles and Applications,* vol..1, pp. 135-140, 1982.

[14] W. Herzog, B. Nigg, L. Read, and E. Olsson, "Asymmetries in Ground Reaction Force Patterns in Normal Human Gait," *Medicine in Science and Sports and Exercise*, vol. 21, pp. 110-114, 1982.

[15] C. Gabbard, "Coming to Terms with Laterality," *Journal of Psychology*, vol. 131, pp. 561-564, 1997.

[16] J. T. Cutting, D. R. Proffitt, and L. T. Kozlowski, "A Biomechanical Invariant for Gait Perception," *Journal of Experimental Psychology: Human Perception and Performance*, pp. 357-372, 1978.

[17] H. Sadeghi, P. Allard, F. Prince, and H. Labelle, "Symmetry and Limb Dominance in Able Bodied Gait: a Review," *Gait and Posture*, vol. 12, pp. 34-45, 2000.

On a Large Sequence-Based Human Gait Database

J. D. Shutler[1], M. G. Grant[2], M. S. Nixon[2], and J. N. Carter[2]

[1] Plymouth Marine Laboratory, Plymouth, United Kingdom jams@pml.ac.uk
[2] University Of Southampton, Department of Electronics and Computer Science, Southampton, United Kingdom {mgg, msn, jnc}@ecs.soton.ac.uk

Abstract. Biometrics today include recognition by characteristic and by behaviour. Of these, face recognition is the most established with databases having evolved from small single shot single view databases, through multi-shot multi-view and on to current video-sequence databases. Results and potential of a new biometric are revealed primarily by the database on which new techniques are evaluated. Clearly, to ascertain the potential of gait as a biometric, a sequence-based database consisting of many subjects with multiple samples is needed. A large database enables the study of inter-subject variation. Further, issues concerning scene noise (or non-ideal conditions) need to be studied, ideally with a link between ground truth and application based analysis. Thus, we have designed and built a large human gait database, providing a large multi-purpose dataset enabling the investigation of gait as a biometric. In addition, it is also a useful database for many still and sequence based vision applications.

Keywords. human gait, biometric, sequence based database, database design, video database.

1 Introduction

Gait is an emergent biometric with increasing interest (e.g. [1, 2, 3] - primarily through DARPA's Human ID at a Distance program). Early gait databases contained very few subjects (primarily due to technological issues) and analyses resulting in high classification rates have prompted further research. For example, the UCSD (University of California San Diego) database consisted of six subjects with seven sequences of each. Several different publications have reported close to 100% classification on that data eg. [4, 5, 6]. Here we describe the design and capture of a new large sequence-based human gait database, detailing the ideas behind its structure through to areas of human psychology used to ensure good quality, consistent data.

2 Database design

A large gait database has been designed and built, providing both ideal laboratory data (indoor, controlled lighting) as an approximation to ground truth, and (non-ideal) outdoor data for application based analysis. Data was also derived using treadmills, to analyse the potential of capturing gait information in a confined space. High resolution still images were taken to enable face classification and to estimate a subject's body and size. Therefore, subjects extracted from indoor and outdoor scenes can be compared, both in terms of classification and extraction. The database consists of fronto-parallel and obliquely viewed imagery (allowing orientation independent analysis), extracted silhouette data, still images (side on and frontal) and ancillary data including subject specific information, camera setups and extraction parameters [7]. Further, a considerable amount of information redundancy was designed into the acquisition, storage and final structure of the database. This allows for cross-referencing within the data itself, and eased the (manual and automatic) validation and consistency checking of the data.

3 Filming Issues

Human psychology plays a large part in collecting data of this type and magnitude. Firstly, to avoid affecting the subject's walking patterns, the treadmill training and filming took place after each subject had first walked outdoors, and then inside on the laboratory track. There is much debate in the use of treadmills for gait analysis concerning their suitability, speed and inclination. The speed and inclination were set at constant values derived by evaluation, however, it is worth noting that treadmills allow for capture of long high resolution continuous gait sequences. Further issues [7] included not informing subjects when the cameras were filming (reducing shyness issues by switching the cameras on prior to the subjects entering the laboratory), not talking to subjects as they walked (as invariably humans will turn their head to address the person), using a talk-only radio station for background noise (to reduce the impulse need of a human to break the talk-silence), removing the need for a person to control the cameras reduced the camera-shyness and talking issues further and placing a mirror in front of the treadmill aided balance and stopped the subject from looking at their feet and/or the treadmill control panel.

4 Imagery

A sequence-based database is rather a massive one. Our target was for over 100 subjects, with approximately 8 sequences of 1.5 steps per subject (as this is periodic data, this is enough of the gait cycle) together with background and supporting data. Given that digital video (DV) is now an established technology, we acquired the imagery via good quality off-the-shelf DV camcorders, having to experimentally examine the differences between cameras due to incomplete published specifications, as the units are designed for domestic-use. This analysis involved capturing data and looking specifically at colour and optical quality, frame rates and picture scan methods. To allow for 25 fps progressive-scan capture and good quality optical data, two different types of cameras were used, as at that time no single camera met the requirements (not all DV camcorders allow for 25 fps progressive scan data). Three Canon MV30i's were used for the progressive scan data while three Sony DCR TRV900E PAL's allowed for interlaced data with improved optical and colour quality (the Sony uses a three chip system, one for each colour, in comparison to the single-chip technology of the Canon). Two cameras filmed each scene (one progressive scan and one interlaced) allowing for analysis of viewpoint-independence. The progressive scan technology provides high resolution imagery (suitable for ground truth) whilst security video often uses interlaced data. The resulting database consists of over 100 subjects, filmed in 3 scenarios each from 2 different views, with accompanying high resolution still images. The subjects walked around a continuous track and were filmed walking in both directions, the track layout can be seen in Figure 1. The outdoor data presents a similar scenario, however with a greater distance between subject and camera, allowing scale independent analysis and an increased number of in-view gait cycles.

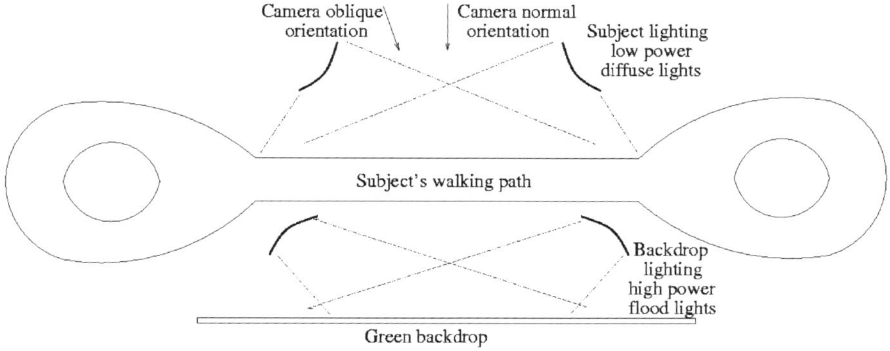

Fig. 1. Laboratory lighting arrangement, enabling the separation of the two lighting schemes.

5 Data quality and post-processing

Blue screening (or chroma-keying) is the process of filming an object, or subject in front of an evenly lit, bright pure coloured backdrop. Object or subject extraction from the images is easily achieved through background removal, allowing an alternative background colour or scene to be used in its place. Here a bright green backdrop was used, mainly as it is an unlikely colour for the subjects to wear, although any pure colour could have been used. Thus, a purpose built chroma-key laboratory was constructed to allow controlled lighting conditions for the inside track and treadmill scenarios. This near perfect subject extraction provides an approximation to ground truth. Further ground truth includes the still photos and subject information. Iterative optimisation of the laboratory and equipment setups produced blur free data and significantly reduced shadows. This iterative process used Canny and Sobel edge detectors and a statistical subject extraction technique [8] enabling the assessment of the quality of the lighting and camera positions. A software implementation of chroma-key extraction allows for easy monitoring of the data quality between filming sessions, as invariably (over time) equipment can be knocked and the output from lights can change. Figure 2 shows example images and a chroma-keyed result (actual data is in colour). The complete database (captured in one month) includes details of all settings and session specifics, recorded on a per-filming session basis. To increase the available information for the ground truth (and to ease use), the data from the fronto-parallel track camera has been significantly described and labelled using XML, an example fragment of XML can be seen in Figure 3. Further information includes: less detailed labeling for the remaining viewpoints, camera sync information (between views) and parameters enabling radial distortion correction.

Analysis

To date, three different recognition approaches have been applied to the data, all with encouraging results. This analysis of the database suggests that it has indeed met its design objectives. First, high gait recognition performances have been achieved on the largest yet number of subjects for gait, an overview of these results can be seen in Table 1. The progression of these results reflects the gradual construction of the database and detailed explanations of these results can be found in [9, 10, 11]. The processing of the data used much of the available support material, enabling streamlined and in some cases automated analysis. These results used a selection of binary silhouettes and optical flow descriptions generated from the inside laboratory data. The use of the chroma-keyed binary silhouettes has also provided a test-bed for performance analyses of the various techniques eg. [11]. A preliminary analysis of the outdoor data confirms the increased variance of features extracted from application scenario imagery as opposed to ground truth [11], shown here for

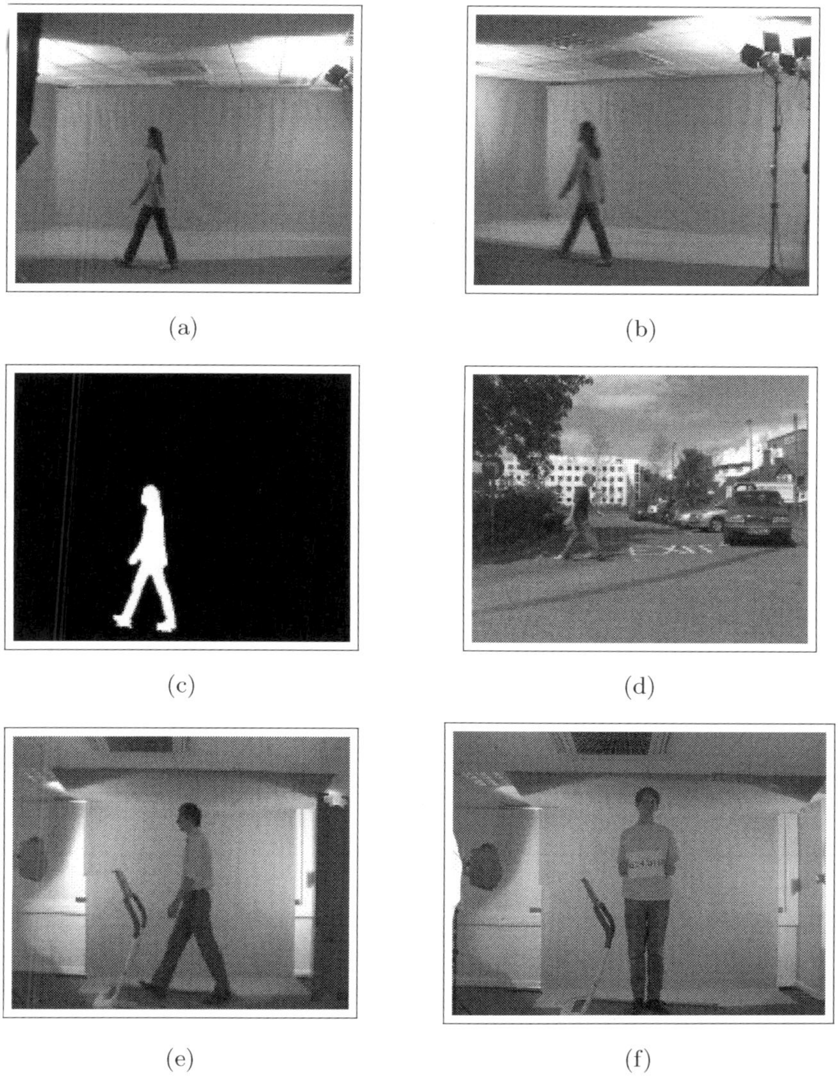

Fig. 2. Example (a) inside data normal (b) inside data oblique, (c) processed silhouette from the inside data normal, (d) outdoor data, (e) treadmill and (f) high resolution still (actual data is in colour).

```
<?xml version="1.0" ?>
<data camera="a" date="23/05/01" direction="left"
filename="008a013s00L.dv" sequence="00L" session="008"
start_frame="2369" start_time="0 : 01 : 34.19" stop_frame="2546"
stop_time="0 : 01 : 41.21" subject_id="013" xmlcreatedby="Time Code
re-writer" xmlcreatedon="Mon Mar 25 11 : 51 : 16 2002">
   <comments>
      No comment
   </comments>
   <heelstrikes automatic="yes" checked="yes">
      <strike leg="left" time="62"/>
      <strike leg="right" time="77"/>
              . . .
```


Fig. 3. Example fragment of XML for the normal-view camera.

two subjects in Figure 4. In each plot the tight Gaussian represents the variance of the inside data, whereas the larger Gaussians (greater variance) are the outdoor data, the mean of each shows the mean feature point drift from the baseline (inside data). Finally, only data-quality checks have been applied to the inside treadmill data.

Table 1. Early published results using the SOTON HiD database.

Algorithm Employed	No. of Subjects	No. of Sequences per Subject	Classifier Result $k = 1$	Classifier Result $k = 3$
Symmetry [9]	28	4	97.30 %	96.40 %
Velocity Moments [11]	50	4	97.00 %	96.00 %
Moment Based [10]	114	8	78.00+ %	-

6 Beyond gait research

The use of the database can be extended past that of gait-driven biometric research. In terms of technique development, the database allows for the evaluation of background removal techniques, high level moving-feature analysis techniques (both rigid and non-rigid) and face recognition analysis. As such, our new database allows not just for evaluation of the potential of gait as a biometric but also the evaluation of many other sequence based vision and image processing problems. The inside laboratory data enables the performance of an extraction technique to be compared with that of the near-ideal (baseline) chroma-keyed result. The outdoor data can be used to assess the performance of subject extraction techniques, providing data of single subjects, multiple subjects (both in the foreground and background) and rigid

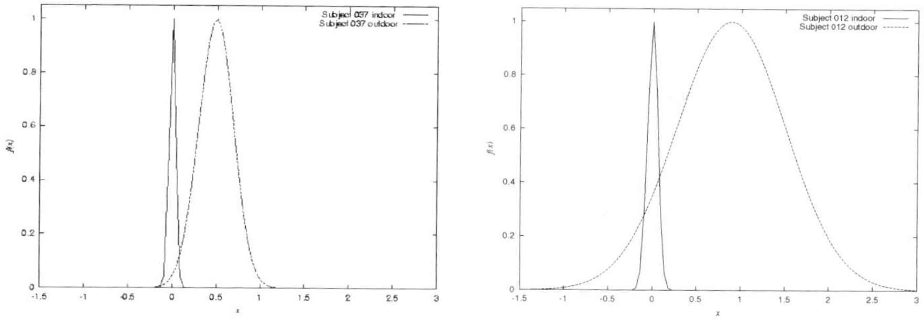

(a) Subject 037 (b) Subject 012

Fig. 4. Gaussian representation of the within-subject distributions for indoor ('tight' Gaussians) and outdoor data ('loose' Gaussians) for two different subjects.

objects in the background eg. cars, buses, bicycles etc. The outdoor data allows for the analysis of these techniques under various weather environments as it includes data taken on bright cloudless days through to overcast, dull conditions. Further, the subjects themselves are non-rigid allowing the application of high-level feature analysis techniques. Finally, the high resolution stills are applicable to face recognition research, not only stand alone, but also allowing for research into the buttressing of gait and face biometrics, as face data exists for all gait subjects.

7 Conclusions

High recognition rates have already been observed on 100+ subjects using a variety of techniques, highlighting gait's potential as a biometric and justifying the detail behind the design and construction of the database. The chroma-key extraction does indeed allow for a basic ground truth that can be achieved with comparable ease, and the results can be compared with extraction by computer vision approaches. The new DV technology allows for ease in digitisation and has helped to mitigate the large storage requirements. This new database allows for estimation of inter- and (limited) intra-subject variability on a large number of subjects, together with assessment of the efficacy of computer vision techniques that extract and describe moving objects, as the evaluation of gait covers many pragmatic factors. As such the data is applicable to many areas of interest within the vision community including moving object extraction algorithms, face recognition, and the buttressing of face and gait biometrics. Our future research will seek to capitalise the variability and extent offered by this new gait database and includes the construction of a new database designed specifically to enable the study of intra-subject variability, studying issues including, but not limited to, variations over time, effects of alternative footwear and the carrying of objects - this database is nearing completion. Early analysis of the outdoor data has confirmed the increased variance of features extracted from application scenario imagery as opposed to ground truth. Finally, further information and access to the database can be obtained at: *http://www.gait.ecs.soton.ac.uk/* and the database is available through contact with the authors.

Acknowledgements

We gratefully acknowledge support from the European Research Office of the US Army, Contract No.N68171-01-C-9002. The input from Peter Myerscough, Robert Boston, Chew Yean Yam, Jeff Foster, James Hayfron-Acquah, Layla Gordon and Nick Spencer is gratefully appreciated.

References

1. C. BenAbdelkader, R. Cutler, H. Nanda, and L. Davis. EigenGait: Motion-Based Recognition of People Using Image Self-Similarity. *Proc. AVBPA01*, :pp. 284–294, 2001.
2. A. Y. Johnson and A. F. Bobick. A multi-view method for gait recognition using static body parameters. *Proc. AVBPA01*, :pp. 301–311, 2001.
3. P. J. Phillips, S. Sarkar, I. Robledo, P. Grother, and K. Bowyer. Baseline results for the challenge problem of human ID using gait analysis. *Proc. FGR02*, :pp. 137–142, 2002.
4. P. S. Huang, C. J. Harris, and M. S. Nixon. Recognising humans by gait via parametric canonical space. *Artificial Intelligence in Engineering*, **13**:pp. 93–100, 1999.
5. J. J. Little and J. E. Boyd. Recognising people by their gait: the shape of motion. *Videre*, **1**(2):pp. 2–32, 1998.
6. J. D. Shutler and M. S. Nixon. Zernike velocity moments for the description and recognition of moving shapes. *Proc. BMVC01*, **2**:pp. 705–714, 2001.
7. J. D. Shutler and M. G. Grant. The large Southampton HiD human gait database. Technical report, ISIS, University of Southampton, U.K., 2002.
8. S. Jabri, Z. Duric, H. Wechsler, and A. Rosenfeld. Detection and location of people in video images using adaptive fusion of color and edge information. *Proc. ICPR00*, **4**:pp. 627–630, 2000.
9. J. B. Hayfron-Acquah, M. S. Nixon, and J. N. Carter. Automatic gait recognition by symmetry analysis. *Proc. AVBPA01*, :pp. 272–277, 2001.
10. J. P. Foster, M. S. Nixon, and A. Prugel-Bennet. Gait recognition by moment based descriptors. *Proc. RASC02*, :accepted for publication, 2002.
11. J. D. Shutler. Velocity Moments for Holistic Shape Description of Temporal Features. *PhD Thesis, University of Southampton, U.K.*, :, 2002.